RESTORING ACID WATERS

RESTORING ACID WATERS: LOCH FLEET 1984–1990

Edited by

G. HOWELLS
Department of Zoology, University of Cambridge
Cambridge, UK

and

T.R.K. DALZIEL
PowerGen plc, Ratcliffe Technology Centre
Nottingham, UK

ELSEVIER APPLIED SCIENCE
LONDON and NEW YORK

ELSEVIER SCIENCE PUBLISHERS LTD
Crown House, Linton Road, Barking, Essex IG11 8JU, England

Sole distributor in the USA and Canada
ELSEVIER SCIENCE PUBLISHING CO., INC.
655 Avenue of the Americas, New York, NY 10010, USA

WITH 52 TABLES AND 101 ILLUSTRATIONS

British Library Cataloguing in Publication Data

Restoring acid waters: Loch Fleet 1984–1990.
I. Howells, Gwyneth, 1925– II. Dalziel, Thomas Ralph Kennedy, 1959–
T. R. K. (Thomas Ralph Kennedy)
363.7386

ISBN 1-85166-663-X

Library of Congress Cataloging-in-Publication Data

Restoring acid waters: Loch Fleet, 1984–1990/edited by G. Howells
and T.R.K. Dalziel.
p. cm.
Includes bibliographical references and index.
ISBN 1-85166-663-X
1. Acid pollution of rivers, lakes, etc.—Scotland—Fleet, Loch
(Dumfries and Galloway) 2. Water quality management—Scotland—
Fleet, Loch (Dumfries and Galloway) 3. Lake renewal—Scotland—
Fleet, Loch (Dumfries and Galloway) 4. Lake ecology—Scotland—
Fleet, Loch (Dumfries and Galloway) 5. Neutralization (Chemistry)
6. Lime. I. Howells, Parry, Gwyneth II. Dalziel, Thomas Ralph Kennedy
TD427.A27R48 1991
628.1′68—dc20 91-4107
CIP

Note: The views expressed in the following text are those of the authors and should not be taken to represent the views or policy of the agencies who employ them or have supported their research.

The funding agencies of the Loch Fleet project have supported the presentation of the project findings in this book, for which permission to publish has been granted.

Photoset by Alden Multimedia Ltd.

Foreword

Even today, after more than a decade of practical experience in Sweden and a number of trials in Scandinavia and in the UK, the restoration of acid waters by "liming" is still viewed with suspicion by environmental purists. Tolerated as a stopgap until agreement was reached on emission reductions, the addition of limestone to acidified soils and waters was, at one time, condemned as an unethical diversion of attention and resources from that supreme objective. More recently has come the realisation that the acidification accumulated in soil by 150 years of industrialisation will not respond overnight to emission reductions, however swingeing.

The Loch Fleet project was born of environmental pragmatism. To get fish back soon, you have to do something to restore the quality of the water or, for a longer-lasting effect, of the soil. What you do has to be scientifically based to ensure that it will be effective and without unacceptable side-effects. This volume marks the successful achievement of these objectives at Loch Fleet and will surely be a pathfinder for success elsewhere. Not least, it makes a fitting testimonial to the foresight and persistence of the scientific team led by Dr Howells and to the practical and financial assistance of the coal and electricity industries in a project well ahead of its time.

Dr Peter Chester
Executive Director, Technology and Environment,
National Power plc

December 1990

Preface

As the effects of acid rain in the aquatic environment become better documented, there is growing recognition that, while acidification is undoubtedly linked with sulphur and nitrogen emissions, a major reduction of those emissions is unlikely to bring benefit to aquatic systems in the short term. Indeed, the short-term effect of reducing the acidity of rain will be to reduce the yield of base cations from soils and parent material. Without sufficient dissolved calcium in natural oligotrophic waters, the fauna and flora are restricted in variety, and in particular, membrane physiological processes in fish will be impaired, and they will be exposed to toxicity from dissolved aluminium. Without supplementary calcium, fish and other organisms will be at risk, even where the acidity is at a moderate level.

As with agriculture, where the need for supplementary calcium (and magnesium) has been recognised for at least three thousand years, materials and methods for making good this deficiency have long been developed. Most studies have focused on direct treatment of the acidified waters, but in Britain, where upland oligotrophic waters draining unmanaged terrain are acidic, principal interest lies in protecting migratory salmonids which spawn in these low-order streams and develop in montane lakes over their first two years. The hydrological regime of these waters is highly variable so that direct water treatment improves water quality only while the water is retained. This reasoning led to the selection of Loch Fleet, in south-west Galloway, for an experimental trial of catchment liming to improve both the major tributary streams and the loch itself. This book relates the outcome of those experiments.

Major programmes of lake and river liming have developed in Sweden and Norway, and more recently in North America. For the most part direct liming of water bodies has been employed, coincident with the more stable hydrological regimes of those waters, and the in-lake habitat of many of the fish species of concern. However, there are a handful of catchment liming experiments in these countries to which the following chapters will refer. For completeness some well-documented lake and stream liming projects are also included for comparison, and some reference made to other methods to provide acceptable water quality for acidified areas: these include hydrological management, land-use management, fisheries management, as well as some novel water additives to counter acidity.

The Loch Fleet project developed through a planned sequence of preliminary, pre-treatment observations, and detailed testing of liming materials and fish response. Following the lime applications, detailed and careful monitoring of the changing water quality and the fishery status was undertaken. In addition, to meet those sometimes speculative claims that the environment created does not return to that assumed to have prevailed prior to acidification, a continuing appraisal of non-target components of the terrestrial and aquatic environment is in hand. The characteristics of a "pristine" environment, moreover, are difficult to define and in the case of the area of Cairnsmore of Fleet NNR unlikely to have been maintained beyond pre-Bronze Age times. The richest reported assemblage of wildlife in the area corresponds to the scrub/heathland conditions of the fifteenth and sixteenth centuries, subsequently changed by land and wildlife management rather than the influence of industrial pollution.

We believe that this endeavour is unique in its complexity and completeness, and that those engaged in liming projects, and in understanding the basic scientific aspects of acidification, will find this account of interest.

Many have contributed to the following text. Some chapters are contributed by one or two named authors, but in some cases (such as Chapter 14 on aquatic components other than fish, and Chapter 15 on the effects of liming on the terrestrial fauna and flora), the diversity of material is reflected in many sectional contributions. The chapters listed (p. ix) identify by section all those who contributed. Their names and addresses in full are given on p. xviii. Our grateful thanks are given for their readiness to write their part of the story of Loch Fleet, and their patience and forebearance with editorial direction.

The funding and in-house support of the Central Electricity Generating Board (subsequently National Power and PowerGen), the South of Scotland Electricity Board (now ScottishPower), the North of Scotland Hydro-Electric Board (now Scottish Hydro-Electric), and British Coal are gratefully acknowledged. The logistic support of the South of Scotland Electricity Board and the maintenance of the field sampling and analysis programme have been an essential component.

Many other people, who cannot all be listed here, have been involved with the Loch Fleet Project. In particular, we wish to thank our field assistants, first Meikle MacKay and Karen Coxon, later replaced by Donald MacMillan and Jessie MacKay. They exhibited great fortitude, perseverance, and determination in following the programmes set down for them, often in the face of most taxing conditions on site. There are also many contracting scientists who contributed to the scientific programme, some contributors to this account, but many who are not, since we have drawn out only the main thread of the story. However, their reports are included in three reports of the project to which the interested reader may refer: the Preliming Phase (1), the Intervention Phase (2), and the Report to Sponsors (3) at the close of the 5-year period of intensive work. It is expected that much of this work will be reported in due course in the open professional literature.

Finally, we wish to thank Mrs Barbara Leonard who efficiently and enthusiastically undertook the task of typing up the manuscript, along with its inevitable changes, to meet a short time-scale for completion.

Gwyneth Howells
Tom Dalziel

25 October 1990

Contents

List of Contributing Authors

R.J. Aston

PowerGen plc, Ratcliffe Technology Centre, Ratcliffe-on-Soar, Nottingham NG11 0EE, UK

R.W. Battarbee

Department of Geography, University College London, 26 Bedford Way, London WC1H 0AP, UK

R.M. Bell

Environmental Advisory Unit, University of Liverpool, 131 Mount Pleasant, Liverpool L3 5TF, UK

K.H. Bishop

Swedish University of Agriculture Science, Department of Forest Site Research, Svartberget, 5119, 92200 Vindeln, Sweden

D.J.A. Brown

National Power plc, Senator House, 85 Queen Victoria Street, London EC4V 4DP, UK

R.S. Clymo

School of Biological Sciences, Queen Mary/Westfield College University of London, Mile End Road, London E1 4NS, UK

T.R.K. Dalziel

PowerGen plc, Ratcliffe Technology Centre, Ratcliffe-on-Soar, Nottingham NG11 0EE, UK

A. Dickson
National Power plc, Senator House, 85 Queen Victoria Street, London EC4V 4DP, UK

G.N. Foster
Scottish Agricultural College, Auchincruive, Ayr KA6 5HW, UK

G. Howells
Department of Zoology, University of Cambridge, Downing Street, Cambridge CB2 3EJ, UK

G. Hudson
The Macaulay Land Use Research Institute, Craigiebuckler, Aberdeen AB9 2QJ, UK

N.A. Logan
Department of Biological Sciences, Glasgow College of Technology, Glasgow G4 0BA, UK

J. MacKay
c/o Glenlee Power Station, New Galloway, Kirendbrightshire, DG7 3SF, UK

K.J. Murphy
Botany Department, University of Glasgow, Glasgow G12 8QQ, UK

A.F. Nisbet (nèe Leech)
National Radiological Protection Board, Chilton, Didcot, Oxon. OX11 0RQ, UK

T.R. Nisbet
Forestry Commission Research Station, Alice Holt Lodge, Wrecclesham, Farnham, Surrey GU10 4LH, UK

K. Paterson
ScottishPower plc, Technology Centre, 45/47 Hawbank Road, College Milton North, East Kilbride, Glasgow G74 5EG, UK

M.V. Proctor
National Power plc, Technology and Environmental Centre, Kelvin Avenue, Leatherhead, Surrey KT22 7SE, UK

P.J. RAVEN
National Rivers Authority, Rivers House, Waterside Drive, Almondsbury, Bristol BS12 4UD, UK

K. RICHARDS
Department of Geography, University of Cambridge, Downing Place, Cambridge CB2 3EN, UK

J. ROBERTSON
The Macaulay Land Use Research Institute, Craigiebuckler, Aberdeen AB9 2QJ, UK

R. SHORE
Institute of Terrestrial Ecology, Monkswood Laboratory, Abbots Ripton, Huntingdon, Cambs. PE17 2LS, UK

R.A. SKEFFINGTON
National Power plc, Technology & Environmental Centre, Kelvin Avenue, Leatherhead, Surrey KT22 7SE, UK

D.I. SKIDMORE
c/o CERC Research Consultants, Castle Hill, Cambridge CB3 0AX, UK

B.F.L. SMITH
The Macaulay Land Use Research Institute, Craigiebuckler, Aberdeen AB9 2QJ, UK

B.R. STEWART
ScottishPower plc, Technology Centre, 45/47 Hawbank Road, College Milton North, East Kilbride, Glasgow G74 5EG, UK

A.W.H. TURNPENNY
National Power plc, Marine Biology Laboratory, Fawley Power Station, Nr. Southampton SO4 1TW, UK

P. WARFVINGE
University of Lund, Institute of Technology, Department of Chemical Engineering II, PO Box 124, 8-22100 Lund, Sweden

E.J. WILSON
National Power plc, Technology & Environmental Centre, Kelvin Avenue, Leatherhead, Surrey KT22 7SE, UK

R.N. VIZARD
c/o Mr J. Beeley, British Coal Corporation, HQ Scientific Control, Ashby Road, Burton-on-Trent, Staffs. DE15 0QD, UK

USEFUL CONVERSION FACTORS

Major Ions

Chloride	Cl^-	1 mg litre^{-1} × 28·2 = × μeq litre^{-1}
Sulphate	SO_4^{2-}	1 mg litre^{-1} × 20·8 = × μeq litre^{-1}
Nitrate	NO_3^-	1 mg litre^{-1} × 16·1 = × μeq litre^{-1}
Bicarbonate (alkalinity)	HCO_3^-	1 mg litre^{-1} × 16·4 = × μeq litre^{-1}
Sodium	Na^+	1 mg litre^{-1} × 43·5 = × μeq litre^{-1}
Ammonium	NH_4^+	1 mg litre^{-1} × 55·6 = × μeq litre^{-1}
Potassium	K^+	1 mg litre^{-1} × 25·6 = × μeq litre^{-1}
Calcium	Ca^{2+}	1 mg litre^{-1} × 50·0 = × μeq litre^{-1}
Magnesium	Mg^{2+}	1 mg litre^{-1} × 82·0 = × μeq litre^{-1}
Aluminium	Al^{3+}	1 mg litre^{-1} × 111·3 = × μeq litre^{-1}

pH Values

Antilog (6·0 − pH) = μeq H^+ litre^{-1}

pH	μeq H^+ litre^{-1}	pH	μeq H^+ litre^{-1}	pH	μeq H^+ litre^{-1}
3·50	316	4·55	28	5·60	3
3·55	282	4·60	25	5·65	2
3·60	251	4·65	22	5·70	2
3·65	224	4·70	20	5·75	2
3·70	200	4·75	18	5·80	2
3·75	178	4·80	16	5·85	1
3·80	158	4·85	14	5·90	1
3·85	141	4·90	13	5·95	1
3·90	126	4·95	11	6·00	1
3·95	112	5·00	10	6·05	<1
4·00	100	5·05	9	6·10	<1
4·05	89	5·10	8	6·15	<1
4·10	79	5·15	7	6·20	<1
4·15	71	5·20	6	6·25	<1
4·20	63	5·25	6	6·30	<1
4·25	56	5·30	5	6·35	<1
4·30	50	5·35	4	6·40	<1
4·35	45	5·40	4	6·45	<1
4·40	40	5·45	4	6·50	<1
4·45	35	5·50	3	· · ·	<1
4·50	32	5·55	3	7·00	0

Concentrations in Precipitation

Sulphate:

$$1\ \mu\text{mol litre}^{-1} = 2\ \mu\text{eq litre}^{-1} = 0{\cdot}032\ \text{mg S litre}^{-1}$$
$$= 0{\cdot}096\ \text{mg}\ SO_4^{2-}\ \text{litre}^{-1}$$

Nitrate:

$$1\ \mu\text{mol litre}^{-1} = 1\ \mu\text{eq litre}^{-1} = 0{\cdot}014\ \text{mg N litre}^{-1}$$
$$= 0{\cdot}062\ \text{mg}\ NO_3^{-}\ \text{litre}^{-1}$$

Ammonium:

$$1\ \mu\text{mol litre}^{-1} = 1\ \mu\text{eq litre}^{-1} = 0{\cdot}014\ \text{mg N litre}^{-1}$$
$$= 0{\cdot}018\ \text{mg}\ NH_4^{+}\ \text{litre}^{-1}$$

Sea Salt Corrections

Non-sea salt sulphate (SO_4^*) = $[SO_4] - (0{\cdot}0135 \times [Cl])$
Non-sea salt calcium (Ca*) = $[Ca] - (0{\cdot}0375 \times [Cl])$
Non-sea salt magnesium (Mg*) = $[Mg] - (0{\cdot}1989 \times [Cl])$

These corrections are based on the assumption that all chloride (as μeq litre^{-1}) in rain is derived from sea salt.

Calculation of Wet Deposition

$$D_M\ (\text{g M m}^{-2}) = [\text{M}]\ (\text{mg litre}^{-1}) \times \text{rainfall (mm)} \times 10^{-3}$$

where D_M is the deposition of ion M and [M] is its concentration.

$$1\ \text{mm rain} = 1\ \text{kg m}^{-2} = 1\ \text{litre m}^{-2}$$

$$\text{and, mm rain} \times 10^4 = \text{rain per hectare}$$

$$\text{or, } 1\ \text{g m}^{-2} = 10\ \text{kg ha}^{-1}$$

PART I

INTRODUCTION, GENERAL PRINCIPLES, AND OBJECTIVES

CHAPTER 1

Acidification and Liming in Context

G. Howells
University of Cambridge, Cambridge, UK

D.J.A. Brown
National Power plc, London, UK

1.1 INTRODUCTION

The underlying principles of limnology, the fruit of much scientific endeavour in the latter part of the nineteenth century and in the early decades of the twentieth, reach beyond the purely scientific or academic pursuits to the development of management strategies to maintain or improve water quality of acid surface waters and their dependent biological communities. The more recent advances in understanding acidified waters have been applied similarly to practical methods of alleviating their unwelcome effects. This book reviews a number of these techniques and their applications and, through the experience of the individual contributors to the Loch Fleet project, 1984 to 1989, considers the results gained there in the wider context of activities with similar objectives elsewhere.

Concern about the real and potential effects of acid rain, especially within the aquatic ecosystem, has been a major scientific and political issue for nearly three decades. An understanding of the extent and nature of acidifed waters, their progressive development, and the mechanisms of acidification has grown as research and field observations have confirmed or refuted earlier claims. Recent international milestones include:

- 1972: UNEP Stockholm Conference, which argued that the acidity of rain in northwest Europe had increased substantially; that this was matched by observations of an increased acidity of lakes and rivers; and that a parallel decrease was evident in the growth of trees and in fish species or populations in acid lakes and rivers.
- 1977, 1979: The OECD developed emissions/deposition budgets from which "blame tables" and national costs of sulphur pollution could be estimated.
- 1980 to present: The UNECE established a European sampling grid (EMEP) for atmosphere and rain, further developing trajectory models and some predictive capacity, reviewing effects and remedies.
- 1982: UNEP Stockholm Conference analysed the need for and extent of sulphur emission control, and recognised the role of nitrogen emissions.
- Further international programmes (sulphur and nitrogen loading, forest damage) were initiated through national and European agencies.

In view of a forthcoming UNEP conference in 1992, further attention is likely to be given to the effectiveness and costs of sulphur and nitrogen emission control, as well as to alternative remedial measures.

In parallel with the improvement in understanding the extent and degree of acidification in sensitive regions and its effects on soils, forests, and aquatic ecosystems, a number of large-scale field experiments have been initiated, both to demonstrate the course and timescale of acidification and to test its reversibility. While simulation of emission control has received most attention, other possible techniques include chemical treatment of water quality, hydrological management, land-use and management changes, and fisheries management.

In particular, considerable effort has been directed to the treatment of acid waters, or their catchments, to improve water quality for fisheries. For the most part this involves the application of limestone in some form or "dose", either directly to lakes and rivers or to the soils of their catchments. The focus in this book is a liming project at Loch Fleet (south-west Scotland), undertaken during the years 1984 to 1990, but reference is made to other liming or alternative projects in northern Europe and North America, seeking their common elements as well as the contrasts between them.

1.2 FISHERIES AND ACIDIFICATION

Even before the end of the last century, a hatchery for salmon operated in southern Norway encountered difficulties which could not then be explained.

At the time, a disease affecting the salmon ovaries was favoured as the cause of a progressive decline in the fish stocks in many lakes and rivers in Norway, but this did not lead to catastrophic fish kills. By the second decade of the twentieth century, a number of reports in Scandinavia documented the mass death of fish in a number of lakes and rivers, particularly in waters near the Baltic coast, in association with drought conditions. The cause of death was attributed to "vitriol" (sulphuric acid), which was generated in soils and sediments exposed to oxygenating conditions as the water table receded. It was also recognised that lakes with peaty sediments and adjacent peat bogs (i.e. inland and at higher elevation) were also prone to vitriol poisoning of their fish populations. These conditions were thought to be the cause of the loss of fish populations and were described as "not unusual" at that time in the moorlands of southern Norway (Huitfeldt-Kaas, 1922) where, indeed, a number of water bodies were classified as "fishless".

Progress was made when a British chemist, Atkins, developed methods for the field measurement of acidity and alkalinity of natural waters (Atkins, 1922), and a Norwegian fishery expert, Knut Dahl, demonstrated the toxic effect of low pH waters on trout and salmon fry (Dahl, 1921, 1927). This was followed by attempts at Kristiansand (southern Norway) to reduce the acidity of the water in a local hatchery by passing it through a limestone filter (Sunde, 1926). The alkalinity of the water increased, with the pH increasing from 5·2 to 6·4, and as Dahl had demonstrated, the mortality of the fry was eliminated.

In Scotland, contemporary records of spate conditions at Loch Dhu, Caithness, following heavy rains, reduced the normal pH of 6·5 to 5·5 and the poor fishery in lakes of the area was attributed to acid drainage from peat (Berry, 1932). At the same time, the sparse fisheries of lakes and streams in the English Lake District were notable, and one correspondent described the Lake District "as thousands of acres of water almost wasted" for want of suitable management. Indeed, the reputation of the river Duddon was derived "more from the beauty of the valley down which it flows, and its association with Wordsworth's sonnets, than for the fish it contains" (Watson, 1899). This was attributed to local physical and chemical characteristics, including the bare rocky nature of the terrain and the great purity of the water; only in the lower reaches was fishing for trout and salmon good, although salmon migrate to the source to spawn. Improvement of the fishery there was thought in the 1920s, however, to lie with limiting the activities of netsmen in the estuary.

Further understanding resulted from recognition of fisheries damage as a result of acid mine drainage. This had already been seen in drainage from metalliferous mines in west Wales, where fisheries' problems were clearly linked with high levels of potentially toxic trace metals such as lead, zinc, and copper (Jones, 1939), stimulating some of the first studies of heavy metal toxicity in low conductivity waters. At the same time, adverse effects on fisheries also became evident in mining areas in the United States of America, especially in Pennsylvania (Zurbuch, 1984). This led to the development of river "lime-drums", initially rather unsophisticated systems activated by stream flow, to correct the acidity and calcium-poor quality of the stream water.

Half a century later, varied techniques have developed in response to the recognition that acidity alone can be responsible for the loss or lack of freshwater fish, and that acid conditions are exacerbated by trace metal contamination. Some recent measures to mitigate surface water acidity relate to lakes adjacent to the Canadian smelter at Sudbury, once the world's largest emission source of sulphur, as well as a major source of potentially toxic trace metals such as nickel, copper, manganese, and zinc. In this case there is some uncertainty whether the loss of fish populations in lakes near to the primary source was primarily attributable to acidity (and associated high sulphate concentrations) or to the excessive levels of toxic metal species. It is established, however, that the reduction of emissions there (by more than 350%) has improved water quality in less than 10 years, both with regard to acidity and trace metal concentrations (Hutchinson and Havas, 1986), and with biological recovery. Fish, benthic invertebrates, and macrozooplankton communities in some lakes have returned to levels of similar lakes in unpolluted areas (Gunn and Keller, 1990).

Areas distant from industrial emissions in Scandinavia and North America were also known to have been affected. The decline of fisheries (Atlantic salmon) in Nova Scotia (Canada) was recognised in the 1950s, and was generally attributed to the progressive acidification of 10 southern rivers (Watt *et al.*, 1983), although those in the adjacent states of Maine and Connecticut (USA) and in northern rivers of Nova Scotia were not affected. This led to the use of "lime weirs" in some rivers to protect emerging salmon fry (Watt *et al.*, 1984) and to maintain smolt production.

Liming of agricultural lands, especially in Britain, was practised by the Romans 2000 years ago and was described in some detail at the time by Pliny; it was probably a common practice even in pre-Roman agrarian cultures, evidently to encourage the vigour and production of favoured crops. But measures to "lime" soft surface waters, some of them acid, no doubt awaited the ability to analyse natural waters, and particularly to characterise their acidity (as "pH", defined by a Danish chemist, Sorenson, in 1909), and the commercial development of colour reagents that could be used in the field in a practical and quick way. The concept of "acidity" is attributed to Arrhenius and its effects on solution chemistry had developed in the nineteenth century and provided a sound chemical basis for these practical developments.

Taking account of this history, it is evident that interest has often stemmed from concern for fisheries, both commercial and recreational. The acknowledged causes of perceived decline are various – they include climatic variations, drainage from peat and boggy areas, changes in hydrological regime, widespread infectious agents, and drainage from mining operations. In recent decades the effects of acid deposition have been given prime importance.

1.3 REVERSING ACIDIFICATION

It is often argued that the only "ethical" way to remedy acidification caused by acid deposition is to reduce sulphur emissions at source. This position, while unarguable in a strict sense, ignores the acidifying role of natural rain (about pH 5), and of the progressive consequences in the catchment soils of long-term growth of vegetation such as forests, and runoff from naturally acidic *Sphagnum* wetlands. Reduction of sulphate deposition in rain must also be ineffective as a means to restore the depleted levels of calcium and alkalinity in drainage waters without which the lost ecosystem components – such as fish, invertebrates, or epiphytic diatoms, all organisms dependent on sufficient calcium – cannot be re-established.

The difficulties of an uncompromising strategy of sulphur emission control are evident in current observations that rain appears to remain as acid notwithstanding the substantial decreases in sulphur emissions (about 25% for western Europe and about 40% for the UK; Barrett *et al.*, 1987) since peak values in about 1965. In spite of emission control over the past 20 years, rivers in southern Sweden and Norway are still as acid as in the 1960s and 1970s, and trout populations are still being lost. Further reductions, even targeted to $>70\%$ (Fig. 1.1), cannot ensure that all lakes in southern Norway, for example, will contain fish (Henriksen *et al.*, 1987). Further direct evidence comes from deposition exclusion in the field, notably the RAIN project in Norway, where five years' application of "cleaned" rain has resulted in only a limited reduction in the drainage water sulphate, possibly because some dry deposition component still

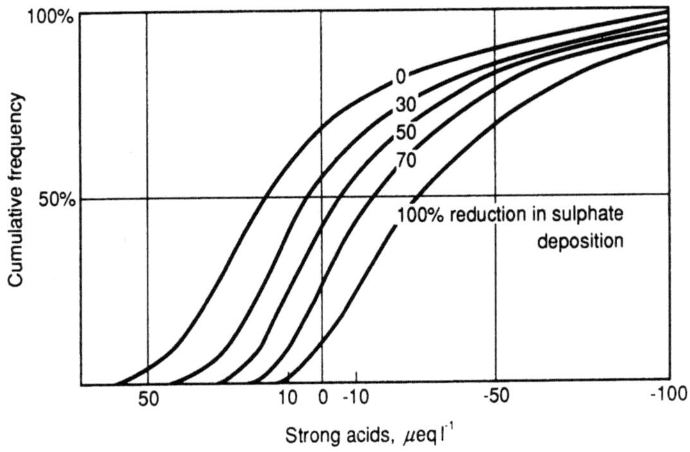

FIG. 1.1 Acidity predicted in southern Norway lakes with different levels of reduction of sulphate deposition (after Henriksen *et al.*, 1987).

contributes, as well as that from the soil reserve (Wright *et al.*, 1988). The resulting runoff waters are still inadequate to support trout (Rosseland *et al.*, 1990).

One possible approach to the mitigation of acidity has been to promote a programme of liming of surface waters. Most interventions of this kind were directed initially to reducing the acidity of lakes by limestone applications direct to surface waters or their shores. It has also become evident that acid deposition over the long term, in addition to vegetative accumulation, has led to the accumulation of acidity in the catchment soils, reducing their base reserve. As the role of soil processes and hydrological influences on stream chemistry was appreciated, more attention has been given to liming applications to catchments or to wetland source areas. In the conditions prevailing in Britain, where acid and organic soils overlie granitic or other base-poor materials, and heavy rainfall and topography result in rapid hydrological turnover, it will be more effective to improve water quality by using limestone or other possible treatments to the catchment rather than direct to lakes and streams. Furthermore, in Britain, the resource most at risk is the migratory salmonid fishery, dependent on the water quality in upland streams which provide nursery areas for spawning and for juvenile growth.

With the growth of understanding, given that reduction of acidifying emissions will be difficult and costly, especially on a regional or hemispheric scale, other options for improving or restoring acid waters appear more realistic. Projects, both scientific and practical, for testing these alternatives and their effective time-scales have been undertaken in western Europe and Scandinavia, and in North America. This book will review some of these projects, but is based on the experience gained by the intensive investigation of restoration techniques at Loch Fleet, south-west Scotland, undertaken in the years from 1984 to 1990.

Considerable debate (see Chapter 2) attaches to the question whether acidification can be reversed, over what time scale, and to what degree historic or pristine conditions will be achieved. It is claimed that the chemistry of drainage waters following liming can never be "pristine" since calcium concentrations are inevitably somewhat higher than before liming, and possibly higher than those characteristic of oligotrophic upland waters. However, the evolution of these waters has presumably been one reflecting a slow but progressive leaching of base materials from their catchment soils; lime additions will initially reverse that condition, but a similar progressive depletion will inevitably follow. Thus, there is no stable or "pristine" quality finite for these upland waters. Moreover, concentrations of calcium in Loch Fleet 3 years after lime treatment are about 150 μeq litre^{-1}, only three times greater than that prior to liming, possibly similar to conditions about 100 years ago (Cook *et al.*, 1990), and characteristic of many oligotrophic waters now.

Predictive modelling of acidification on a catchment scale (MAGIC) has also demonstrated that a substantial reduction (of at least 60%) of acid emissions is needed even to halt the natural and progressive decline in drainage water alkalinity, and that any change will take many decades, perhaps even a century (Fig. 1.2) (Whitehead, 1989). Changes in land use (e.g. coniferous afforestation or agricultural management) may thus be seen to be more influential in this time scale than a change in deposition acidity.

1.4 LIMING ACTIVITIES WORLD WIDE

In several European countries liming of acidified waters has been used to counter adverse water quality restricting salmonid survival. These are summarised briefly below providing a context in which the Loch Fleet experience can be assessed. The examples given (Table 1.1) will be found in more detail in the following chapters which provide a wider spectrum of conditions and experience on lake and stream restoration.

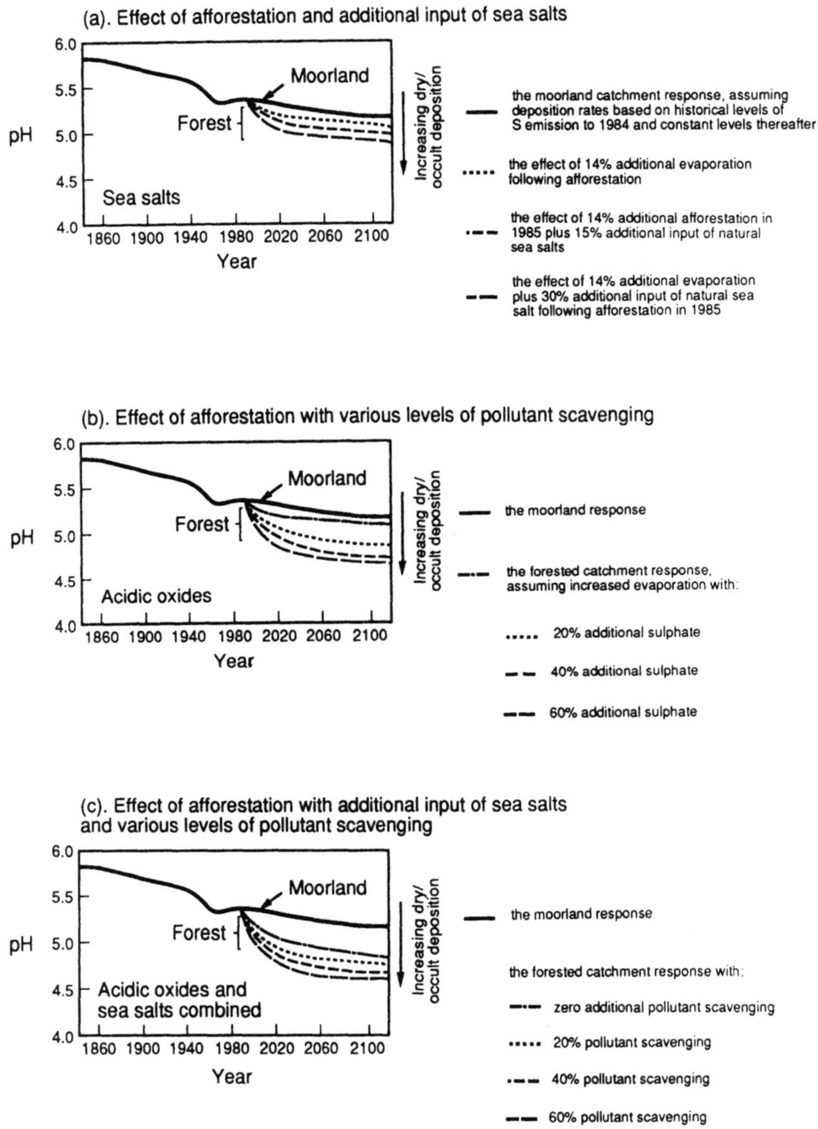

FIG. 1.2 Predicted pH for stream runoff at Loch Dee (Galloway, Scotland) with different pollutant and marine salt inputs, and alternative land-use (after Whitehead, 1989).

1.4.1 Sweden

Starting in 1977, liming of acid waters for a trial period was undertaken in Sweden with government financial support. On the basis of successful applications, a national operational liming programme was initiated in 1982 (Nyberg and Thornelof, 1988). This programme continues, with a government subsidy (about 110 m SEK*, in 1988) for all waters of pH less than 6·0 or alkalinity less than 50 μeq litre^{-1}, in total comprising about 5000 lakes. It is estimated that about one-quarter of Sweden's lakes and streams are affected or threatened by acidification and liming of these lakes would cost about 200 m SEK per year; to lime accessible streams would cost about 400 m SEK per year. Several thousand kilometres of streams and rivers in Sweden have been limed in the past decade with sufficient improvement of water quality to restore salmonid and other fisheries.

In all waters in Sweden where liming has resulted in a sufficient and durable pH and alkalinity increase, fish started to reproduce again and population age structure has returned to the normal pattern. Eight major rivers in Sweden have been treated successfully. In the river Atran (Edman *et al.*, 1988) and its tributary river, Hogvadsan, Atlantic salmon smolt migrations and adult spawning runs decreased during the 1970s. In this system, lake liming and river dosing silos were found to be the most effective measures, and further liming of the river Hogvadsan was provided up to 1986. The pH and alkalinity values in the river waters have reached target values, although not always maintained in high flow conditions, so that the pH may fall transiently to around 5·0.

The improved water quality resulted in increased survival of salmon fry and parr, with a 15-fold increase in abundance in 1984/86 compared with that in 1978. The spawning migration of salmon has also increased, with higher numbers in 1986 than in previous years (Appelberg *et al.*, 1989).

Liming of the acid Lake Gårdsjön (pH = 4·7) was undertaken in April 1982 by application of limestone powder direct to the lake surface. Fish species in the lake were historically roach, tench, and Crucian carp (introduced between 1910 and 1938) and these species spawned in the lake until 1949. However, all fish species were lost by 1973 (Hultberg, 1985). Acidification was also associated with changes in benthic invertebrate communities, benthic algae, and macrophyte flora (Hultberg and Grähn, 1975). Lake turnover is about 1·2 years, and limestone addition in these conditions was calculated to last for a period of three or more years,

*SEK = Swedish kronor.

TABLE 1.1
Some examples of lake, stream, and catchment liming

Site/location	*Conditions*	*Liming objectives*	*Applications* (*date, method, and dose*)	*Reference*
Loch Fleet, UK	Upland catchment, small lake (0·17 km^2) with turnover ~6 months. Rain 2215 mm yr^{-1}, pH 4·8. Fish absent before liming.	1. To provide water quality suitable for brown trout. 2. To re-establish the trout fishery.	1986, 350 t to catchment at 10–30 t ha^{-1}; 1987, 100 t to catchment, 5–7·5 t ha^{-1}.	Howells (1986); Howells and Dalziel (1988, 1989); Howells *et al.* (1992 and this volume)
Loch Dee, UK	Large catchment (> 150 km^2) and shallow lake (1 km^2), turnover 40 days. Rain 2232 mm yr^{-1}, pH 4·8. Brown trout present before and after liming.	1. To assess effects of acid rain and afforestation.	1983, 75 t to sub-catchment and streamside.	Burns *et al.* (1984)
		2. To improve potential of trout fishery, and develop management.	1985, 110 t to improve to lower sub-catchment	Welsh and Burns (1987)
R. Esk, UK	River (26·8 km) draining volcanic and granitic rocks, soils podsols, and brown earths. Land used for sheep grazing. Migratory salmon and sea trout and resident brown trout.	To restore water quality to conditions pertaining prior to withdrawal of liming practices in 1976.	Powdered limestone 3200 t to 640 ha (5 t ha^{-1}) on stream-side fields, July 1987 and July 1988.	Diamond *et al.* (1992)
Llyn Brianne, UK	Upland catchments (n = 17, ~21 km^2), moorland and forested. Variety of sites, areas, treatments. Rain 1800 mm yr^{-1}, pH 4·7. Stream studies. Brown trout present, but few in forest streams.	To test alternative strategies (lime, land improvement, tree acidification).	1986, 1987, 1988 at 9–30 t ha^{-1}, areas up to 34 ha treated.	Welsh Water Authority (1987, 1988)
Gårdsjön, Sweden	Lowland, spruce forested catchment (2·1 km^2), lake 0·31 km^2, turnover 1·4 yr. Rain 1100 mm yr^{-1}, pH 4·7. Mixed fish species lost by 1973. Lake pH raised from 4·5 to 7·5.	To improve WQ (to pH 7 & Ca 250 μeq $litre^{-1}$) over 3+ yr.	1982, liming (slurry) lake surface, 110 t on 31 ha.	Dickson (1988)
R. Atran, Sweden	Mixed coniferous forest in western Sweden on gneissic geology with mires. Rain 1100 mm yr^{-1}, pH 4·3–4·5.	To restore Atlantic salmon runs.	1978–81, > 10 000 t applied to lakes (55%), wetlands (17%), catchment (13%), and to river (15%). 1986, 11 500 t to tributary river.	Edman *et al.* (1988)

L. Lysevatten, L. Bredvatten Sweden	Lakes in south-west Sweden, underlying bedrock of igneous and metamorphic rocks, including granites. Soils shallow. Rain 1100 mm yr^{-1}, pH 4·7 Lysevatten (0·41 km^2) max. depth 18 m, retention 2·5 yr. Bredvatten is smaller and shallower, retention 2 yr.	Mitigation of acidic conditions for restoration or protection of fisheries and biota.	Both lakes limed 1974 with Ca-silicate slag (52% CaO, 11.5% MgO), dose 60 and 40 t, respectively, directly to lakes. Later treatment employed ground agricultural limestone (76 and 51 t) in a slurry, and at Lysevatten dose was later increased (98 t) to counter acidic runoff from catchment.	Hultberg and Andersson (1982); Alenas *et al.* (1991)
St. Hovvatn, Norway	Montane lake (1·1 km^2), shallow soils in catchment. Lake turnover 0·8 yr. Rain 1500 mm yr^{-1}, pH 4·2. Lake pH raised from 4·5 to 5·7. Trout restored after liming.	To demonstrate lime treatment can be effective measure for deacidification.	1981, 200 t to ice cover and along shore; 1987, 91 t.	Rosseland and Hindar (1988); Wright (1985)
Tjonnstrond, Norway	Catchment (0·25 km^2), two lakes connected by a stream, turnover ~2 months. Rain pH 4·2. Brown trout populations absent, restocking unsuccessful prior to liming. Trout restored after liming, but did not reproduce or migrate. Lake pH changed from 4·5 to a maximum of pH 7·1; pH > 5·8 4 months after liming.	1. To document chemical and biological responses. 2. To improve liming strategies, for most cost efficient activities.	1983, 75 t (3 t ha^{-1}) to whole catchment, estimated sufficient for 30 yr.	Rosseland and Hindar (1988); Baalsrud (1985/87)
R. Audna, Norway	Catchment 450 km^2, flow 22 $m^3 s^{-1}$. Rain pH 4·2. Atlantic salmon lost in early 1970s, but sea trout remained plus other fish species.	As Tjonnstrond.	1985, 3320 t as limestone slurry to a large lake upstream and also, 330 t *via* limestone doser at lake inlet, and at river mouth.	Rosseland and Hindar (1988)
Woods Lake, USA	Small headwater lake (0·23 km^2) in forested (broadleaved) area (2·07 ha), turnover 6 months. Rainfall 1268 mm yr^{-1}, pH > 4·5. Lake pH raised from 5·0 to 9·0 in the short term (lake liming).	To follow chemical and biological changes in water and soils post-liming. To re-establish a brook trout population.	1985–86, limestone slurry direct to lake, 23 and 35 t; 1989, catchment liming, 1000 t limestone pellets at 10 t ha^{-1} on two sub-catchments.	Olem (1990); Porcella (1989; pers comm, 1990)
Bowland Lake, Canada	Acid-sensitive lake (1·1 km^2) in forest area (108 ha), turnover 2·0 yr. Lake pH > 4·76, increased to pH 5·8. Yellow perch (*Perca flavescens*) present, lake trout lost.	To document chemistry and biological responses.	1983 and 1984 direct limestone addition by aircraft.	Dodge *et al.* (1988)
Nova Scotia rivers	Rain 1200 mm yr^{-1}, pH 4·6. Granite and quartzite geology, peaty terrain, humic acid rivers, fast flowing.	To restore Atlantic salmon production in rivers.	Upstream lake liming. Lime dosers in spawning reaches. Limed substrates, 180 t crushed lime in 300 m reach of stream.	Watt *et al.* (1984); Watt (1986); White *et al.* (1984); Lacroix (1992)

providing a calcium concentration of 250 μeq litre^{-1}, a pH of 7·0, and an alkalinity of 200 μeq litre^{-1}. Treatment with 110 t of limestone in 1982 resulted in most of the limestone (74%) dissolving in the first year, 18% in the second year, and 7% in the third, but acid surges in April 1985 and high discharges in the early period after liming led to an unexpected reacidification within about 5 years. Changes in the phytoplankton, zooplankton, and macrophyte communities are reported (Dickson, 1988); recovery of fish populations has not yet been reported.

1.4.2 Norway

Research and operational liming schemes in Norway have been supported by government funding since 1983. Liming all acid waters in Norway is estimated to require 300 m NOK* per year (Hindar and Rosseland, 1988); in 1988, government funds provided 14 m NOK for research and liming, increasing to 20 m NOK in 1989. Two salmon rivers are included in the national programme – the rivers Audna and Storelva – while another – the river Vikedal – is limed with local funding. Lake Hovvatn (littoral and lake liming), and Lake Tjonnstrand (catchment liming) are further documented examples.

In the river Audna the native salmon stock was lost early in the 1970s (Leivestad *et al.*, 1976) and by the 1980s only a few sea trout were spawning in this river. In autumn 1985, a large lake upstream of the main river was limed and two stream-side dosers were installed on the main inlet stream to the lake and in the salmon stretch of the main river. Water quality response has been favourable (Johannessen and Skogheim, 1987; Rosseland and Hindar, 1988) and stocking experiments with juvenile salmon have been promising (Matzow *et al.*, 1985).

River water quality has been restored upstream (Baalsrud, 1985/87) principally by a lime doser restoring spawning conditions which led to a satisfactory hatch of brown trout. The effects of liming on downstream reaches accessible to Atlantic salmon are not yet reported.

The river Vikedal had a record of episodic mortality of salmon smolts, especially during the spring snow melt (Henriksen *et al.*, 1984). Since 1987, the river has been limed during the spring melt period (March–May) using a stream-side doser upstream. Water quality during the treatment periods is now satisfactory (Hindar and Lien, 1988) but numbers of salmon fry and parr are still poor.

The shores around Lake Hovvatn were limed (200 t) in 1981 and again

*NOK = Norwegian kronor.

in 1987. Acidity was reduced (pH > 6·7 in the first year) and calcium level raised (to greater than 100 μeq litre^{-1}) but acid events (pH < 5) associated with snow-melt continued. About 50% of the limestone was dissolved over three and a half years but reacidification is slow (pH 4·9 in 1984) presumably because of further dissolution of the littoral limestone (Wright, 1984).

The whole catchment of Lake Tjonnstrand was limed (3 t ha^{-1}) in 1983; acidity was reduced and calcium levels raised as limestone dissolved in the runoff water. Although some deterioration in pH and calcium concentration was evident during storm rainfalls, reasonable conditions have been maintained over more than four years. Brook trout and brown trout were stocked at the end of 1981, and successful survival and growth are reported.

1.4.3 North America

Within the United States, until recently most liming efforts were directed towards neutralisation of acid mine drainage waters. However, many of the techniques used have been adapted or modified for use in mitigating oligotrophic acid waters. Approximately 150 surface waters in the USA have now been limed with the emphasis again being largely on research and, as in early work in Sweden, directed towards direct lime additions to lakes (Brocksen and Emler, 1988; Schreiber, 1988).

Studies have been reported on 10 lakes in the Adirondacks (NY) treated with limestone additions direct to the waters (Schofield *et al.*, 1986). At Woods Lake (pH ~ 4·5) direct lake liming led to short-term improvements but did not lead to sustained water quality or biological improvement, and catchment liming has now been undertaken.

Canada's liming programme is also fairly restricted, with an experimental focus (Dodge *et al.*, 1988). Lake liming projects have included six lakes near to the Sudbury smelter, Ontario, but high residues of heavy metals in these locally polluted lakes are a compounding factor; the lake responses are not uniform. Salmon rivers in Nova Scotia have been limed (by lime addition to headwater lakes) with short-term benefits not maintained because of high flushing rates of the lakes and heavy winter rainfall (Watt *et al.*, 1984).

1.4.4 Other United Kingdom Projects

In England and Wales, several liming projects of a research or pilot nature have been undertaken to improve conditions in acid waters; they are mostly lime applications to lakes or their tributaries, or to their catchments.

In Cumbria, lands adjacent to the river Esk (10% of the catchment) were limed in 1986/87 with 3200 t of powdered limestone. Surprisingly, a decrease in the calcium concentration of the river water downstream was seen, although the pH was somewhat higher than before (Diamond *et al.*, 1990), possibly because of spate flows during sampling. An increase in the run of spawning salmonids in the adjacent R. Esk and even in streams above the limed areas is reported, and it is difficult to attribute changes to liming. However, no repetition has occurred of the kill of spawning fish seen in the lower river spate conditions in spring 1981 and 1984.

In several small catchments at Llyn Brianne on the Tywi in west Wales, restricted riparian liming within an afforested area (at $30\,t\,ha^{-1}$) was of limited value in improving water quality (Edwards *et al.*, 1990). In another area, upland "improvement" (draining, liming, ploughing) improved grazing productivity but had little effect on the chemistry of the stream draining the treated area. In another treatment, however, application of limestone to a wetland source area (at $10\,t\,ha^{-1}$) led to a rapid rise in the pH of stream water, although this improved water quality has not since been maintained (Edwards *et al.*, 1990).

Elsewhere in west Wales a number of stocked lakes have been treated with direct application of limestone; this has improved water quality, and the value of the "put-and-take" fisheries of brown trout there, although on time-scales determined by time of water replacement (Underwood *et al.*, 1987).

In addition to the Loch Fleet project, another scheme in this acid-sensitive area of Galloway, south-west Scotland, has been undertaken at Loch Dee, where liming of the riparian zone of the lower White Laggan tributary and direct to that stream has led to the apparent "loss" of most of the applied limestone to the lake, mostly to the sediments discharged by the White Laggan (Burns *et al.*, 1984; Welsh and Burns, 1987). In this lake, average turnover is only 40 days, so that the loss of calcium carbonate in solution in the outflow is also a significant factor. In the event, fishery (brown trout) production, thought to be declining in 1980, seems to have recovered in this not very acid lake (pH = 5·6), possibly by improved fishery management.

1.5 THE LOCH FLEET LIMING PROJECT

The Loch Fleet project was set up in 1984 for an initial period of 5 years, with the joint commitment and participation of the Central Electricity Generating Board (now National Power and PowerGen), the South of Scotland Electricity Board (now ScottishPower), the North of Scotland Hydro-Electric Board (now Scottish Hydro-Electric), and the National Coal Board (now British Coal). Its aim was to investigate and correct the acidity of Loch Fleet, a small acid lake in the uplands of south-west Scotland where the lake water had become progressively more acid over the previous two decades, and where the fishery for brown trout (*Salmo trutta*) had been lost. The objectives specified for the project were:

- to demonstrate that the water chemistry of a lake can be brought into a range suitable for trout by one or more of several treatments to the catchment or waters, including the addition of basic minerals to soils and/or waters, manipulation of hydraulic contact with natural or added minerals, manipulation of the ion-exchange system, e.g. by burning, and
- once suitable chemical conditions have been achieved in portions of the lake, to demonstrate the suitability of the water for a self-sustaining trout population.

As a result of this work, it was expected that practical land, lake, or stream management procedures could be developed, and would be applicable elsewhere.

The project work was phased: first, to collect and collate existing information while necessary field installations were put in place; secondly, to apply sectoral treatments to the catchment and to document changes in water quality; and then to re-establish a trout population and to monitor its survival or success as a self-sustaining community.

Many of the components needed for the investigation were commissioned with scientists in universities, colleges, and research institutes; their results have been freely available as informal reports and in the professional literature.

Within the time and resources available to the Loch Fleet project, priority was given to developing effective treatments for sub-catchments (sectors) of the whole catchment and to follow the detailed time-course of changes in water quality. It was recognised that a target water quality for re-establishing a trout population should be specified, and when trout were reintroduced that their fate should be followed. The treatments applied were designed to cover a range of "doses" and formulations, bearing in mind the nature of the soils and the acidity of deposition, following the resulting yield of calcium and alkalinity. Attention has also been given to the wider consequences, in both the short and the long term, on non-target species in the aquatic and terrestrial ecosystems.

In the first two years the project management completed the design, installation, and testing of field equipment, and established the pre-intervention conditions of the aquatic ecosystem and of the surrounding terrestrial environment. Data gathered included atmospheric deposition, characteristics of the soils and vegetation at the site, hydrology and water quality in the loch and inflow/outflow streams, and the variety and abundance of the aquatic fauna and flora. Information on these topics, both before the intervention programme and in the years following, are summarised and the findings assessed in the following chapters of this book. Emphasis is given to methods for the application of limestone to the lake catchment, although some other procedures were also tested.

As a result of preliminary laboratory and small-scale field studies, applications of fine particulate calcium carbonate powder in a variety of modes and doses were made in the spring of 1986 and 1987. These treatments, and others tested, are set out in some detail in Chapter 10.

It was evident that the lake water prior to treatment was unable to support trout (or salmon). With limestone addition, fish survival was demonstrated initially in laboratory and tank trials, then in the improved water quality following lime applications in the tributary flows, and in due course in the lake as a whole (see Chapter 13). Suitable conditions for trout have been maintained over the subsequent period of observation (more than 5 years). Estimates from the continuing water quality observations, and chemical budget calculations, suggest that satisfactory conditions will be maintained for a further period of 5 to 15 years without further lime applications (see Chapter 16).

Fish were the focus of the biological studies undertaken. It is recognised, however, that other important components of the upland environment will be affected by the limestone treatments. Fishery success is in part dependent on meeting the nutritional needs of fish – to this end, primary and secondary production in the aquatic ecosystem has been monitored, as well as documentation of the relative importance of aquatic and terrestrial fauna to the trout diet (see Chapter 13). In addition, some effects of liming on the natural terrestrial fauna and flora, possibly unwelcome, are acknowledged, and ancillary studies have been undertaken by independent scientists to establish the nature and scale of any such effects (see Chapters 14 and 15).

It is evident that the liming treatments applied at Loch Fleet have been successful in meeting the declared objectives. It is also considered reasonable to expect that the findings could be applied to other sites, depending on their characteristic hydrology, soil conditions, and vegetation. Work is in hand to explore these wider perspectives, to identify critical components, and to develop guidelines for similar treatments in a wider range of conditions.

This book will highlight the major findings of the Loch Fleet project, and by comparison with other projects will try to identify the components leading to success or failure in this approach to mitigating surface water acidification.

1.6 SUMMARY

Thus, the greater part of liming experience available worldwide is with direct lake application, and has been primarily concerned with fishery restoration or protection rather than biological restoration as a whole. Generally, the effects that have been observed following liming have been positive. If, however, the retention time of the system is short (less than 4 months) and/or the restoration (as opposed to the protection) of a side-stream spawning fish species is involved, it is becoming recognised that catchment application of the limestone is the only feasible route for sustained improvements and is a necessary adjunct to any programme of emission control of acid-generating gases.

While this book gives an opportunity to relate the experience gained in catchment liming at Loch Fleet, it will also set this into a wider context of other liming projects elsewhere, including direct liming of lakes and streams, and will also refer to alternative hydrological and land-use management techniques.

REFERENCES

Alenas, I., Andersson, I.B., Hultberg, H. and Romarin, A. (1990). Liming and reacidification reactions of a forest lake ecosystem, Lake Lysevatten, in SW Sweden, *Water Air Soil Poll.*, **59**: 55–77.

Appelberg, M., Degerman, E., Karlsson, K. and Johlander, A. (1989). Liming increases the catches of Atlantic salmon of the west coast of Sweden. *Nordic J. Freshw. Res.*, No 65, 44–53.

Atkins, W.R. (1922). Measurements of acidity and alkalinity of natural waters and their biological relationships. *Salmon & Trout Mag.* (as cited in Dahl, 1927).

Baalsrud, K. (editor) (1985/87). *Liming of Acid Water: Final Report*. The Norwegian Liming Project. Dept of Environment, Trondheim, Norway, 177 pp.

Barrett, C.F., Atkins, D.H.F., Cape, J.N., Crabtree, J., Davies, T.D., Derwent, R.G., Fisher, B.E.A., Fowler, D., Kallend, A.S., Martin, A., Scriven, R.A. and Irwin, J.G. (1987). *Acid Deposition in the United Kingdom 1981–1985*. Dept Environment UK, WSL, Stevenage, 124 pp.

Berry, J. (1932). The trapping of spawning trout: an account of some recent experiments in the Altnabreac district of Caithness. *Salmon & Trout Mag.* **67**: 119–24.

Brocksen, R.W. and Emler, P.W. Jr (1988). Living Lakes: an aquatic liming and fish restoration demonstration program. *Water Air Soil Poll.* **41**: 85–94.

Burns, J.C., Coy, J.S., Tervet, D.J., Harriman, R., Morrison, B.R.S. and Quine, C.P. (1984). The Loch Dee Project: a study of the ecological effects of acid precipitation and forest management on an upland catchment in southwest Scotland. *Fish Manag.* **15**: 145–67.

Cook, J.M., Edmunds, W.M. and Robins, N.S. (1991). Groundwater contribution to an acid upland lake (Loch Fleet, Scotland) and the possibilities for amelioration. *J. Hydrol.*, **125**: 111–128.

Dahl, K. (1921). Investigations into the extinction of trout in the mountain lakes of south-west Norway. *NJ & FF Tidsskrift* **50**: 249–67.

Dahl, K. (1927). The effects of acid water on trout fry. *Salmon & Trout Mag.* **46**: 35–43.

Diamond, M., Hirst, D., Winder, L., Crawshaw, D.H. and Prigg, R.F. (1992). The effect of liming agricultural land on the chemistry and biology of the river Esk, NW England. Internat. Conference on Acidic Deposition – its Nature and Impacts, Glasgow, UK. *Env. Poll.*, in press.

Dickson, W. (editor) (1988). *Liming of Lake Gårdsjön, An Acidified Lake in SW Sweden*, Rep. No. 3426. Nat. Swedish Environmental Protection Board, 327 pp.

Dodge, D.P., Booth, G.M., Richman, L.A., Keller, W. and Tomassini, F.D. (1988). An overview of lake neutralization experiments in Ontario. *Water Air Soil Poll.* **41**: 75–84.

Edman, G., Fleischer, S., Fritz, O. and Stibe, L. (1988). Hövadsan 1978–86, Förorad-Kalkad-Panyttfodd. Lansstyrelsen: Hallands län, Medd. 1988: **4**, 90 pp. (In Swedish, summary in English.)

Edwards, R.W., Stoner, J.H. and Gee, A. (1990). *Acid Waters in Wales*. Kluwer, Dordrecht, Netherlands, 337 pp.

Gunn, J.M. and Keller, W. (1990). Biological recovery of an acid lake after reductions in industrial emissions of sulphur. *Nature* **345**: 431–33.

Henriksen, A., Lien, L., Traaen, T.S. and Sevaldrud, I. (1987). *1000 Lake Survey, 1986*. Norwegian State Pollution Control Authority, and Norwegian Institute for Water Research, 33 pp.

Henriksen, A., Skogheim, O.K. and Rosseland, B.O. (1984). Episodic changes in pH and aluminium speciation kill fish in a Norwegian salmon river. Vatten **40**: 225–60.

Hindar, A. and Lien, L. (1988). Overvaking ar Vikedalseleva i forbindelse med kalking 1987. Kalkingsrapp. 1/88 NIVA 0-870-87, 33 pp. (in Norwegian).

Hindar, A. and Rosseland, B.O. (1988). Liming acidic waters in Norway: national policy and research and development. *Water Air Soil Poll.* **41**: 17–24.

Howells, G. (1986). *The Loch Fleet Project: A Report of the Pre-intervention Phase* (1) 1984–1986. CEGB, SSEB & NSHEB, British Coal, 74 pp (+ annexes).

Howells, G. and Dalziel, T.R.K. (1988). *The Loch Fleet Project: A Report of the Intervention Phase* (2) 1986–1987. CEGB, SSEB & NSHEB, British Coal, 102 pp.

Howells, G. and Dalziel, T.R.K. (1989). *The Loch Fleet Project: A Report to the Sponsors*, 1984–1989. CEGB, SSEB & NSHEB, British Coal, 64 pp.

Howells, G., Dalziel, T.R.K. and Turnpenny, A.W.H. (1992). Loch Fleet: liming to restore a brown trout fishery. Internat. Conference on Acid Deposition: Its Nature and Impacts, Glasgow, UK. *Env. Poll.*, in press.

Huitfeldt-Kaas, H. (1922). On the cause of mass death of salmon and trout in Frafjord, Helle and Dirdal rivers, Ry County (Norway), autumn, 1920. *NJ & FF Tidsskrift* **51**: 37–44.

Hultberg, H. (1985). Changes in fish populations and water chemistry in Lake Gardsjon and neighbouring lakes during the last century. *Ecol. Bull.* **37**: 64–72.

Hultberg, H. and Andersson, I.A. (1982). Liming acidified lakes: induced long term changes. *Water Air Soil Poll.* **18**, 311–21.

Hultberg, H. and Grähn, O. (1975). Effects of acid precipitation on macrophytes in oligotrophic Swedish lakes, pp. 208–17 in *Proc. First Symp. on Chemistry of Acid Waters*. Great Lakes Research.

Hutchinson, T.C. and Havas, M. (1986). Recovery of previously acidified lakes near Coniston, Canada, following reduction of atmospheric sulphur and metal emissions. *Wat. Air Soil Poll.* **28**: 319–33.

Johannessen, M. and Skogheim, O.K. (1987). Hydrological effects of lime treatment, pp. 57–78 in *Liming of Acid Water: Final Report* (ed. K. Baalsrud) (1985). Department Environment and Directorate for Nature Management, Trondheim, Norway, 147 pp.

Jones, J.E. (1939). The relation between the electrolytic solution pressures of the metals and their toxicity to the stickleback (*Gasterosteus aculeatus* L.). *J. Exp. Biol.* **16**: 425–37.

Lacroix, G. (1992). Mitigation of low pH by substrate manipulation in a stream and its effects on salmonids. International Conference on Acidic Deposition: Its Nature and Impacts, Glasgow, 1990. *Env. Poll.*, in press.

Leivestad, H., Hendrey, G., Muniz, I.P. and Snekvik, E. (1976). Effects of acid precipitation on freshwater organisms, pp. 87–111 in *Impact of Acid Precipitation on Forest & Freshwater Ecosystems* (ed. F. Brakke). SNSF Project IR 76/6, 111 pp.

Motzau, D., Rosseland, B.O. and Skogheim, O.K. (1985). Effects of lime treatment on fish. In *Liming of Acid Water*, ed. K. Baaldsrud, Dept. of Env. (Norway), Dir. Nature Management (Norway), (English version, 1987) pp. 110–30.

Nyberg, P. and Thornelof, E. (1988). Operational liming of surface waters in Sweden. *Water Air Soil Poll.* **41**: 3–16.

Olem, H. (1990) *Liming Acidic Surface Waters*. Lewis Publishers Inc., Chelsea Michigan, USA, 331 pp.

Porcella, D.B. (1989). Mitigation of acidic conditions in lakes: an overview of an ecosystem perturbation experiment. *Can. J. Fish Aquat. Sci.* **46**: 246–48.

Rosseland, B.O. and Hindar, A. (1988). Liming of lakes, rivers and catchments in Norway. *Water Air Soil Poll.* **41**, 165–88.

Rosseland, B.O., Lien, L., Krogland, F., Sadler, K. and Dalziel, T.R.K. (1990). *Strains of Brown Trout (S. trutta L.) Stocking and Test Fishing 1988 and 1989;*

Field and Laboratory Toxicity Experiments. NIVA Report No. 0-87178, 26 pp.
Schofield, C.L., Gloss, S.P. and Josephson, D. (1986). *Extensive Evaluation of Lake Liming, Restocking Strategies, and Fish Population Response in Acidic Lakes Following Neutralization by Liming*. EELUT, Interim Progress Report, NEC - 86/18. USFWS, 117 pp.
Schreiber, R.K. (1988). Cooperative Federal-State liming research on surface waters impacted by acid deposition. *Water Air Soil Poll.* **41**: 53–74.
Sorenson, S.L.P. (1909). Uber die Messung und die Bedeutung der Wasserstoff konzentration bei enzymatischen Prozessen. *Biochem. Zeits.* **21**: 131–39.
Sunde, S.E. (1926). Acid water kills off trout and trout fry. *NJ & FF Tidsskrift* **2**: 1–4.
Underwood, J., Donald, A.P. and Stoner, J.H. (1987). Investigations into the use of limestone to combat acidification in two lakes. *J. Environ. Manag.* **24**: 29–40.
Watson, J. (1899). *English Lake Fisheries*. Laurence and Bullen, London, 271 pp.
Watt, R.D. (1986). The case for liming some Nova Scotia salmon rivers. *Wat. Air Soil Poll.* **31**: 775–89.
Watt, W.D., Scott, C.D. and White, W.J. (1983). Evidence of acidification of some Nova Scotian rivers and its impact on Atlantic salmon, *Salmo salar*. *Can. J. Fish Aquat. Sci.* **40**: 462–73.
Watt, W.D., Farmer, G.J. and White, W.J. (1984). Studies on the use of limestone to restore Atlantic salmon habitat in acidified rivers, pp. 374–79 in *Lake and Reservoir Management*, EPA 440/5 84-001. USEPA, Washington, DC.
Welsh, W.T. and Burns, J.C. (1987). The Loch Dee Project: runoff and surface water quality in an area subject to acid precipitation and afforestation in southwest Scotland. *Trans. Roy. Soc. Edin. Earth Sci.* **78**: 249–60.
Welsh Water Authority (1987). *Llyn Brianne Acid Waters Project*, First Technical Summary Report. Department Environment & Welsh Office UK, 94 pp.
Welsh Water Authority (1988). *Llyn Brianne: Acid Water Project*, Second Technical Summary Report, Nov. 1988, 32 pp.
White, W.J., Watt, W.D. and Scott, C.D. (1984). An experiment on the feasibility of rehabilitating acidified Atlantic salmon habitat in Nova Scotia by the addition of lime. *Fisheries* **9**: 25–30.
Whitehead, P. (1989). Future trends in acidification. In *Acidification in Scotland*, Scotl. Devel. Dept., Edinburgh, pp. 114–21.
Wright, R.F. (1984). Chemistry of Lake Hovvatn, Norway, following liming and reacidification. *Can. J. Fish. Aquat. Sci.* **42**: 1103–13.
Wright, R.F., Lotse, E. and Semb, A. (1988). Reversibility of acidification shown by whole catchment experiments. *Nature* **334**: 670–5.
Zurbuch, P. (1984). Neutralization of acidified streams in West Virginia. *Fisheries* **9**: 42–47.

CHAPTER 2

Soil/Water Acidification and the Potential for Reversibility

R.A. Skeffington

National Power plc, Leatherhead, Surrey, UK

2.1 INTRODUCTION

To put the Loch Fleet project into scientific perspective, some theoretical background is needed. Loch Fleet is a recently acidified system. It once sustained a small trout population which had died out by the 1970s, and it is known that water quality at that time was not suitable for trout (see Chapter 3). Diatom stratigraphy has also demonstrated that the loch has become more acid in recent years (see Chapter 14). What processes caused it to become more acid? How has acidification there been related to acid deposition and change in land use? Why is catchment manipulation necessary – will the lake not simply recover as acid deposition declines under the influence of present emission control legislation? This chapter sets out to answer these questions and to give a basic understanding of acidification and recovery mechanisms.

2.2 NATURE OF ACIDIFIED SYSTEMS

Early studies of acidified systems tended to concentrate on study of surface waters and the response of aquatic organisms to acidity. Acidification of water was assumed to be a straightforward response to acid precipitation, akin to a laboratory titration (e.g. Henriksen, 1979). It soon became apparent that this approach was inadequate, because in a typical catchment area most of the water present in streams and lakes has interacted with soil in some way. The nature of acid soils, and the influences on them, is thus very relevant to the acidification problem. These influences include land use and management as well as pollution.

Much research on soils of acidified systems has focused on one group of soils, the podzols (Spodosols) and their relatives. These soils are characterised by a subsurface horizon with an accumulation of organic matter and oxides of aluminium and iron (Fig. 2.1). This illuvial "B" horizon is formed by deposition of material leached from the soil above, leaving an eluvial "E" horizon which is very highly weathered. On top of the eluvial horizon is typically a layer of very acid (mor) humus. Podzols tend to form in sandy, freely-draining parent material under coniferous forest or ericaceous vegetation. They are normally very acid, but pH tends to increase down the profile, as does base saturation. This has implications for the chemistry of surface waters in that water draining through the different layers will have different chemical compositions. Water from the surface ("O" and "A") horizons will be acid and have high concentrations of organic matter, but in most cases little inorganic aluminium. Water from the "E" and "B" horizons will be acid but less so than the "O" horizon water, will contain very low concentrations of base cations, but can have high inorganic aluminium concentrations. Water from the "C" horizon at the base of the profile will typically have a higher pH, higher base cation concentrations, and lower inorganic aluminium concentrations than water from the "A" and "B" horizons. Because the "B" horizon tends to have high concentrations of sesquioxides, which have considerable sulphate-adsorbing capacity, water from below the "B" horizon may have a lower sulphate concentration. Podzols form through natural pedogenic processes, the exact nature of which is still controversial. Runoff water from the upper horizons of podzols will tend to be acid, irrespective of whether the rain is acid or not, but higher concentrations of the mobile sulphate ion in acid precipitation imply that these differences will be expressed more clearly. The implication is that any manipulation of the soil that alters flow paths, such as ditching or afforestation, will also tend to alter surface water composition.

The above principles also apply to soils related to the podzols, such as brown earths and brown forest soils, though to a lesser extent. At Loch Fleet, however, only about 1% of the catchment soils are podzolic (see

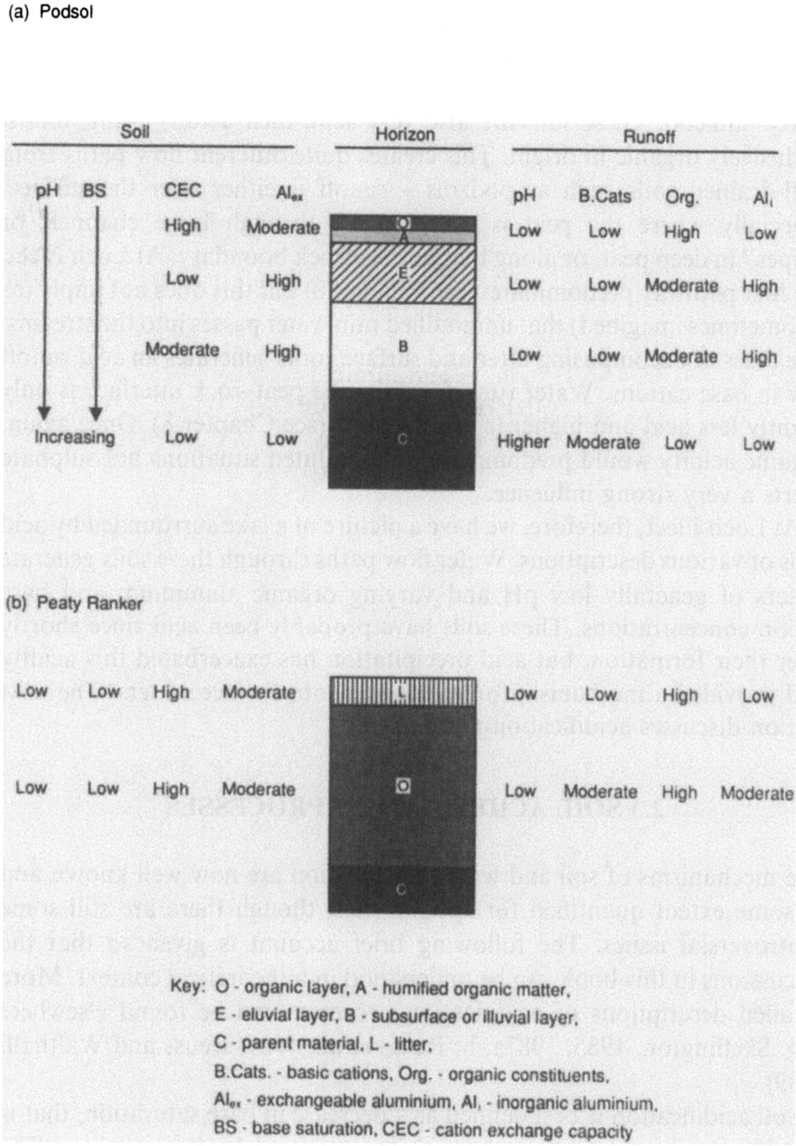

FIG. 2.1 Simplified soil profiles for a typical podsol and peaty ranker soils, showing physico-chemical characteristics.

Chapter 11); most are rankers of various types consisting largely of peat sitting directly on bedrock with little or no mineral content. Peaty rankers and dystrophic peat of this type occupy 64% of the Loch Fleet catchment, the distinction between the two relating largely to the depth of the "O" horizon (more than 50 cm in the case of deep peat; less than 50 cm for peaty rankers). These soils are also very acid, their acidity being almost exclusively organic in origin. This creates quite different flow paths from well-drained soils such as podzols – runoff is either over the surface, especially where the peat is waterlogged, through large channels or "pipes" in deep peat, or along the peat–bedrock boundary. At Loch Fleet, the first pathway predominates (see Chapter 8) but this does not imply (as is sometimes imagined) that unmodified rain water passes into the streams. The mass of decomposing litter and surface roots generates an acid runoff low in base cations. Water running along the peat–rock interface is only slightly less acid and higher in base cations (see Chapter 8). Once again, organic acidity would predominate in unpolluted situations but sulphate exerts a very strong influence.

At Loch Fleet, therefore, we have a picture of a lake surrounded by acid soils of various descriptions. Water flow paths through these soils generate waters of generally low pH and varying organic aluminium and base cation concentrations. These soils have probably been acid since shortly after their formation, but acid precipitation has exacerbated this acidity and provided a mechanism for its transfer into surface waters. The next section discusses acidification mechanisms.

2.3 SOIL ACIDIFICATION PROCESSES

The mechanisms of soil and water acidification are now well known and to some extent quantified for specific sites, though there are still some controversial issues. The following brief account is given so that the discussions in this book can be understood in a theoretical context. More detailed descriptions of acidification processes can be found elsewhere (e.g. Skeffington, 1985, 1987a, b; Reuss *et al.*, 1987; Reuss and Walthall, 1989).

Soil acidification is best defined as a decrease in base saturation, that is the percentage of cation exchange sites occupied by base cations. The usual operational definition is

$$\frac{Ca^{2+} + Mg^{2+} + Na^{+} + K^{+}}{Ca^{2+} + Mg^{2+} + Na^{+} + K^{+} + H^{+} + Al^{3+}} \times 100$$

where the element symbols represent the amounts of each ion which can be displaced from the exchanger. Base saturation at Loch Fleet is fairly low – ranging from only 3% to 19% if measured at pH 7 (Hudson *et al.*, 1986) or 32% to 53% if measured at soil pH, which is probably more appropriate (Wilson *et al.*, 1988). The equivalent sum of H^+ and Al^{3+} is known as the exchangeable acidity. To decrease base saturation, two conditions are necessary:

(a) a source of hydrogen ions to exchange for the exchangeable bases, and
(b) some means of removing the displaced base cations.

Acid precipitation provides both of these (Fig. 2.2). Hydrogen ions are contributed from rain, and the base cations are leached from the soil in conjunction with the mobile sulphate (or nitrate, chloride) anion. It is this leaching process, in which passage of mobile anions drives cation loss, that is the key to acidification of soils and water by acid precipitation (Johnson and Cole, 1980). The resultant lowering of soil base saturation implies an increased concentration of H^+ and aluminium in leachate water, because the relative concentrations of H^+, aluminium, and base cations are dependent on the composition of the soil cation exchanger. When the soil is acidified, sulphate and other mobile anions will leach H^+ and Al^{3+} into surface waters as well as base cations, leading to conditions potentially unsuitable for fish and other aquatic life.

Figure 2.2 also shows the acidification mechanisms that operate independently of acid deposition. Most of these mechanisms increase in intensity in areas that have high rainfall, such as in the catchment of Loch Fleet. High rainfall leads to a high water flux through the soil, which in turn enhances the fluxes of mobile anions. It can also lead to waterlogging, which inhibits the decomposition of dead plant residues and gives rise to organic matter accumulation (mechanism (f) – hence many such catchment soils are peaty (see Chapter 11)). At Loch Fleet, a special case of mechanism (f) also applies: *Sphagnum* moss is a very effective cation exchanger and if actively growing it has an acidifying influence comparable to that of acid deposition (Clymo, 1984). In this case the living biomass, as well as its decay products, acts as a cation exchanger. Consideration of all these mechanisms leads to the conclusion that the soils of the Loch Fleet catchment have probably always been acid. However, the additional acidification pressure exerted by acid deposition may have led to further soil acidification, so that the soil has less exchangeable calcium than before the Industrial Revolution. It is also reasonable to support the view that the increased flux of sulphate has increased H^+ and Al^{3+} concen-

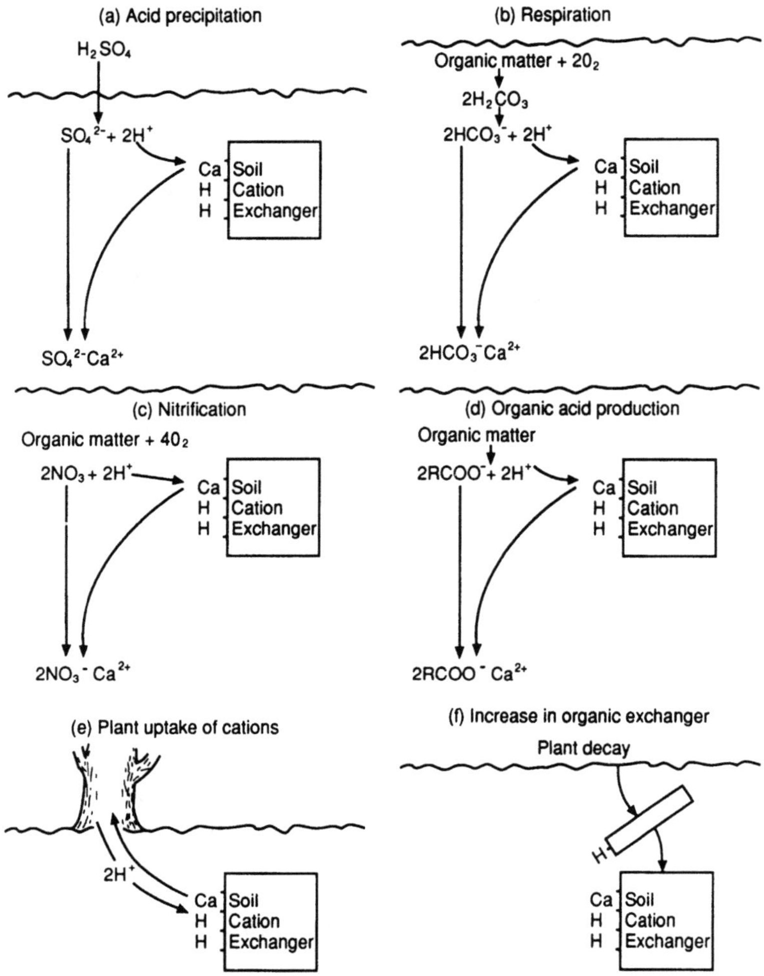

FIG. 2.2 Principal mechanisms of soil acidification.

trations in the loch waters, although there is no hard evidence on either of these points.

2.4 ACIDIFICATION AND LAND USE

There has been much controversy in the literature over the extent to which changes in land use can explain observations of surface water acidification and loss of fisheries. Rosenqvist (1978), in particular, has pointed out that the potential production of exchangeable acidity due to the accumulation of an organic-rich surface soil far exceeds acidity deposited from the atmosphere. This situation would occur if farmland was replaced by forest: a common situation in Norway during the last century. Krug and Frink (1983) and Krug (1989) extended these arguments to conditions in the USA. It is therefore worth considering to what extent land-use change in the Loch Fleet catchment is likely to have affected water chemistry.

The land-use history of the Loch Fleet catchment itself is not well known, although that of the adjacent Cairnsmore of Fleet region is known to some extent (see Chapter 4). In pre-historic times it may have supported a deciduous forest, with the addition possibly of some Scots pine, although judging from the performance of the recently planted Sitka spruce on the hillside ridges, this would have been confined to the more sheltered parts of the catchment. It seems unlikely that such an area – high, wet, infertile, exposed, and remote – would ever have supported a permanent settlement within the catchment. In the seventeenth and eighteenth centuries subsistence agriculture in this area was replaced by estate management for sheep and later for grouse. Moorland management involved regular burning (about every 7 years) which may well have had an effect on lake chemistry. This is because burning reduces exchangeable acidity by destroying exchange sites and liberating base cations otherwise unavailable. The effect will depend greatly on the intensity and extent of the fire. The normal aim in moorland burning was to remove the old *Calluna* growth without affecting their root mass and the peat below – in this case the effect on acidity of runoff water would probably be small. However, the process is difficult to control and the area was also, in the past, subject to wildfires from the steam railway a few kilometres to the south. These wildfires could be much more intense, and Mr Jack Henderson, the former gamekeeper on the Cally Estate, noticed that fishing in the area always improved after one of these wildfires. This agrees with observations in Norway by Rosenqvist (see Rosenqvist and Seip, 1986) who observed reduced acidity in a small lake which had been subjected to an intense forest fire. It therefore seemed worthwhile to burn part of a catchment sector as an experimental treatment (see Chapter 9) to establish whether this effect could be reproduced.

The biggest land-use change in the Loch Fleet catchment has been the plantation of Sitka spruce and lodgepole pine over about 10% of the catchment in 1963 (see Chapter 4, Table 4.2). The effect of afforestation on water quality is currently a controversial issue in the UK (Egglishaw *et al.*, 1986; Nisbet, 1989; Edwards *et al.*, 1990) but trees are known to enhance pollutant capture, to take up cations in excess of anions from the soil, and to produce a very acidic litter layer (see Chapter 8), thus enhancing acidification by mechanisms (a), (d), (e) and (f) in Fig. 2.2. Diatom stratigraphy (Chapter 4) shows acidification of the Loch about 10 years after afforestation: there is speculation that this was due to sediment from catchment ploughing sealing off an in-lake groundwater source of alkalinity, rather than the effect of forestry *per se*. Fishery decline, however, pre-dated the estimated acidification by about two decades.

2.5 REVERSAL OF ACIDIFICATION

2.5.1 Introduction

Currently, the European Community Large Combustion Plant Directive requires the UK to reduce emissions of sulphur dioxide from large boiler plant to 40% of 1980 values by 2003. A reduction of sulphate deposited in precipitation and in lake water compared with 1970s levels has already been reported for Galloway (Battarbee *et al.*, 1988). There is no direct relationship between emissions and wet deposition; the latter depends on many other factors, including rainfall amount, weather patterns, and supply of atmospheric oxidants. The UK Review Group on Acid Rain (Irwin *et al.*, 1990) found evidence for declining trends in H^+ and non-marine sulphate deposition at some Scottish sites, including at Eskdalemuir 70 km from Loch Fleet, but not at others. It is worth considering whether sulphur emission reduction would have been sufficient to restore Loch Fleet to a chemical status suitable for fish without liming treatment. Consideration of the theory of acidification outlined above shows that reduction in sulphate concentration passing through the soil into stream water will reduce H^+ and aluminium, which will improve conditions for fish, but will also reduce calcium, which will make conditions worse. Loch Fleet already had calcium levels marginal for fish survival (see Chapter 3) and therefore any reduction was likely to be deleterious provided the pH was low.

2.5.2 Calcium Status of Waters

Early attempts to mitigate the effects of surface water acidity concentrated on liming surface waters directly. This has in many cases been effective, particularly in Sweden. However, it leaves untouched the acidity within the catchment area which continues to feed acid water into a limed lake.

Lakes have thus periodically to be re-limed to maintain the target pH. Lakes with short hydrological turnover times will need liming more frequently. In western Britain, rainfall is high and the ratio of lake volume to catchment area tends to be low, which implies very short turnover times, of the order of a few months in many cases. Moreover, one resource at risk is the migratory salmonid fishery which is dependent on a satisfactory water quality in the upland streams to provide nursery areas for spawning and juvenile growth. Brown trout also usually spawn in streams. Under these circumstances lake liming will not alleviate the problems of acidity, and catchment liming is indicated.

2.5.3 Calcium Status of Soils

Some of the calcium which has been lost from the catchment over the centuries must be replaced to restore the fishery. Are "natural" processes adequate for this task? There are only two "natural" sources of calcium – atmospheric deposition and mineral weathering. A rough calculation can be made to show that these sources will not restore the calcium to the soil ion exchange system in a reasonable time-scale measured at least in decades. Using rather conservative assumptions, the soil at Loch Fleet can be regarded as an organic layer 0·35 m thick with a cation exchange capacity of 50 meq $100\,g^{-1}$ (see Chapter 11). Assuming a density of $1\,g\,cm^{-3}$, this equates to an exchange capacity of 175 000 meq m^{-2}. The bedrock of Fell of Fleet is a biotite–muscovite granite: 1 mm thickness contains 5400 meq m^{-2} base cations of which 570 meq m^{-2} is calcium (Parslow, 1968). The weathering rate of a granite with such a composition would not be expected to exceed 50 meq $m^{-2}\,yr^{-1}$ base cations (Sverdrup and Warfvinge, 1988). This equates to the release of about 5 meq $m^{-2}\,yr^{-1}$ of calcium. To increase calcium saturation of the exchanger by only 10% would therefore take 3500 years (175 000/10 $\times$ 5), even assuming all the calcium stays on the exchange surface and does not leach into the drainage water. Atmospheric deposition supplies 20–30 meq $m^{-2}\,yr^{-1}$ of calcium (see Chapter 5): even at this rate it would take 760 years to increase the calcium saturation of the exchanger by 10%, assuming no calcium is leached to the drainage water. In contrast, liming increases the calcium saturation of the exchanger by more than 50% within a year (Wilson, 1989). It must therefore be concluded that even if atmospheric acid deposition were eliminated entirely, chemical conditions in the Loch would remain unsuitable for fish for an unacceptably long time. Indeed, deposition of calcium (and other cations) shows evidence of a decline in Britain over the past decade, including at Eskadalemuir (Irwin *et al.*, 1990).

2.5.4 Sulphur Retention in Soils

There is another reason why recovery of the lake chemistry due to reduction of deposition will be slow. The soils of the catchment probably contain a large store of organic sulphur which has accumulated from both natural and anthropogenic sources. As sulphate deposition declines, this store will tend to release stored sulphate, buffering the decline, and slowing recovery. Potentially, the sulphur store could maintain present sulphate release to drainage waters for many hundreds of years. In practice most of this sulphur is not available for oxidation (Brown, 1982), but experience so far (e.g. in the RAIN project, Wright *et al.*, 1988) suggests that some catchment sulphate mineralisation does occur and is likely to delay the fall in sulphate concentrations in lakes and streams at least for some years.

Many soils also have a store of adsorbed inorganic sulphate. This sulphate will tend to desorb if sulphate concentrations in the soil solution decline owing to reduced deposition. It is not known whether the adsorption–desorption process is fully reversible, but this is also potentially a mechanism which will delay the fall in sulphate concentrations for some years. This mechanism applies particularly to soils with large amounts of sesquioxides in the lower horizons (such as podzols). At Loch Fleet, where organic soils predominate, the organic sulphur store is probably more important.

2.5.5 Base Restoration of Soils

Some form of restoration of calcium to the catchment soils will therefore be necessary to restore fisheries to Loch Fleet, whatever the time-scale and amount of reduction of sulphur emissions. The simplest way to do this is to add calcium carbonate – the calcium displaces H^+ and aluminium from exchange sites and raises the base saturation, the displaced H^+ combining with the carbonate (which is the conjugate base of a weak acid) to remove the acidity from the system in the form of carbon dioxide and water. It would be equally possible to use the calcium salt of a strong acid (such as $CaCl_2$) to replenish calcium on the exchanger, but the result would be a massive flush of toxic aluminium and H^+ into the aqueous system. The interesting questions relate to how long the increased calcium will last (see Chapter 16) and thus how long the fishery will be sustained without re-treatment. Other methods involve increasing the relative calcium concentration by burning off the organic exchange capacity and releasing unavailable calcium (see above); at Loch Fleet, however, this does not seem to have been very successful (see Chapter 12).

2.6 MODELS AND REVERSIBILITY EXPERIMENTS

There is now some evidence from experimental, observational, and modelling studies that bears on the likely effects of a reduction of acid deposition on water chemistry and hence fisheries. This is discussed in more detail by Skeffington and Brown (1992). The most direct evidence is provided by manipulation experiments in which acid deposition is reduced by means of a roof over a catchment area. The longest established of these is the RAIN (Reversing Acidification in Norway) Project (Wright *et al.*, 1988). Here ambient acid deposition has been excluded from a small catchment and replaced by cleaned rain, while an adjacent control catchment is used for comparison. The results are mostly (though not entirely) those predicted by theory (Section 2.3). The sulphate concentration in the output water declined to 40% of that in the control after 4·5 years. About 25% of the sulphate in the output is being supplied by mineralisation from the catchment. In spite of this, the pH of the output water is much the same as in the control catchment. The reducing sulphate concentration is being compensated for by an increasing organic anion concentration, as predicted by Krug and Frink (1983). Calcium is declining as predicted and inorganic aluminium is decreasing (Fig. 2.3). After 4·5 years of complete removal of wet acid deposition, the water is still unsuitable for fish (Rosseland *et al.*, 1990). Fish have probably never existed in water from such catchments where there is no groundwater input to neutralise the acidic runoff from the surface soil layers. However, even when mixed with groundwater in constant amount, treatments in the RAIN project so far show no sign of improving conditions for fish and bear out predictions that recovery from acidification will be very slow (Skeffington and Brown, 1992).

Observations at acidified sites in areas where deposition has been declining present a less consistent picture. In Norway, a resurvey of 305 lakes in 1986, which were previously sampled in 1974–75, showed reduced sulphate concentrations, reduced calcium and magnesium, slightly increased pH, and a continued deterioration in fishery status (Henriksen *et al.*, 1988). This picture is consistent with the results of the RAIN project, namely that there is no sign of recovery in spite of reduced sulphur deposition. The evidence is not unequivocal, however, because there was also an increase in (total) aluminium and an increase in nitrate. Moreover, fish populations might be expected to take time to respond to changes in water chemistry. In contrast to the Norwegian situation, some lakes near to the Sudbury smelter in Canada appear to be recovering quite quickly

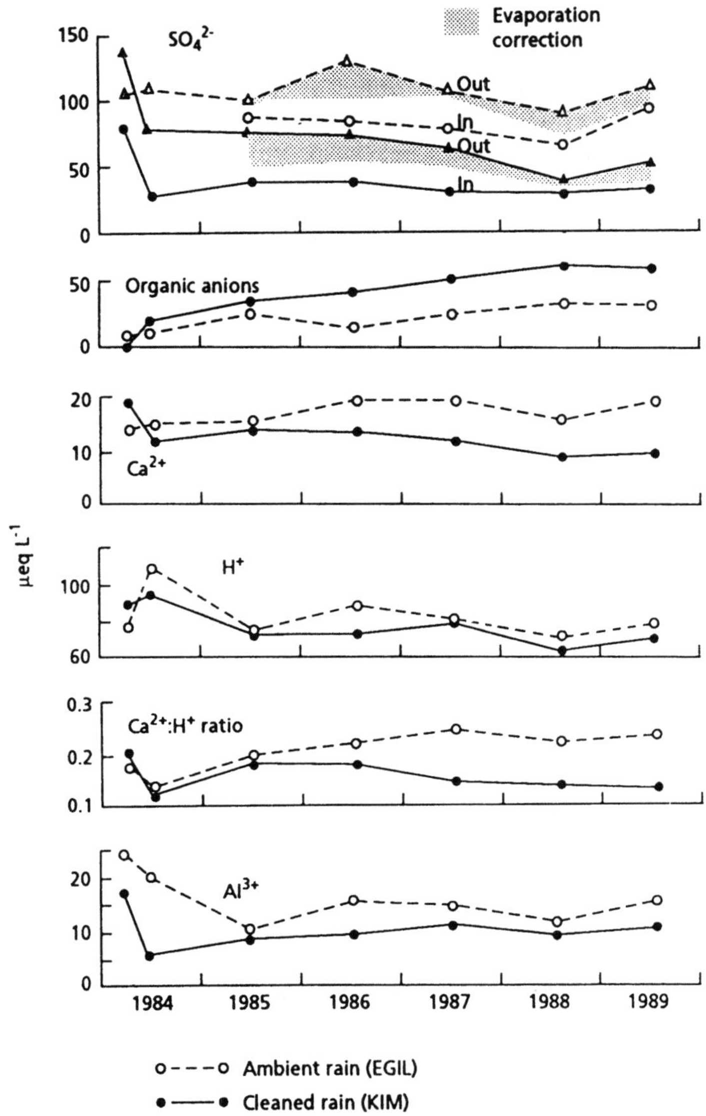

FIG. 2.3 Progressive changes in the mean concentrations of some major constituents and in the equivalent ratio of calcium and hydrogen and of toxic aluminium runoff from the acid rain exclusion experiment at Risdalsheia (RAIN Project). The sulphate graph includes inputs: the shaded area represents the production of SO_4^{2-} from the catchment (output less input corrected for evapotranspiration).

(Dillon *et al.*, 1986; Gunn and Keller, 1990). The smelter reduced its sulphur emissions very considerably (by 350%), however, and the weathering rate is clearly higher than in Norway (Skeffington and Brown, 1992). Moreover, the Sudbury lakes had high concentrations of toxic trace metals, possibly an influence on fish decline in acid conditions prior to emission control. Modelling studies using MAGIC (Cosby *et al.*, 1985) demonstrate that weathering rate is one of the primary determinants of recovery rate (Skeffington and Brown, 1992). The implication is that, in the slow-weathering areas that have been most affected by acid deposition, recovery also will be slow, measured in decades or centuries. This will be true with any conceivable degree of emission reduction. If a more rapid recovery is desired, then liming to restore lost calcium and increase soil base saturation is the only solution.

Acidification experiments also give information that is pertinent to recovery rates, because they highlight acidification and deacidification mechanisms and allow their quantification. Schindler and Turner (1982) added sulphuric acid to a Canadian lake (ELA223) and found that this lake had a considerable capacity to buffer the added acid through sulphate reduction mechanisms. Skeffington and Brown (1986) quantified significant soil buffering mechanisms even in a base-poor sandy soil in a lysimeter experiment. In another component of the RAIN Project, at Sogndal in western Norway, where in contrast, acid addition was made to a relatively pristine catchment, the resulting quality of runoff water was similar to that from the acid deposition-impacted areas of southern Norway (Wright *et al.*, 1988). These experiments all bear out the general picture of acidification and its reversibility presented in this chapter.

2.7 SUMMARY AND CONCLUSIONS

Most processes governing acidification and reversibility are now reasonably well understood and some can be quantified, though some questions remain unanswered. Recovery can be expected to be slow (decades to centuries) in the slow-weathering areas most impacted by acid deposition, given any conceivable degree of emission reduction. An obvious way to restore fisheries to these acidified areas more quickly is to add the lost calcium and alkalinity back to the ecosystem. This is the focus of the Loch Fleet project.

REFERENCES

Battarbee, R.W., Flower, R.J., Stevenson, A.C., Jones, V.J., Harriman, R. and Appleby, P.G. (1988). Diatom and chemical evidence for reversibility of acidification in Scottish Lochs. *Nature* **332**: 530–32.

Brown, K.A. (1982). Sulphur in the environment: a review. *Env. Poll.* **B3**: 47–80.

Clymo, R.S. (1984). *Sphagnum*-dominated peat-bog: a naturally-acid ecosystem. *Phil. Trans. Roy. Soc.* **B305**: 487–99.

Cosby, B.J., Hornberger, G.M., Galloway, J.N. and Wright, R.F. (1985). Time scales of catchment acidification. *Env. Sci. Technol.* **19**: 1144–49.

Dillon, P.J., Reid, R.A. and Girard, R. (1986). Changes in chemistry of lakes near Sudbury, Ontario following reductions of SO_2 emissions. *Water, Air Soil Poll.* **31**: 59–65.

Edwards, R.W., Stoner, J.H. and Gee, A.S. (1990). *Acid Waters in Wales*. Kluwer, Dordrecht, Netherlands, 337 pp.

Egglishaw, H., Gardiner, J.R. and Foster, J. (1986). Salmon catch decline and forestry in Scotland. *Scott. Geogr. Mag.* **102**: 57–61.

Gunn, K.M. and Keller, W. (1990). Biological recovery of an acid lake after reductions in acid emissions of sulphur. *Nature* **345**: 431–33.

Henriksen, A. (1979). A simple approach for identifying and measuring acidification of freshwater. *Nature* **278**: 542–45.

Henriksen, A., Lien, L., Traaen, T.S., Sevaldrud, I.S. and Brakke, D.F. (1988). Lake acidification in Norway – present and predicted chemical status. *Ambio* **17**: 259–66.

Hudson, G., Robertson, J.S., Campbell, C.G.B. and Henderson, D.J. (1986). *Soils and Vegetation of the Loch Fleet Catchment: Baseline Studies*. Macaulay Institute Report, Aberdeen, 137 pp.

Irwin, J.G., Campbell, G.W., Cape, J.N., Clark, P.A., Davies, T.D., Derwent, R.G., Fisher, B.E.A., Fowler, D., Kallend, A.S., Longhurst, J.W.S., Martin, A., Smith, F.B. and Warrilow, D.A. (1990). *Acid Deposition in the United Kingdom, 1986–1988*. Dept. of Environment, Warren Spring Laboratory, 124 pp.

Johnson, D.W. and Cole, D.W. (1980). Anion mobility in soils: relevance to nutrient transport from forest ecosystems. *Env. Internat.* **3**: 79–90.

Krug, E.C. (1989). *Assessment of the Theory and Hypotheses of the Acidification of Watersheds*, Illinois State Water Survey Contract Report 457. Champaign, Illinois, USA, 217 pp.

Krug, E.C. and Frink, C.R. (1983). Acid rain on acid soil: a new perspective. *Science* **221**: 520–25.

Nisbet, T.R. (1989). *Forests and Surface Water Acidification. Forestry Commission Bulletin* **86**. HMSO, London, 8 pp.

Parslow, G.R. (1968). The physical and structural features of the Cairnsmore of Fleet granite and its aureole. *Scott. J. Geol.* **4**: 91–108.

Reuss, J.O. and Walthall, P.M. (1989). Soil reaction and acidic deposition, pp. 1–33 in *Acidic Precipitation, Vol. 4: Soils, Aquatic Processes, and Lake Acidification* (eds. S.A. Norton, S.E. Lindberg and A.L. Page). Springer-Verlag, New York, Berlin, Heidelberg.

Reuss, J.O., Cosby, B.H. and Wright, R.F. (1987). Chemical processes governing soil and water acidification. *Nature* **329**: 27–32.

Rosenqvist, I. Th. (1978). Alternative sources for acidification of river water in Norway. *Sci. Total Env.* **10**: 39–49.

Rosenqvist, I. Th. and Seip, H.M. (1986). Acidification of watercourses – how great is the disagreement? *Kjemi* **46**: 13–17. CEGB Translation T16668.

Rosseland, B.O., Lien, L., Krogland, F., Sadler, K. and Dalziel, T.R.K. (1990). *Strains of Brown Trout* (S. trutta L.) *Stocking and Test Fishing 1988 and 1989; Field and Laboratory Toxicity Experiments.* NIVA Report No. 0-87178, 26 pp.

Schindler, D.W. and Turner, M.A. (1982). Biological, chemical and physical responses of lakes to experimental acidification. *Water, Air Soil Poll.* **18**: 259–71.

Skeffington, R.A. (1985). Effect of acid deposition and natural acidification processes on soil: some studies in the Tillingbourne catchment, Leith Hill, Surrey. *Seesoil* **2**: 18–34.

Skeffington, R.A. (1987a). The transport of acidity through ecosystems, pp. 139–54 in *Pollutant Transport and Fate in Ecosystems* (eds. P.J. Coughtrey, M.H. Martin and M. Unsworth). British Ecological Society Special Publications, Blackwell, Oxford, 414 pp.

Skeffington, R.A. (1987b). Soil and its responses to acid deposition, pp. 16–29 in *CEGB Res.* **20**, 64 pp.

Skeffington, R.A. and Brown, K.A. (1986). The effect of five years acid treatment on leaching, soil chemistry and weathering of a humo-ferric podzol. *Water, Air Soil Poll.* **31**: 891–900.

Skeffington, R.A. and Brown, D.J.A. (1992). Timescales of recovery from acidification: implications of current knowledge for aquatic organisms. *Env. Poll.*, in press.

Sverdrup, H. and Warfvinge, P. (1988). Assessment of critical loads of acid deposition on forest soils, pp. 81–130 in *Critical Loads for Sulphur and Nitrogen* (eds. J. Nilsson and P. Grennfelt). Nordic Council of Ministers, Copenhagen.

Wilson, E.J. (1989). *The Effect of Lime Application on Soils at Loch Fleet – 1987 Sampling.* CERL Report ESTD/L/0010/R89, 81 pp.

Wilson, E.J., Roberts, D.J. and Skeffington, R.A. (1988). *The Effect of Lime Application to the Loch Fleet Catchment on Tree and Soil Chemical Properties – An Initial Report.* CERL Report TPRD/L/3283/R88, 25 pp.

Wright, R.F., Lotse, E. and Semb, A. (1988). Reversibility of acidification shown by whole-catchment experiments. *Nature* **334**: 670–75.

CHAPTER 3

The History of the Loch Fleet Fishery and Fresh Water Quality for Fish and Other Aquatic Fauna

A.W.H. Turnpenny
National Power plc, Southampton, UK

3.1 INTRODUCTION

This chapter provides information about the fishery status at Loch Fleet prior to liming and on the preliminary laboratory and field bioassay tests from which a target water quality was derived. Some comparisons with other upland waters in the UK are also included.

3.2 FISHERY STATUS AT LOCH FLEET BEFORE LIMING

3.2.1 The Historical Fishery

Prior to its decline in recent years, the fishery for brown trout (*Salmo trutta*) at Loch Fleet provided sport for a small number of anglers each year who visited the loch on foot, since at that time it was not served by the present network of forestry tracks. The fishery was owned and run by the Cally Estate (Gatehouse of Fleet). Visits by anglers were few, but sport was deemed to be good, with bags of up to 40 trout in a single day according to Estate records, and a total annual catch of about 150 fish per year (Fig. 3.1(a)). The average weight of fish from the loch was around

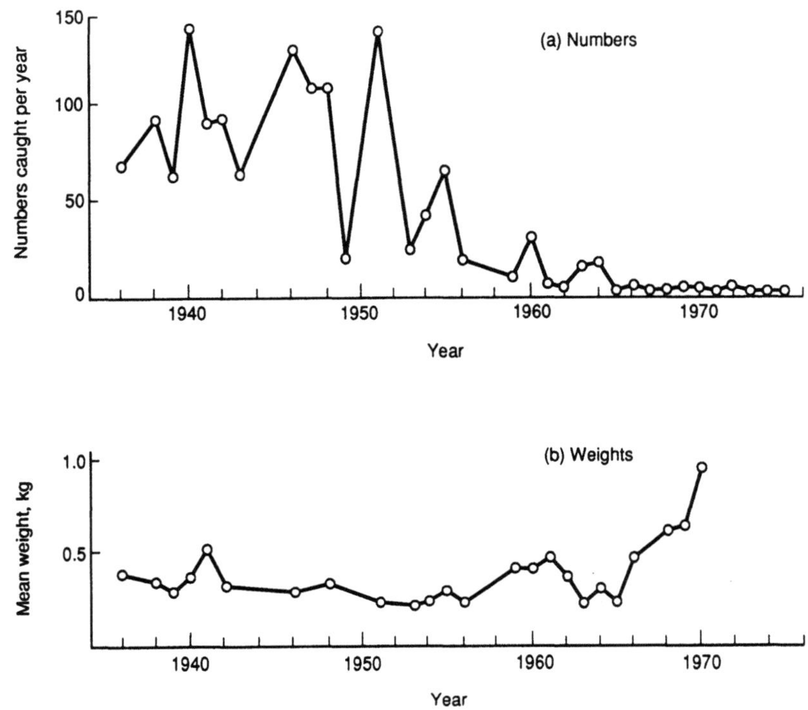

FIG. 3.1 (a) Numbers of brown trout caught by rod and line and (b) their mean weight, from Cally Estate records 1936–75.

0·4–0·5 kg, until the last years of the fishery when increasingly larger specimens were caught (Fig. 3.1(b)).

The data shown in Fig. 3.1 are not corrected for fishing effort (for which no data are available) and thus can be taken only as an indication of the pace of decline. However, the overall pattern, taken with average weight of landings, suggests recruitment failure from the 1950s, leading to an ageing population of larger individuals. Reduced competition under these circumstances may even lead to enhanced growth rates, a feature observed in other fish-containing Scottish lochs which are thought currently to be becoming acidified (Campbell, 1987).

There are few further facts known about the Loch Fleet fishery. Spawning is believed to have centred on the main feeder stream, the Altiwhat, where suitable gravels exist, though it is not clear whether these were the only spawning grounds accessible to loch trout. Other suitable gravels exist more than 2 km downstream in the loch outlet stream known as the Little

Water of Fleet (Fig. 3.2) (Turnpenny *et al.*, 1988), but the substrate immediately adjacent to the loch outlet is solid bedrock and would not have afforded any spawning potential. The steep gradient (greater than 12·5%) between the downstream spawning grounds and the loch suggests that most fish spawning in these waters would not return to the loch. There are anecdotal reports that the gravel beds of drainage ditches feeding into the loch at one time might have provided additional spawning beds for trout. In the earlier decades of this century such ditches were maintained regularly to improve sheep grazing on the catchment.

The Little Water of Fleet itself runs for a length of some 13 km before joining the larger Big Water of Fleet at Akiehill, and in its lower reaches supports a mixed population of brown trout and anadromous sea trout (*S. trutta*). The Craigie Linn, a 5 m waterfall 7 km below the Loch Fleet outlet, marks the geological boundary between the Fleet granite (upstream) and sedimentary greywacke beds (downstream). It creates an impassable barrier to the upstream movement of salmonids, so that in recent geological time Loch Fleet will not have supported an anadromous salmonid population. Both the Little and Big Waters of Fleet still support trout angling in their lower reaches, administered by the Cally Estate.

The decline of the Loch Fleet fishery can be viewed within the context of northern European, especially Scottish, fisheries as a whole. The decline of salmonid fisheries in Scandinavia widely reported in the early 1970s (Jensen and Snekvik, 1972) prompted investigations in other upland areas with ultra-soft waters, and by the late 1970s it was becoming apparent that the decline of salmonids in some upland waters in Scotland might be associated with acidification. In 1978 the Department of Agriculture and Fisheries for Scotland (Harriman *et al.*, 1987) undertook a survey of some 22 upland lochs and associated streams in south-west Scotland. They obtained information on fishery status of these waters from the literature, from angling records, and by interview with commercial fishermen. The results indicated a decline of fish in a number of lochs (including Fleet) and streams overlying granitic geology. As a result of these findings the Institute of Terrestrial Ecology was commissioned to undertake a larger-scale programme of field surveys over the whole of Scotland. Their report (Maitland *et al.*, 1987) concluded that the great majority of fisheries affected or lost due to acidification were on the Doon and Cairnsmore of Fleet granites of the Galloway region. There, 6 of 11 lochs sampled were found to be fishless, compared with 4 of 38 examined in the rest of Scotland. Also in Scotland a widespread decline of Atlantic salmon (*S. salar*) fisheries has been associated with the recent expansion of coni-

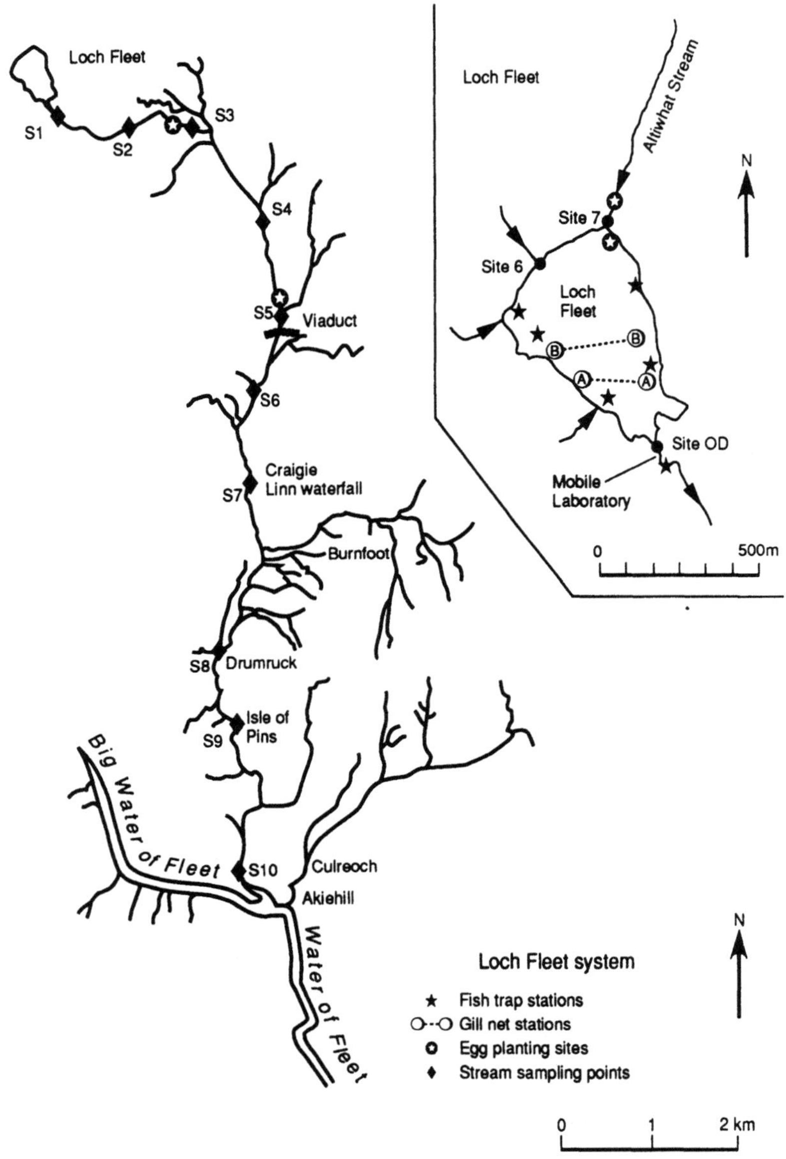

FIG. 3.2 The River Fleet system, showing sampling and experimental sites (from Turnpenny *et al.*, 1988).

TABLE 3.1
Record of sea trout catch on River Fleet, 1937–79

Year	*No.*	*Year*	*No.*	*Year*	*No.*
1937	667	1951	–	1965	1178
1938	487	1952	329	1966	972
1939	514	1953	317	1967	981
1940	404	1954	460	1968	567
1941	252	1955	764	1969	312
1942	341	1956	461	1970	514
1943	414	1957	–	1971	230
1944	351	1958	–	1972	70
1945	–	1959	520	1973	98
1946	–	1960	529	1974	99
1947	274	1961	517	1975	86
1948	374	1962	883	1976	87
1949	317	1963	716	1977	30
1950	–	1964	786	1978	65
				1979	93

Source: T.M. Drakeford Report to Forestry Commission, 1979.

ferous afforestation (Egglishaw *et al.*, 1986). Elsewhere in Britain a decline of salmonid fisheries attributed to acidification and afforestation has been reported in Cumbria and central and western Wales since the early 1980s (see the review by Turnpenny, 1989).

Prior to the start of the Loch Fleet project in 1984, fisheries surveys had been undertaken within the Fleet catchment on behalf of the Forestry Commission (T.M. Drakeford, unpublished reports to the Forestry Commission, 1979) and by Harriman *et al.* (1987). The reports to the Forestry Commission, undertaken as a result of declining angling catches of sea trout from the lower Fleet following a peak in 1965 (Table 3.1), highlighted factors associated with afforestation as being responsible for the deterioration of the fishery. These factors are well known from investigations elsewhere and include: (1) soil erosion into stream channels and consequent siltation of spawning gravels (Cordone and Kelley, 1961); (2) overplanting of stream margins with conifers thereby shading light from streams and reducing water temperature and primary production (Smith, 1980); (3) elimination of marginal broadleaf species which support diverse insect food sources for fish; (4) lowering of the water table due to enhanced evapotranspiration (Gash *et al.*, 1978); and (5) acidification of the drainage waters due to soil oxidation (Harriman and Morrison, 1982; Stoner and Gee, 1985). Of these, siltation and acid water quality were considered to be the major factors in the Fleet. The report records pH values from the granitic reaches of the Little Water of Fleet in the autumn of 1979 in the range 3·9–4·0, with values as low as 3·5 in some forest ditches draining into the main channel. The stream was also electrofished in 1979, and whilst sea trout were present in the Little Water and its tributaries below the Craigie Linn, no salmonids were recorded at the (now demolished) railway viaduct, some 2 km upstream of the Craigie Linn (Fig. 3.2).

Harriman *et al.* (1987), who sampled the Fleet system between 1978 and 1984, similarly found salmonids (*S. trutta*) present in the lower streams where the Little Water passes over sedimentary rocks, but found none in the upper reaches above Craigie Linn, including Loch Fleet itself. Atlantic salmon (*S. salar*) are present in the Big Water of Fleet but not the Little Water. This species is more sensitive than *S. trutta* to acid conditions especially during spawning migrations.

3.2.2 Fish Surveys 1984–85

A more thorough survey for fish in Loch Fleet, its feeder streams, and the Little Water of Fleet was carried out for the Loch Fleet project between 1984 and 1985 (Turnpenny *et al.*, 1988). The aim was to confirm and update previous data, with a larger number of sampling points.

The first stage of this survey was conducted in June–July 1984, when three sampling methods were employed: gill-netting, trapping, and electrofishing. All three methods were used in Loch Fleet (of which only the margins were electrofished); at stream sites, electrofishing only was used. Sampling locations are shown in Fig. 3.2. Gill-nets were of nylon monofilament, 25 m long × 1·5 m deep, with buoyant headlines and weighted footlines, and deployed in banks of four to eight. Within each bank a random series with respect to mesh size (19–54 mm half-mesh) was selected. These were fished in June 1984, with a total of 13 500 gill-net $m^2 h^{-1}$ fishing effort. No fish were caught by this method. By comparison, Harriman *et al.* (1987) used the same technique in other Scottish lochs and caught up to 99 fish for the same effort in some lochs.

Fish traps were of two types. The larger "eel" traps were plastic mesh (5 mm^2 mesh) cylinders, 20 cm diameter × 60 cm long, with re-entrant funnel-type mouths (3 cm dia.), and a second funnel placed half way along the inside to improve fish retention. The smaller "minnow" traps comprised a clear plastic cylinder 10 cm dia. × 20 cm long with a single re-entrant mouth (2 cm dia.). Both types were baited with offal and deployed on the bed of the Loch at points shown in Fig. 3.2. The traps were left in place for a period of 50 h during July 1984, and were inspected periodically during that time. Using this method, a single eel (*Anguilla anguilla*) was captured at the sampling point close to the loch outlet.

The electrofishing survey carried out in June 1984 used a pulsed-DC apparatus at sites shown on Fig. 3.2. The stream reaches fished were not isolated by stop-nets on this survey and therefore no population estimates were made, the aim being to establish the presence or absence of fish.

Again, no fish were found in the loch or its feeder streams. A few eels were found in the Little Water of Fleet close to the loch and these became increasingly common downstream to the Craigie Linn. Below this waterfall, eels were present, along with 0- and 1-group juvenile trout (*Salmo trutta*), presumed to be a mixture of brown and sea trout but indistinguishable in the field at that age. The overall fish distribution pattern revealed by the 1984 survey is shown in Fig. 3.3. Spot pH readings made in June 1984 at electrofishing sites ranged from pH 4·1 at the loch outlet to 7·0 at Akiehill.

A second fish survey was carried out in May 1985 when the three sampling methods were again applied at the same sampling locations. This time, stop-nets (1 cm half-mesh) were used for electrofishing, and quantitative population estimates were made. The results confirmed the distribution pattern of the previous year, but showed population biomass and density values to be low, with no more than 5 small trout per 100 m^2, even in the richer habitat of the more alkaline lower reaches below Craigie Linn (cf. values for other British upland waters in Table 3.2; see also Fig. 3.4). This was considered surprising until a few days after the survey when there was a period of heavy rainfall (50 mm in two days) and water chemistry was monitored before, during, and following spate flows, along the course of the Little Water of Fleet. At the peak of the spate, the pH fell by up to 1 unit (i.e. a tenfold increase in $[H^+]$), calcium levels fell four to eightfold, and aluminium levels rose four to ninefold. Pre-spate levels had not been restored within 3 days. Although the ameliorating effect on water chemistry of the sedimentary geology can be seen from Fig. 3.5 as the water passes over the Craigie Linn, transient conditions severe enough to be acutely toxic to fish extended to the lower limits of the Little Water during the spate.

No dead fish were observed during this event (although they would have been difficult to spot). An alternative possibility is that fish were washed down into the more alkaline reaches of the Big Water of Fleet and survived.

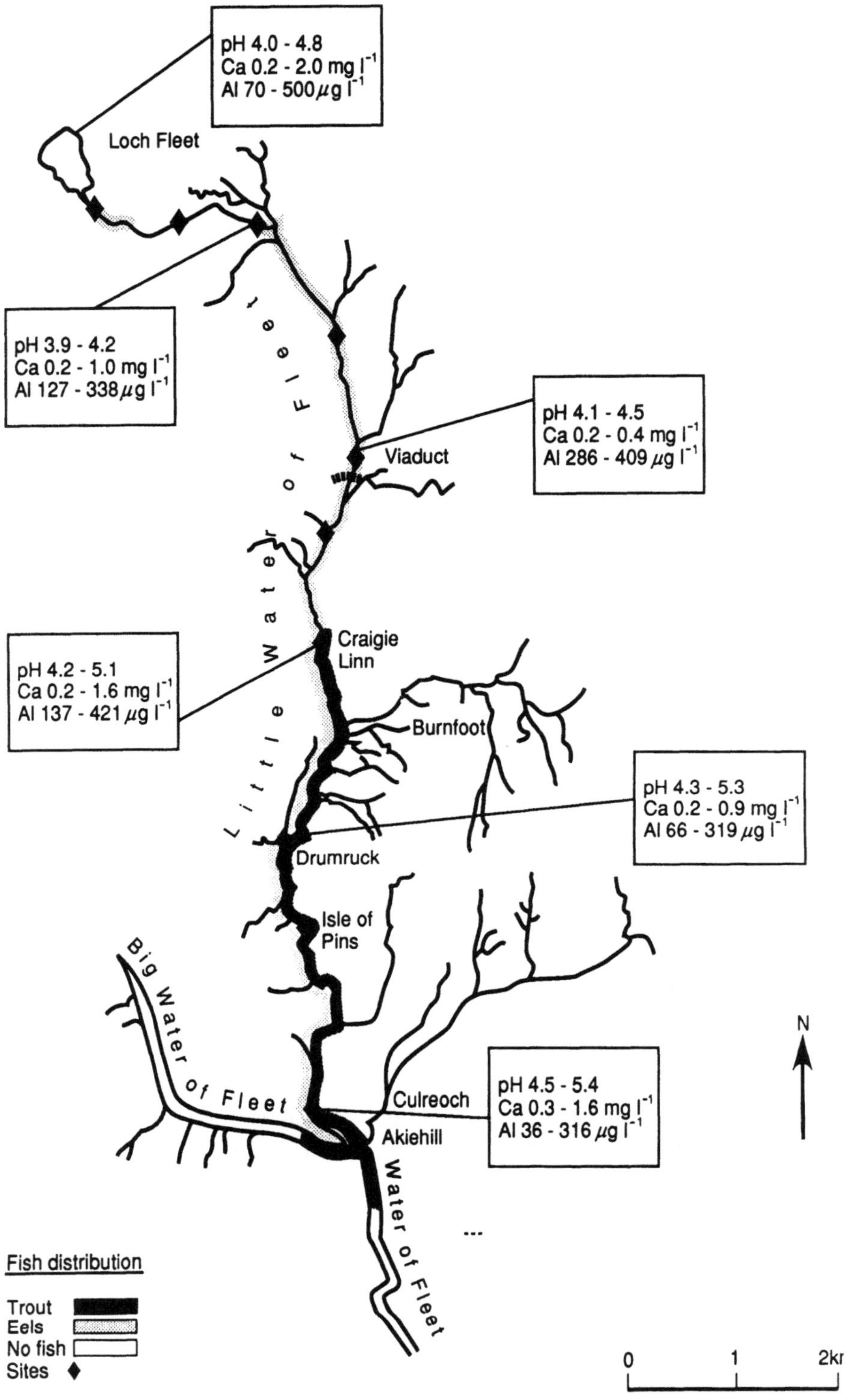

FIG. 3.3 The distribution of fish and water quality in 1984 in the Little Water of Fleet prior to liming. (After May, 1985; Turnpenny *et al.*, 1988).

TABLE 3.2

Brown trout and eel population density and biomass values, Little Water of Fleet, May 1985. Values for other circumneutral British waters are shown for comparison

Site	*Brown trout*		*Eel*	
	Density ($n\,m^{-2}$)	*Biomass* ($g\,m^{-2}$)	*Density* ($n\,m^{-2}$)	*Biomass* ($g\,m^{-2}$)
LF outflow 0·05 km	0	0	0	0
Lime tanks 1·0 km	0	0	0	0
1st bridge 2·1 km	0	0	0·04	0·8
Viaduct 4·6 km	0	0	0·05	2·0
C. Linn above 7·1 km	0	0	0·03	1·5
C. Linn below 7·2 km	0·01	0·09	0·05	1·0
Drumruck 9·1 km	0·05	0·47	0·07	1·2
Culreoch 11·5 km	0·02	0·08	0·06	0·98
British upland waters[a]	0·33 (0·25–0·45)	4·5 (3·5–6·2)	0·075 (0·02–0·17)	3·3 (1·5–6·5)

[a]Turnpenny (1989): mean values for waters of pH $>6·5$; 95% confidence limits shown in parentheses.

3.3 TARGET WATER QUALITY FOR FISH AND OTHER AQUATIC FAUNA

3.3.1 Ecological Objectives of Liming

The objectives of any liming programme will vary according to the species to be protected or re-established, and the type of habitat to be restored (spawning streams, rivers, lakes, hatcheries, fish farms). A suitable target water quality must first be defined therefore. Whilst most fish species can flourish in hard waters at the relatively high pH values of 7 to 9 (Alabaster and Lloyd, 1982), such conditions would be difficult to establish in acidified ecosystems and possibly at the expense of other components of the ecosystem, both aquatic and, in the case of catchment liming, terrestrial. To develop cost-effective treatments that meet these objectives we must first have a concept of the minimum water quality requirements, and of the

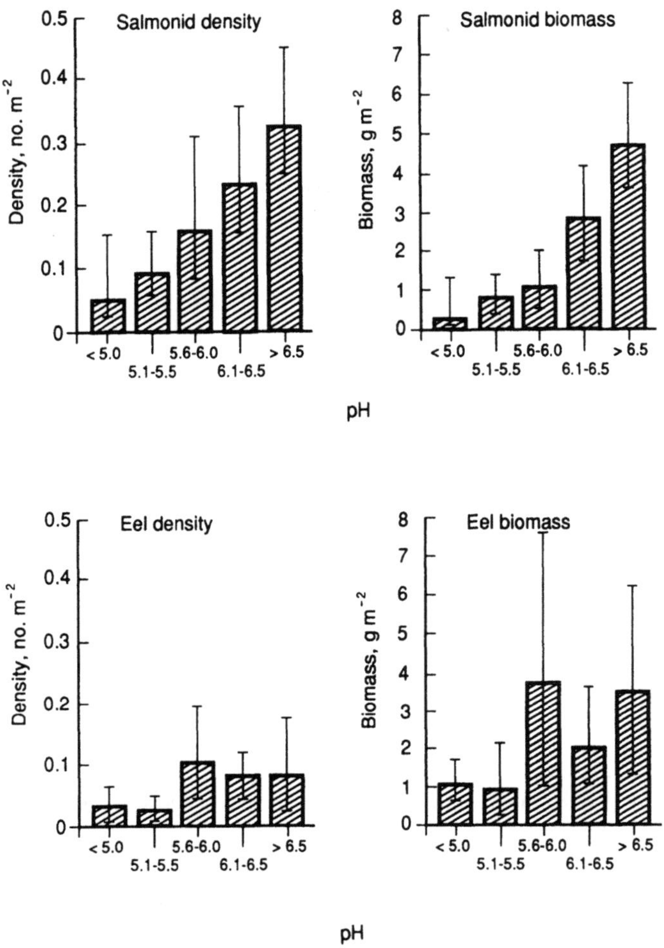

FIG. 3.4 Salmonid (brown trout) and eel population density and biomass in relation to pH in the Little Water of Fleet, 1985 (means and 95% confidence limits). (After Turnpenny, 1989.)

maximum level of liming-induced change that will not radically alter the nature of the ecosystem. The balance between initial dose and repeat frequency for liming treatments should depend primarily on these ecological requirements and only secondarily on economic factors.

The Loch Fleet project was founded on the assumption that the decline and eventual loss of the brown trout fishery from 1950 to the mid-1970s was a direct result of acidification. A principal aim of the project was to restore a self-sustaining fishery after bringing the water quality back into a range suitable for brown trout by the application of one or more treatments to its catchment or waters. During the 2-year (1984–86) pre-intervention phase of the project, it was necessary to test this assumption and to propose target water qualities for the spawning streams and lake that would meet the desired objectives.

In this section the chemical properties of acid waters relevant to fish are reviewed and experimental studies on the toxicity of the waters in the Fleet System before and after liming are described. Finally, critical values of water quality characteristics are identified as a basis for setting target values for liming programmes.

3.3.2 The Toxicity of Acid Waters to Fish

Field surveys of upland waters in Britain and northern Europe indicate that fish populations can exist over a broad range of ambient pH levels, though becoming progressively impoverished with reducing mean pH value (Fig. 3.4).

Both laboratory and field studies have shown that calcium plays a crucial role in moderating hydrogen ion (low pH) toxicity. Laboratory studies (Brown, 1983) have shown that whilst early survival of freshly-fertilised brown trout eggs is poor at pH < 5 in waters of very low calcium content, almost complete protection against $[H^+]$ toxicity is provided by a calcium level $> 100\ \mu eq\ litre^{-1}$ for water of pH greater than 4·5 (Fig. 3.6). Extensive field investigation of 700 lakes in southern Norway where calcium levels were predominantly less than $100\ \mu eq\ litre^{-1}$ showed that the presence or absence of fish was strongly correlated with calcium concentration, and the $Ca^{2+} : H^+$ ratio (expressed as chemical equivalents) provided a useful index of fishery status, with fisheries generally present in waters where $Ca^{2+} : H^+ > 3$ (Chester, 1984). Such correlations have not been established in other synoptic assessments of field data that have included waters with calcium levels predominantly in excess of $100\ \mu eq\ litre^{-1}$ (Turnpenny, 1985; Muniz and Walløe, 1990). Thus field and laboratory data indicate a minimum calcium requirement of $100\ \mu eq\ litre^{-1}$ for waters of pH $> 4·5$ (i.e. the majority of natural waters) ($Ca^{2+} : H^+ > 3·1$) to support a salmonid fishery.

The simple relationships between pH, calcium, and fishery status mask the true complexity of the mechanisms of toxic action in acid waters. Acidity in soil drainage and runoff waters tends to increase the weathering rate of minerals and the leaching rate of trace metals. In addition to releasing basic cations such as calcium, acidity mobilises potentially toxic

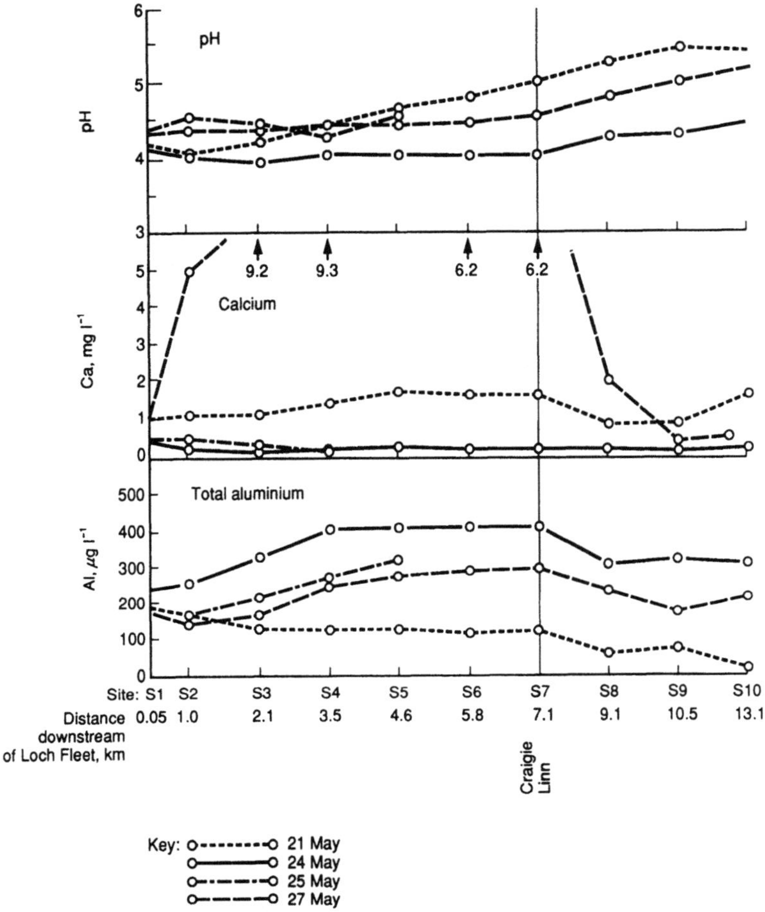

FIG. 3.5 Water quality along the Little Water of Fleet in May 1985, showing pre-spate, peak flow and post-spate conclusions. (Note that the high calcium levels on 27 May were the result of a small liming experiment between sites S1 and S2 on that date.) (After Turnpenny *et al.*, 1988.)

aluminium and heavy metal ions (McDonald *et al.*, 1989). Although low pH appears to be the chief causative agent of toxicity for the freshly-fertilised salmonid egg (Brown and Sadler, 1989), trace metal toxicity is more commonly the cause of mortalities in the post-hatching stages of salmonids. This again is revealed in both field and laboratory investigations. In a study of 60 upland streams in England and Wales, Turnpenny *et al.* (1987) showed significant correlations between biomass of salmonids present and mean concentrations of aluminium and heavy metals (copper,

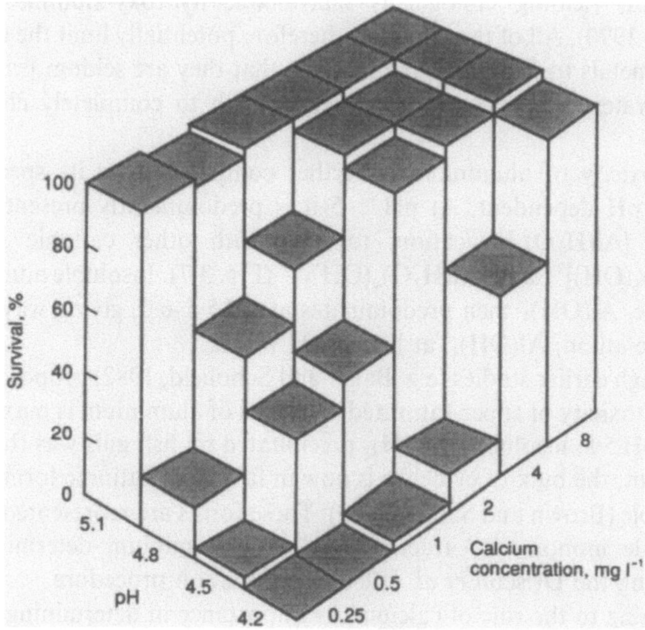

FIG. 3.6 Survival of freshly fertilised brown trout eggs after 8 days in Loch Fleet water adjusted to a range of pH and calcium concentrations (after Brown, 1983).

lead, zinc). Scarcity of fish in this study was related to high levels of aluminium ions in the English Pennine streams, but to heavy metal concentrations in the Welsh streams. This underlines the importance of taking account of regional mineralisation patterns when assessing possible toxic risks to fish and other aquatic biota.

The chemical form of trace metals in natural fresh waters is complex, and determines their degree of toxicity to fish. Ionic forms are generally most toxic (McDonald *et al.*, 1989) but the presence of organic and inorganic ligands which compete with biological surfaces (for example, fish gill membranes) as binding sites for metal ions can reduce the concentrations of biologically available forms, especially of aluminium and copper. Humic acids represent the major category of naturally-occurring organic ligands in upland waters with peat drainage, but inorganic fluoride and sulphate ligands, for which aluminium has a high affinity, are often present at concentrations high enough to limit toxicity. Silicon, present in natural waters ($pH < 9$) as silicic acid ($Si(OH)_4$), also has a very strong affinity for aluminium, yielding biologically unavailable hydroxy-aluminosilicates (Birchall, 1990). All of these ligands therefore potentially limit the toxicity of trace metals to fish, although it seems that they are seldom present in natural waters at concentrations high enough to completely eliminate toxicity.

The toxicity of aluminium is further complicated by its speciation, which is pH dependent. At pH >5 it is predominantly present as the hydrated $[Al(H_2O)_6]^{3+}$ cation, together with other cationic species, $[Al(H_2O)_5(OH)]^{2+}$ and $[Al(H_2O)_4(OH)_2]^{+}$ (Fig. 3.7). Insoluble aluminium hydroxide, $Al(OH)_3$ then predominates at pH 5·5–6·5, giving way to the aluminate anion, $Al(OH)_4^-$ at higher pH values.

Although earlier studies (e.g. Baker and Schofield, 1982) proposed that, since the toxicity of super-saturated solutions of aluminium is maximal at around pH 5·5, insoluble $Al(OH)_3$ precipitated on fish gills was the likely toxic agent, the bulk of evidence is now in favour of cationic forms being responsible (Brown and Sadler, 1989). These forms are represented within the "labile monomeric" fraction (Al_{lab}) of aluminium determinations made using the Driscoll *et al.* (1980) fractionation procedure.

Returning to the role of calcium, its importance in determining fishery status in acid waters can now be stated more clearly. The mechanism of hydrogen ion toxicity to fish is well understood, and derives from its effects on membrane permeability to sodium (Na^+) ions (Potts and McWilliams, 1989). The salt content of fish blood is more concentrated than that of the external medium in fresh waters, and to maintain this gradient the fish must actively take up sodium from the water. Increased $[H^+]$ in the external medium leads to decreased sodium uptake and increased sodium loss, resulting in osmotic stress and ultimately to the death of the fish. Toxic aluminium species act synergistically with hydrogen ions to increase membrane permeability. Calcium ions help to protect fish from these effects. Where heavy metal ions are present, increasing water hardness (calcium plus magnesium) has been shown to limit toxicity, although the biological mechanism is less certain (Alabaster and Lloyd, 1982; Bradley and Sprague, 1985; McDonald *et al.*, 1989). The influential role of calcium in determining fishery status is exerted through amelioration of most or all the naturally-occurring toxic agents present in acid waters.

3.3.3 Significance of Acid Episodes

Streams and rivers draining upland catchments underlain by weathering-resistant bedrock and with thin, acid soils are usually characterised by periodic episodes of acid water, often accompanied by elevated toxic trace

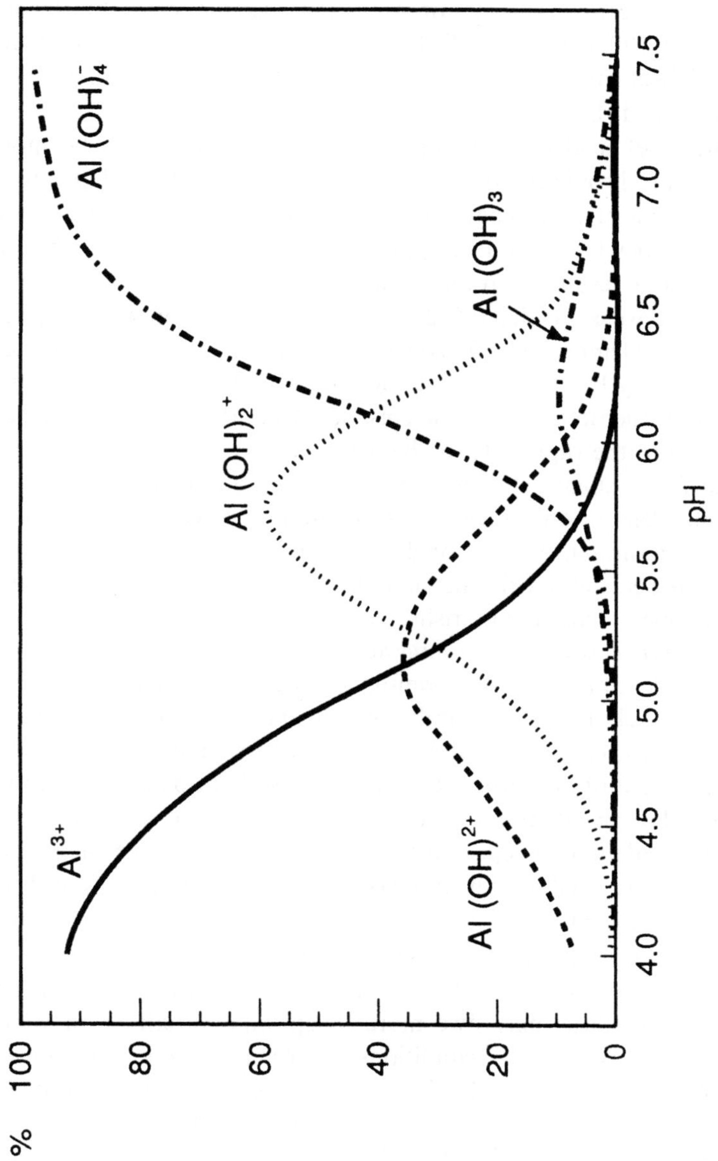

FIG. 3.7 Variation of monomeric aluminium complexes with pH (predicted assuming gibbsite equilibrium) (after Lydersen, 1990).

metal concentrations. These can arise as a result of heavy rainfall or snowmelt leading to rapid surface runoff, or in areas influenced by a maritime climate can be due to deposited sodium ions displacing hydrogen ions from ion exchange surfaces on peaty soils (see the review by Reader and Dempsey, 1989). The episodic behaviour of streams in the Loch Fleet and Fell of Fleet catchments is documented elsewhere in this volume (see Chapters 5 and 11).

The effects of acid episodes on fish may in some waters be more significant than chronic exposure to less acid waters. Reader and Dempsey (1989) provide a brief catalogue of reported fish kills, predominantly of salmonids, associated with acid episodes. The majority of the fish kills reported were a result of prolonged and severe pH depression, i.e. for several days to weeks, with a 0·5–2·5 unit drop in pH. From experimental investigations, Dempsey (1987) concluded that the toxic effects of such episodes are largely the same as toxic effects under constant exposure conditions for the time span of the episode, rather than some heightened response due to the rapid rate of change in conditions.

In small, flashy, upland streams typically used by salmonids for spawning, episodes may last for only a few hours, but may be repeated frequently as a result of intermittent rainfall or diurnal freeze–thaw cycles. In laboratory-simulated episodes with calcium falling from 100 to 0 μeq litre^{-1}, and labile monomeric aluminium rising from 0 to 200 μg litre^{-1} over 6 h, Dempsey (1987) showed that a minimum pH below 3·3 was required to cause mortalities of brown trout swim-up fry in a single episode, conditions which are unlikely to arise naturally. A sequence of such episodes may, however, place a cumulative stress on fish. Morris and Reader (1990) conducted laboratory experiments to measure sodium balance in brown trout exposed to repeated acid episodes. They showed that where a second episode followed an initial episode before homeostasis had been achieved, blood sodium levels fell further. It is easy to envisage how a series of short episodes following in rapid succession could lead to the loss of osmoregulation and eventual death via such a mechanism.

It has been recognised for some time that synoptic relations between low pH and fishery status may owe more to episodic variations in water chemistry than to "average" conditions. Harriman *et al.* (1990) showed that in upland streams in Scotland a good statistical relationship could be shown between mean pH and frequency of acid episodes (defined in their study as periods of pH less than 5·5). Salmonid fry were absent in streams where acid episodes prevailed for more than 30% of the time.

3.3.4 The Toxicity of Fleet System Waters to Fish Before Liming

3.3.4.1 Pre-liming Survival Tests

During the 1984–86 pre-intervention phase, a series of experimental fish survival studies was undertaken (Turnpenny *et al.*, 1988) to confirm the toxicity of waters in the Fleet system to fish. It was established at the start of the project in 1984 that salmonids were absent from the upper 7–8 km of the system, including the loch itself, its inlet streams and the outflow stream. The experiments in turn examined toxic constraints upon each of the life stages of salmonids from the freshly-fertilised ("green") egg, the "eyed" egg, the newly-hatched alevin through to yearling fish, each in its relevant habitat.

3.3.4.2 Brown Trout Eggs

Egg-planting experiments were carried out over the 1984–85 and 1985–86 seasons prior to liming. Freshly-fertilised eggs were obtained from a local source of wild fish in the neighbouring Loch Dee catchment (mean water pH = 5·6). Using an egg box planting technique described by Harris (1973), groups of boxes each containing around a hundred eggs were planted out in streambed gravel at the locations shown in Fig. 3.2. Additional egg boxes were held at a local hatchery operated by the Forestry Commission. These provided an experimental "control", and an indication of developmental progress. Once the "control" eggs reached the "eyed" stage, groups of boxes were lifted at the experimental site, the remainder being lifted after the control eggs had just hatched. Survival rates of "green" eggs through to hatching were less than 20% in both years, mortalities occurring mainly prior to "eyeing" (Fig. 3.8). In a second experiment conducted in the 1984–85 season, batches of eggs were also planted at the just-"eyed" stage to give a better assessment of later development. In this test (Fig. 3.8) more than 95% went on to hatch successfully, confirming the relative sensitivity of the freshly-fertilised egg as shown in laboratory studies.

Water chemistry data, shown only for the 1984–85 season in Fig. 3.9, allow comparison with the laboratory data of Brown (1983) shown in Fig. 3.6. At the average exposure levels of pH 4·5, calcium 50 μeq litre^{-1}, observed mortalities were consistent with laboratory findings. The reported finding that the process of hatching itself is impaired (unsuccessful or delayed) by low pH (Farmer *et al.*, 1980; Muniz and Leivestad, 1980; Peterson *et al.*, 1980; Kugel *et al.*, 1990) was not, however, in evidence at Loch Fleet. Carrick (1979) also has reported normal hatching in Cum-

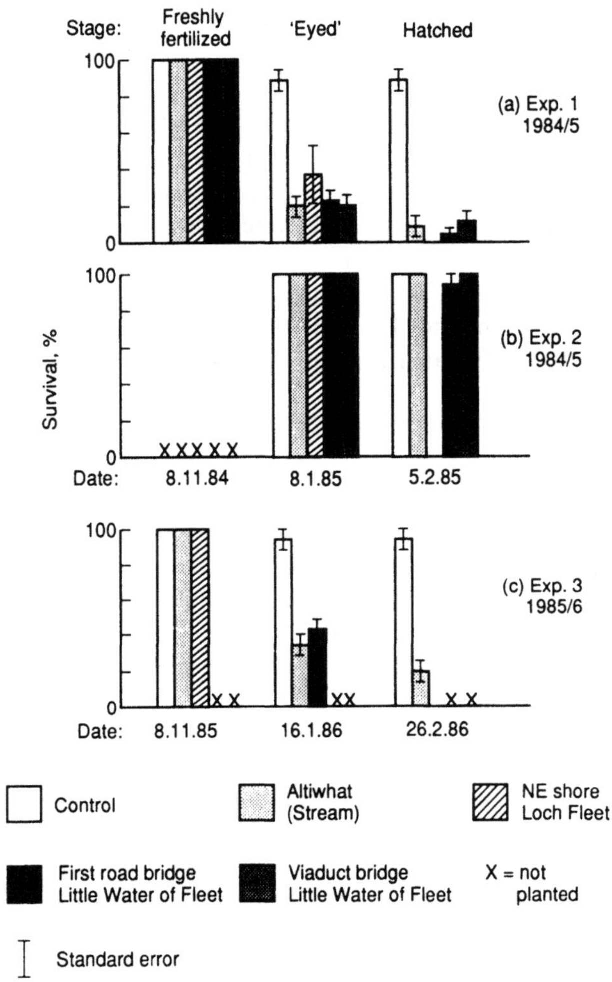

FIG. 3.8 Survival of freshly fertilised and 'eyed' brown trout eggs artificially planted in Loch Fleet and its tributaries in 1984/5 and in 1985/6, prior to liming (after Turnpenny *et al.*, 1988).

brian strains of brown trout and Atlantic salmon eggs at pH values as low as 4·5. It may have been simply that hatching was delayed in our experiments, though we were not able to test this hypothesis.

The possibility that egg mortalities were due to causes other than water quality were investigated but ruled out, for the stream sites at least. Freezing, and the associated drop in loch water levels which exposed egg boxes sited on the loch shore, caused heavy mortalities at this location, but

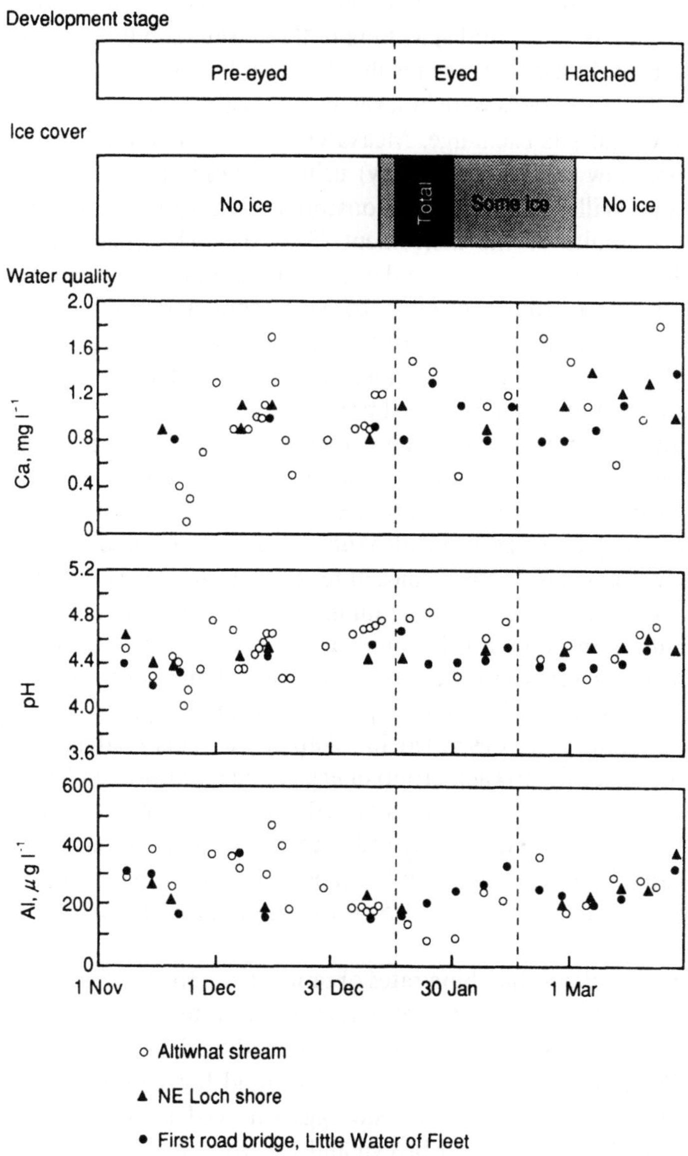

FIG. 3.9 Changes in stream and loch water pH, calcium, and aluminium during egg incubation period, prior to liming 1984/5 (after Turnpenny *et al.*, 1988).

stream water flowed over the gravels throughout incubation. Gravel wash-out resulted in the loss of some egg boxes in both years, and clearly represents a risk to natural spawning in these waters, but data from these boxes were excluded from the results shown in Fig. 3.8. A third possibility was that the gravel beds were blocked by fine sediments, thereby inhibiting water flow and gas exchange. Measurements of streambed permeability and water flow (apparent velocity) using a standpipe technique (Turnpenny and Williams, 1981), demonstrated that values were above the calibration limits of the instrument (8×10^4 cm h^{-1} permeability and 200 cm h^{-1} apparent velocity) and were well in excess of minimum requirements for brown trout eggs given by Turnpenny and Williams (1980).

In parallel with work at Loch Fleet, a series of experiments was carried out on the neighbouring Fell of Fleet site (see Chapter 10) to confirm the water quality requirements of salmonid eggs.

The different intra-gravel water chemistries produced by in-stream lime treatments, together with the untreated "control" conditions, provided the opportunity to assess the requirements not only for the survival to hatching of implanted eggs, but also sub-lethal effects producing low-quality fry which might stand little chance of long-term survival in the wild. A full account of this experimental programme is in preparation (Turnpenny *et al.*, pers. comm.) and only a brief summary of results is given here.

Experiments were carried out over three spawning seasons from 1985 to 1988. Freshly-fertilised eggs were planted out in the egg box/intra-gravel water sampling units described in Chapter 10. This enabled the water quality in proximity to each group of eggs to be monitored independently, and since intra-gravel conditions varied widely, the effects of a range of water chemistries could be assessed. Because of wider interest in species other than brown trout, some of the experiments were extended to include sea trout (anadromous form of *S. trutta*) and Atlantic salmon (*S. salar*) eggs.

Figure 3.10 shows hatching rates obtained from freshly-fertilised brown trout, sea trout, and salmon eggs in relation to water chemistry, with data pooled for the three winter seasons, and including data from the Altiwhat trials. These show a distinct survival threshold for salmon eggs at pH 6 ($Ca^{2+}:H^+ = 100$) below which no eggs survived to hatching. A more gradual trend is apparent for brown and sea trout, reaching an asymptote at around pH 6 ($Ca^{2+}:H^+ = 100$). Also important are the sub-lethal effects: embryos that survived to hatching at pH values below 5–6 were smaller than those hatched at circumneutral values ($Ca^{2+}:H^+ > 100$) (Fig. 3.11) and showed incomplete calcification of the skeleton.

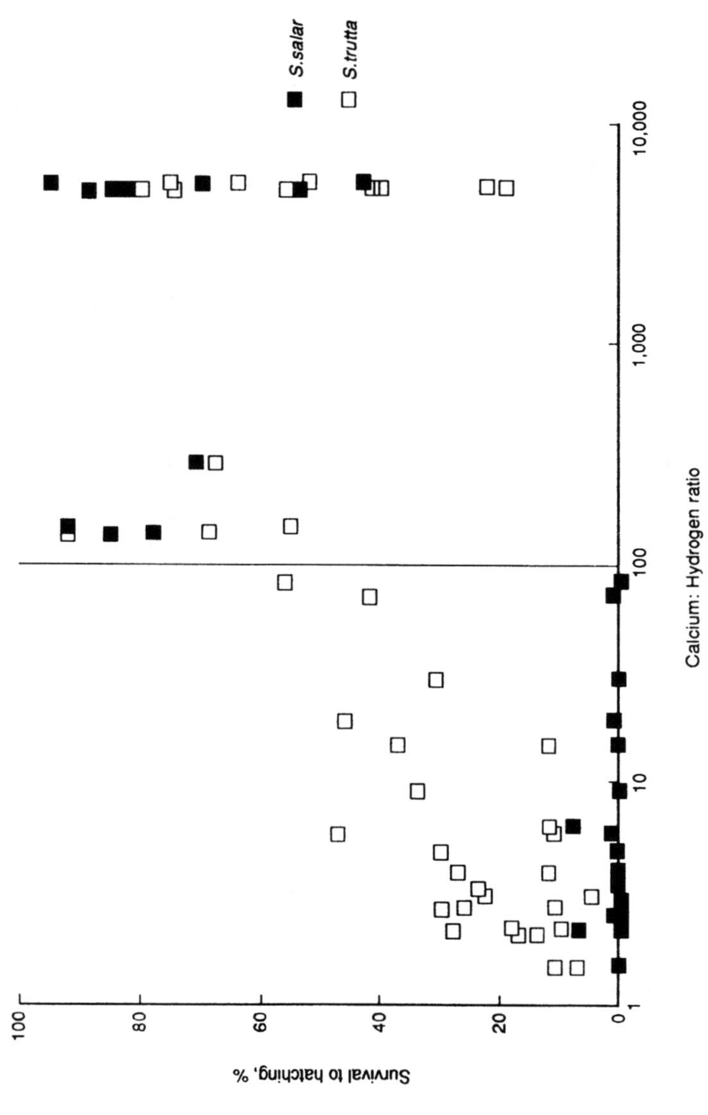

FIG. 3.10 Survival of salmonid eggs in relation to the calcium : hydrogen equivalent ratio in Loch Fleet and Little Water of Fleet waters.

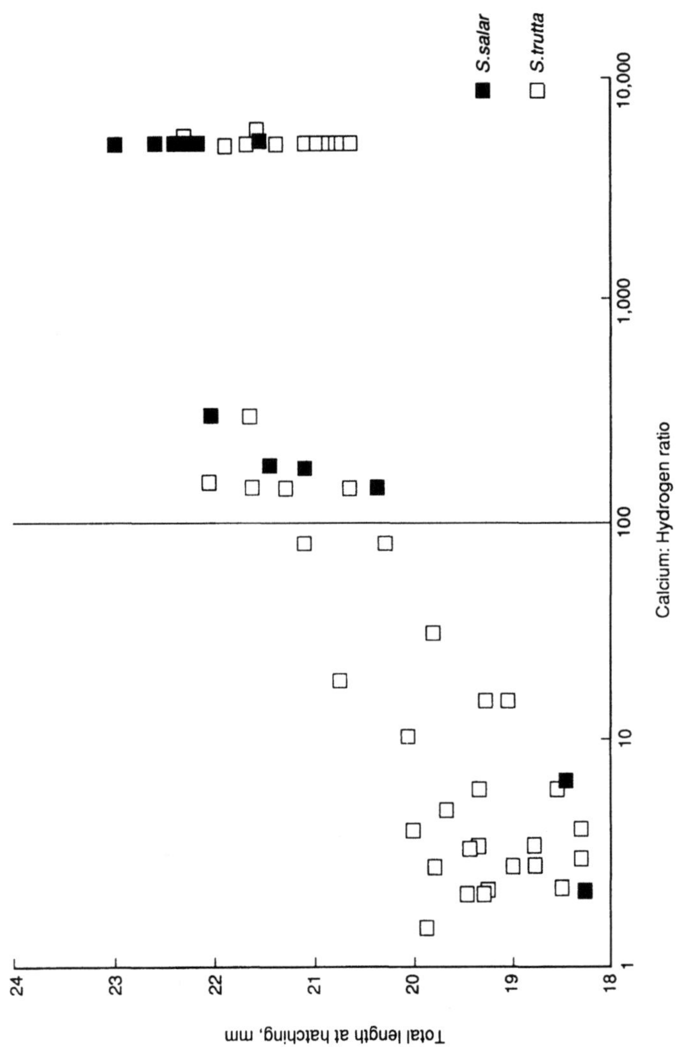

FIG. 3.11 Larval size at hatching of salmonids in relation to calcium : hydrogen equivalent ratio.

Reduced size and incomplete skeletal calcification are likely to limit the chances of survival in the wild. The ecological significance of larval size at hatching in fish has been discussed in general terms by Miller *et al.* (1988) and specifically for *S. trutta* by Elliott (1989). Large size at hatching confers several advantages upon the individual. The point of yolk exhaustion in larvae is considered a critical period for survival as the young fish must find food and learn to feed. Larger individuals can survive longer without food and therefore have increased chances of survival. Secondly, if there is a critical size above which a fish becomes invulnerable to a class of predators, then smaller fish will remain vulnerable for longer. Finally, larger individuals will compete better for available food, and fare better in territorial contests. This last consideration is less important where all the fry are similarly handicapped, but would put fry at a disadvantage where, for example, they migrated downstream to mix with others raised in a better water quality environment.

3.3.4.3 Brown Trout Fry and Fingerlings

Survival trials with brown trout fry and fingerlings were carried out in test aquaria held inside a mobile toxicity testing laboratory (Turnpenny *et al.*, 1988) positioned close to the loch outlet, from which water was drawn. The laboratory has a temperature-controlled aquarium room with an array of 64 test aquaria arranged in 4 columns of 16 (Fig. 3.12). Each column can be pH regulated independently by means of a computer-controlled acid/alkali (H_2SO_4/KOH) dosing system. Separate dosing pumps, one channel per aquarium, introduce additional toxicants or ameliorants to the test tanks as required. The system is most commonly operated in a two-factor design with four levels and four replicates (e.g. four pH levels, four toxicant/ameliorant levels, each replicated four times = 64 aquaria).

For alevin and fingerling tests the following treatment regime was adopted:

treatment 1: untreated loch outlet water, mean pH 4·4–4·5;
treatment 2: loch outlet water adjusted to pH 4·0;
treatment 3: loch outlet water adjusted to pH 5·4;
treatment 4: (control) loch outlet water deionised, carbon filtered, mean pH 5·4, Ca^{2+} = 85 μeq litre^{-1}, with $CaCl_2$ addition, and Na^+ = 23 μeq litre^{-1}, with NaCl addition.

A pH of 4·0 was selected in treatment 1 as the minimal seasonal pH recorded at Loch Fleet. In treatment 3, a pH of 5·4 was chosen to yield

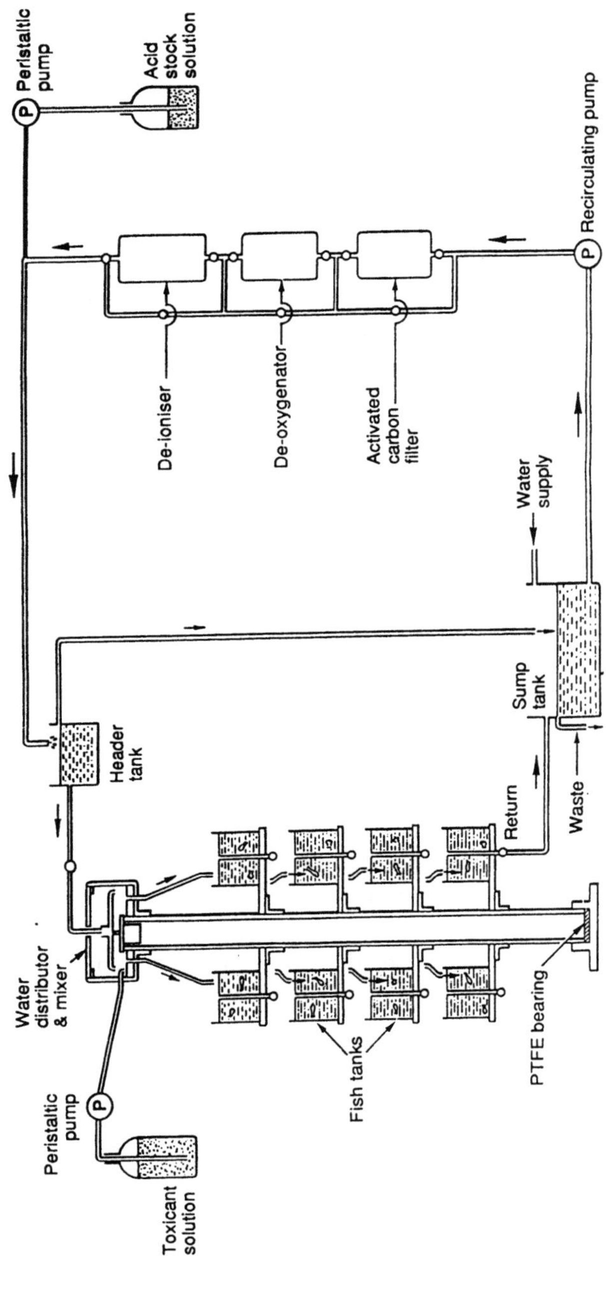

FIG. 3.12 Mobile toxicity laboratory: section of dosing unit exposure tanks and water supply system (after Turnpenny *et al.*, 1988).

maximal aluminium toxicity since any liming treatments applied directly to water, or mixing of acidic loch waters with more alkaline runoff from the land, might lead to enhanced aluminium toxicity owing to changes in speciation.

Trials were carried out in May 1985 using yolk sac fry, and were repeated in October 1985 using fingerlings. The fish tested were from two distinct hatchery-raised stocks: a wild stock from the Dee catchment (as used in egg experiments) raised in the Forestry Commission's Clatteringshaws hatchery, and a reputedly acid-tolerant strain from Stirling University's Bannockburn hatchery. In both tests, trials were carried out over a 9-day period. The results (Fig. 3.13) showed a complete kill of Dee stock exposed as yolk sac fry at pH 4·0 and a 60% kill when exposed to the ambient pH of 4·4. As expected, the Bannockburn stock fared better, but significant mortality (20%–50%) at these pH levels did occur. Raising the pH of the waters to 5·4 increased the 9-day survival to 98%–100%, in line with control values.

The fingerling tests carried out in October 1985, in very similar conditions of water quality (Fig. 3.13), revealed that fish by this stage were much more tolerant of acid conditions. Over 90% of Loch Dee fingerlings survived the 9-day exposure period, with 100% survival in the Bannockburn stock.

More protracted survival tests were carried out between October 1985 and April 1986 with brown trout fingerlings (mean length = 8 cm) held in tanks fed from the loch's two largest feeder streams (sites 6 and 7, Fig. 3.2) and the outlet stream. Dee stock alone were used in these tests. The tanks were heavily lagged to prevent freezing and were divided internally into separate fish-holding compartments. Individual fish could be monitored and replaced on death. The fish were fed initially on a proprietary pelleted trout diet, but it soon became apparent that drift fauna provided an adequate natural food supply and supplementary feeding was abandoned. Median survival times (Table 3.3) of fish in these tests ranged from 8 to 98 days, the longest individual survival being 143 days. No deaths occurred in a control tank held at the Clatteringshaws hatchery over the 180-day test period.

3.3.4.4 Adult Brown Trout

No trials with adult fish were carried out prior to liming, since it was assumed that with high juvenile mortality, few or no fish would reach adulthood in pre-liming water quality.

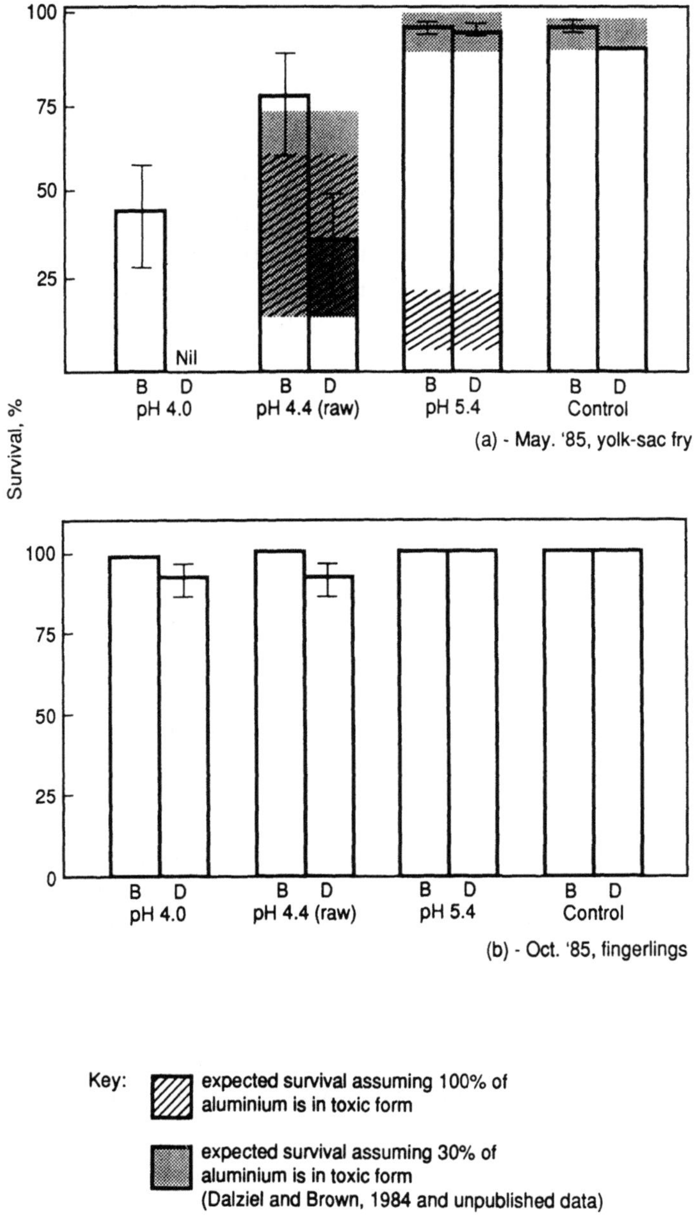

FIG. 3.13 Survival of brown trout yolk-sac fry over a 9 day period from two stocks (B = Bannockburn and D = Loch Dee) in natural and pH adjusted Loch Fleet water, prior to liming. Means and standard errors are shown.

TABLE 3.3
Fish survival and water quality data for Loch Fleet inlet and outlet streams, October 1985 to January 1986 (mean and ranges)

Test site	ET_{50}	*pH*	*Calcium* (μeq *litre*$^{-1}$)	*Aluminium* (*tot*)(μg *litre*$^{-1}$)
Site 6	8 days	4·46	39	28
(n = 54)		4·19–4·68	30–55	140–410
inlet				
Site 7	98 days	4·58	50	22
(n = 58)		4·26–5·03	30–75	140–400
Site OD	41 days	4·59	38	18
(n = 56)		4·29–4·68	30–50	90–230

Notes: ET_{50} is median survival time. "Control" was Clatteringshaws Hatchery where water was circumneutral and ET_{50} was greater than 180 days over the test period.

3.3.4.5 Conditions Responsible for Toxicity at Loch Fleet

Mortality of eggs in the Altiwhat prior to liming can be explained in terms of low pH and calcium levels alone.

Mortality of older fry and fingerling stages appear to have been due to a cocktail of trace metals present acting in concert with low pH and low calcium levels. Dalziel *et al.* (1991) examined levels of iron, copper, manganese, and zinc in Fleet waters pre- and post-liming and showed that pre-liming levels were potentially toxic to trout (Table 3.4). It should be

TABLE 3.4
Changes in some water quality characteristics affecting fish, before and after liming, as measured at Loch Fleet outlet (mean ± SD)

Determinand	*Before liming* *April 1985–March 1986*	*After liming* *April 1987–March 1988*
pH	4·43 ± 0·11	6·60 ± 0·10
Ca^{2+} μeq litre^{-1}	46 ± 19·5	1·84 ± 87·5
Al_{tot} mg litre^{-1}	0·20 ± 0·03	0·08 ± 0·04
Fe mg litre^{-1}	0·15 ± 0·05	0·08 ± 0·03
Cu mg litre^{-1}	0·01 ± 0·01	0 ± 0·02
Zn mg litre^{-1}	0·04 ± 0·03	0·03 ± 0·13
Mn mg litre^{-1}	0·17 ± 0·03	0 ± 0
SiO_2 mg litre^{-1}	0·75 ± 0·32	1·13 ± 0·43
TOC mg litre^{-1}	6·1 ± 1·5	4·8 ± 1·48
F mg litre^{-1}	0	0·08 ± 0·04

noted that the analytical measurements were based on unfiltered samples and may have included non-ionic forms, thereby overestimating their true concentrations.

Aluminium was also present at potentially toxic levels with levels of pH and calcium existing prior to liming. In the 1985 fry/fingerling tests, total aluminium was measured at around 200 μg litre^{-1}. Although no determination of the (toxic) labile monomeric aluminium fraction was made at that time, subsequent analyses (Chapter 12) indicated that about 30% of total aluminium was present in this form, most being organically bound. Thus a concentration of around 60 μg Al_{lab} litre^{-1} can be inferred during these survival tests. Turnpenny *et al.* (1988) showed that on this basis the observed mortality rates were consistent with laboratory findings for brown trout yolk sac fry as reported by Dalziel and Brown (1984).

A separate experiment on the Fell of Fleet site looked at the toxicity of aluminium to brown trout yolk sac fry. The untreated stream waters are characterised by low pH (pH 4·0–4·7), low calcium levels (5–125 μeq litre^{-1}), and high aluminium concentrations (350–600 μg Al_{tot} litre^{-1}).

In the experiment, untreated stream water was fed to the 64-tank array in the mobile laboratory. The water quality was then modified to give a range of four pH levels and four labile monomeric aluminium levels, as shown in Fig. 3.14. Aluminium (Al_{lab}) concentrations were controlled by the addition of a sodium citrate solution, as described by Driscoll *et al.* (1980), so that the highest Al_{lab} concentration was that of the untreated water (425 μg Al_{tot} litre^{-1}; 327 μg Al_{lab} litre^{-1}). Lower concentrations were obtained by adding citrate to achieve 33%, 66%, and 82% complexation, respectively (i.e. 225, 114, and 61 μg Al_{lab} litre^{-1}). Total aluminium levels remained constant. Sodium (as NaCl) was added to the tanks to maintain the same sodium level in all tanks irrespective of citrate level.

The results of percentage survival after 12 days (Fig. 3.14) show a relatively small effect of pH *per se*, as would be expected at the calcium level of 100 μeq litre^{-1}, but substantially increased mortality at the higher aluminium levels. This experiment confirmed that the bulk of the aluminium was present in a biologically available form.

3.4 CAUSES OF FISH LOSS AT LOCH FLEET AND WATER QUALITY: TARGETS FOR LIMING

3.4.1 Decline of the Loch Fleet Fishery

This chapter has identified the principal toxic risks associated with acid

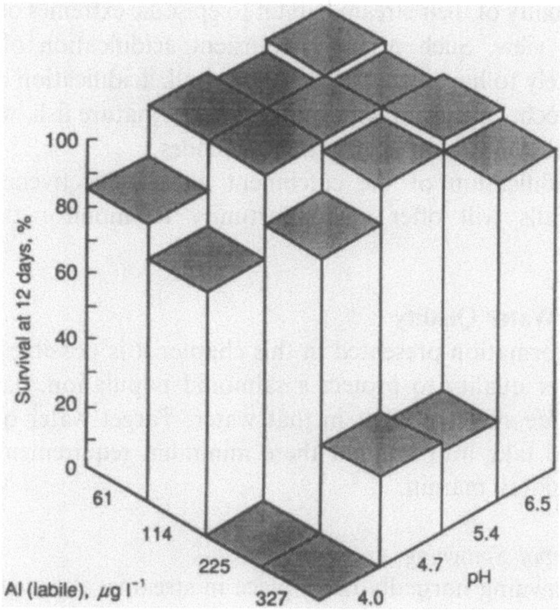

FIG. 3.14 Survival of brown trout in pH-adjusted natural water (Fell of Fleet) with addition of sodium citrate to complex aluminium. Each block represents the mean of 4 replicates of 10 fish.)

waters and a possible explanation for the loss of fish from Loch Fleet between 1950 and 1975.

Although we shall never be certain about the mechanism of fish decline at Loch Fleet, it is quite clear from palaeolimnological evidence (see Chapter 14) and limited historical water quality data (see Chapter 12) that the waters of Loch Fleet underwent acidification from a pH level of around 5·5, or higher recorded in 1965, to a pH of 4·4 measured in 1984. At some point during the course of this rapid acidification process, water quality became critical to fish, leading to their extinction in the mid-1970s.

Fishery records during this period reveal a steady decline in rod returns, accompanied by a rise in average fish weight landed (Fig. 3.1). This pattern is symptomatic of recruitment failure, resulting in a progressively ageing adult population. The particular vulnerability of the early juvenile stages to acid-associated toxicity, demonstrated directly at Loch Fleet, and the high susceptibility of their stream habitat to episodic extremes of chemistry, support this view. Such periodic, transient acidification of spawning streams is likely to have taken place before bulk acidification of the main body of the loch, consistent with the survival of mature fish, which are in any case more tolerant of acidity than juveniles.

The re-acidification of the catchment as the effectiveness of lime treatments falls will offer an opportunity to monitor this process directly.

3.4.2 Target Water Quality

From the information presented in this chapter it is possible to define a minimal water quality to protect a salmonid population, based on the appropriate life stages present in that water. Target water qualities for liming should take into account these minimum requirements, with an appropriate safety margin.

3.4.2.1 Principal Spawning Areas

Salmonid spawning normally takes place in streams, although where no suitable streams exist spawning may occur on lake shores (Frost and Brown, 1967). It should be noted that the intra-gravel waters of spawning beds may differ from the overlying waters.

The freshly-fertilised egg is most sensitive to low pH and calcium but is relatively insensitive to aluminium and trace metal toxicity. As the "eyed" egg develops it becomes less sensitive to water quality, provided that oxygen supply is adequate, but on emergence from the egg membrane the larva becomes more sensitive to low pH, low calcium, and the presence of toxic metal species. Water quality in spawning habitats must be suitable for all these stages.

Whilst a relatively low $Ca^{2+} : H^{+}$ ratio of 3–4 will allow freshly-fertilised brown trout eggs to survive, investigations at Fell of Fleet, adjacent to Loch Fleet, indicate that a ratio of at least 100 is required to promote proper development and skeletal calcification. At a calcium level of $100\,\mu eq\,litre^{-1}$ this would require a minimum pH of 6·0, taking into account the possibility of episodes of lower pH. For Atlantic salmon, a minimum pH of 6·0 is also indicated.

At these calcium and pH levels, laboratory and field exposure tests have shown that brown trout fry are unaffected by aluminium concentrations (as labile monomeric) of less than $100\,\mu g\,litre^{-1}$.

For heavy metals, there is a lack of definitive toxicity data, but data given in Table 3.3 provide some indication of toxic levels. On the basis of field studies in waters with predominantly greater than 100 μeq Ca litre^{-1}, Turnpenny *et al.* (1987) showed that good salmonid populations were present in waters where the total (summed) concentration of heavy metals was less than 0·4 of the threshold LC_{50} for rainbow trout (*S. gairdneri*) (recently renamed *Oncorhynchus mykiss*), as calculated by the method of Brown (1967). In areas of heavy mineralisation involving these metals, calcium levels above 100 μeq litre^{-1} may be necessary to overcome toxicity. However, where the catchment is limed, mobilisation of metals will in any case be decreased by raising the pH of soil waters.

3.4.2.2 Nursery Habitats

Nursery habitats, occupied by fry that have emerged from the gravel, normally overly spawning beds in streams or on lake shores. At this stage of development tolerance to acid waters is increased, and criteria for spawning habitats will therefore protect juveniles.

3.4.2.3 Adult Habitats

Adult salmonids are generally more tolerant of acidic waters than juveniles and water quality criteria for juveniles will therefore normally protect adults. Adult habitats (rivers and lakes) are also better buffered (by their volume) against episodic changes and this provides further protection. In some instances "put-and-take" fisheries based on adult fish, rather than self-sustaining fisheries, are preferred, and can be achieved at lower liming application rates, or with no liming at all.

A special case is presented by the anadromous salmonids returning from salt to fresh water (Potts *et al.*, 1990). At this point in their life cycle the ionoregulatory systems of the fish are already under stress, rendering them more sensitive to acid and aluminium induced sodium losses.

3.4.3 Summary

Target water quality must be related both to the fish species of concern, and to the management objectives of the fisheries. Minimum water quality thresholds for major mitigation programmes are set at pH 6·0–6·5, and ANC at 100 μeq litre^{-1} (Olem, 1990). For reliming, these values are somewhat reduced (pH = 6·0, ANC 10–50 μeq litre^{-1}) because "surplus" alkalinity is not needed to prolong the benefit of one-time treatments. These values will protect sensitive life stages (e.g. spawning and fry development) of salmonid species and lower threshold values are likely to be adequate to ensure growth and health of 2+ year fish (e.g. brook trout; Gloss *et al.*, 1989) stocked for a put-and-take fishery. It must be said, however, that confirmation of these targets awaits longer term intensive studies of fish population response to liming programmes, few of which have so far been reported in the literature.

Since there is strong evidence from both field and laboratory studies that aluminium, in addition to hydrogen ion, concentration is important in fish survival, a target water quality for dissolved inorganic aluminium should also be specified (Howells *et al.*, 1990). In the case of catchment liming, however, where sufficient lime is applied to soils to ensure satisfactory pH and calcium concentrations in runoff, mobilisation of labile aluminium will be reduced and toxic concentrations will not be present in surface waters.

REFERENCES

Alabaster, J.S. and Lloyd, R. (1982). *Water Quality Criteria for Freshwater Fish.* Butterworths, London, 298 pp.

Baker, J.P. and Schofield, C.L. (1982). Aluminium toxicity to fish in acidic waters. *Water Air Soil Poll.* **18**: 289–309.

Birchall, J.D. (1990). The role of silicon in biology. *Chemistry in Britain*, **February 1990**: 141–44.

Bradley, R.W. and Sprague, J.B. (1985). The influence of pH, water hardness and alkalinity on the acute lethality of zinc to rainbow trout (*Salmo gairdneri*). *Can. J. Fish Aquat. Sci.* **42**: 731–36.

Brown, D.J.A. (1983). The effects of calcium and aluminium concentrations on the survival of brown trout, *Salmo trutta*, at low pHs. *J. Fish Biol.* **18**: 31–40.

Brown, D.J.A. and Sadler, K. (1989). Fish survival in acid waters, pp. 31–44 in *Acid Toxicity and Aquatic Animals* (eds. R. Morris, E.W. Taylor, D.J.A. Brown and J.A. Brown), SEB Seminar Series, **34**. Cambridge University Press, Cambridge, 272 pp.

Brown, V.M. (1967). The calculation of the acute toxicity of mixtures of poisons to rainbow trout. *Water Res.* **2**: 723–33.

Campbell, R.N.B. (1987). The status of fish populations in some lochs vulnerable to acid deposition in Scotland, pp. 43–53 in *Acidification in Scottish Lochs* (eds. P.S. Maitland, A.A. Lyle and R.N.B. Campbell). Institute of Terrestrial Ecology, Natural Environment Research Council, 71 pp.

Carrick, T.R. (1979). The effect of acid water on the hatching of salmonid eggs. *J. Fish Biol.* **14**: 165–72.

Chester, P.F. (1984). General discussion, pp. 306–7 in *Ecological Effects of Deposited Sulphur and Nitrogen Compounds.* Royal Society, London, 319 pp.

Cordone, A.J. and Kelley, D.W. (1961). The influences of inorganic sediment on the aquatic life of streams. *Calif. Fish Game* **47**: 189–228.

Dalziel, T.R.K. and Brown, D.J.A. (1984). *Survival of Brown Trout* (Salmo trutta L.) *Fry in Solutions of Aluminium and Calcium at Low pH.* CERL TPRD/L/2562/N83.

Dalziel, T.R.K., Paterson, K. and Proctor, M.V. (1991). Loch Fleet Project – Trace Metal Concentrations before and after liming; significance to fish survival. PowerGen Tech. Mem. TF/91/23113/R.

Dempsey, C.H. (1987). *Investigations into the Causes of Acute Toxicity to Brown Trout of Episodes of Reduced pH.* CERL TPRD/L/3077/R86.

Driscoll, C.D., Baker, J.P., Bisogni, J.J. and Schofield, C.L. (1980). Effects of aluminium speciation on fish in dilute waters. *Nature (London)* **284**: 161–64.

Egglishaw, H., Gardiner, R. and Foster, J. (1986). Salmon catch decline and forestry in Scotland. *Scott. Geogr. Mag.* **102**: 57–61.

Elliott, J.M. (1989). The critical-period concept for juvenile survival and its relevance for population regulation in young sea trout. *Salmo trutta. J. Fish Biol.* **35** (Suppl. A): 91–98.

Farmer, C.J., Goff, T.R., Ashfield, D. and Samant, H.S. (1980). Some effects of acidification of Atlantic salmon rivers in Nova Scotia. *Can. Tech Rep. Fish Aquat. Sci.*, No. 972, 13 pp.

Frost, W.E. and Brown, M.E. (1967). *The Trout.* Collins, London, 316 pp.

Gash, J.H.C., Oliver, H.A., Shuttleworth, W.J. and Stewart, J.P. (1978). Evaporation from forests. *J. Inst. Wat. Eng. Sci.* **32**(2): 104–110.

Gloss, S.P., Schofield, C.L. and Marcus, M.D. (1989). *Liming and Fisheries Management Guidelines for Acidified Lakes in The Adirondack region.* U.S. Fish & Wildlife Service, Biol. Report 80 (40.27). US Dept of Interior, 59 pp.

Harriman, R. and Morrison, B.R.S. (1982). Ecology of streams draining forested and non-forested catchments in an area of Scotland subject to acid precipitation. *Hydrobiologia* **88**: 251–63.

Harriman, R., Morrison, B.R.S., Caines, L.A., Collen, P. and Watt, A.W. (1987). Long-term changes in fish populations of acid streams and lochs in Galloway, south west Scotland. *Water Air Soil Poll.* **32**: 89–112.

Harriman, R., Gillespie, E. and Morrison, B.R.S. (1990). Factors affecting fish survival in Scottish catchments, pp. 343–55 in *The Surface Water Acidification Programme* (ed. B.J. Mason). Cambridge University Press, Cambridge, 522 pp.

Harris, G.S. (1973). A simple egg box planting technique for estimating the survival of eggs deposited in stream gravel. *J. Fish Biol.* **1**: 85–88.

Howells, G., Dalziel, T.R.K., Reader, J.M. and Solbe, J. (1990). EIFAC water quality criteria: aluminium. *Chem. Ecol.* **4**: 117–73.

Jensen, K.W. and Snekvik, E. (1972). Low pH levels wipe out salmon and trout populations in southernmost Norway. *Ambio* **1**: 223–25.

Kugel, B., Hoffman, R.W. and Friess, A. (1990). Effects of low pH on the chorion of rainbow trout, *Onchorhynchus mykiss*, and brown trout, *Salmo trutta v. fario. J. Fish Biol.* **37**(1): 301–10.

Maitland, P.S., Lyle, A.A. and Campbell, R.N.B. (1987). *Acidification and Fish in Scottish Lochs.* Institute of Terrestrial Ecology, Natural Environment Research Council, 71 pp.

McDonald, D.G., Reader, J.P. and Dalziel, T.R.K. (1989). The combined effects of pH and trace metals on fish ionoregulation, pp. 221–42 in *Acid Toxicity and Aquatic Animals* (eds. R. Morris, E.W. Taylor, D.J.A. Brown and J.A. Brown), SEB Seminar Series, **34**. Cambridge University Press, Cambridge, 282 pp.

Miller, T.J., Crowder, L.A., Rice, J.A. and Marschall, E.A. (1988). Larval size and recruitment mechanisms in fishes: toward a conceptual framework. *Can. J. Fish. Aquat. Sci.* **45**: 1657–70.

Morris, R. and Reader, J.P. (1990). The effects of controlled chemical episodes on the survival, sodium balance and respiration of brown trout, *Salmo trutta* L., pp. 357–68 in *The Surface Waters Acidification Programme* (ed. B.J. Mason). Cambridge University Press, Cambridge, 522 pp.

Muniz, I.P. and Leivestad, H. (1980). Toxic effects of aluminium on the brown trout, *Salmo trutta* L., pp. 320–21 in *Proc. Int. Conf. Ecological Impact of Acid Precipitation* (eds.) D. Drablos and A. Tollan). Sandefjord, Norway, SNSF Project, 383 pp.

Muniz, I.P. and Walløe, L. (1990). The influence of water quality and catchment characteristics on the survival of fish populations, pp. 327–38 in *The Surface Water Acidification Programme* (ed. B.J. Mason). Cambridge University Press, Cambridge, 522 pp.

Olem, H. (1990). *Liming acidic surface waters.* Lewis Publishers, Michigan, USA, 331 pp.

Peterson, R.H., Daye, P.G. and Metcalfe, J.L. (1980). Inhibition of Atlantic salmon (*Salmo salar*) hatching at low pH. *Can. J. Fish Aquat. Sci.* **37**: 770–74.

Potts, W.T.W., Talbot, C., Eddy, F.B. and Williams, M. (1990). Sodium balance in adult Atlantic salmon (*Salmo salar* L.) during migration from seawater to freshwater and to acid freshwater, pp. 369–82 in *The Surface Waters Acidification Programme* (ed. B.J. Mason). Cambridge University Press, Cambridge, 522 pp.

Potts, W.T.W. and McWilliams, P.G. (1989). The effects of hydrogen and aluminium ions on fish gills, pp. 201–20 in *Acid Toxicity and Aquatic Animals* (eds. R. Morris, E.W. Taylor, D.J.A. Brown and J.A. Brown), SEB Seminar Series, **34**. Cambridge University Press, Cambridge, 282 pp.

Reader, J.P. and Dempsey, C.H. (1989). Episodic changes in water quality and their effects on fish, pp. 67–83 in *Acid Toxicity and Aquatic Animals* (eds. R. Morris, E.W. Taylor, D.J.A. Brown and J.A. Brown), SEB Seminar Series, **34**. Cambridge University Press, Cambridge, 282 pp.

Smith, B.D. (1980). The effects of afforestation on the trout of a small stream in southern Scotland. *Fish. Manag.* **11**: 39–58.

Stoner, J.H. and Gee, A.S. (1985). Effects of forestry on water quality and fish in Welsh rivers and lakes. *J. Inst. Wat. Eng. Sci.* **39**: 27–45.

Turnpenny, A.W.H. (1985). *The Fish Populations and Water Quality of Some Welsh and Pennine Streams.* CERL, TPRD/L/2859/N85.

Turnpenny, A.W.H. (1989). Field studies on fisheries in acid waters in the United Kingdom, pp. 45–65 in *Acid Toxicity and Aquatic Animals* (eds. R. Morris, E.W. Taylor, D.J.A. Brown and J.A. Brown), SEB Seminar Series, **34**. Cambridge University Press, Cambridge, 282 pp.

Turnpenny, A.W.H. and Williams, R. (1980). Effects of sedimentation on the gravels of an industrial river system. *J. Fish Biol.* **17**: 681–93.

Turnpenny, A.W.H. and Williams, R. (1981). Factors affecting the recovery of fish populations in an industrial river. *Env. Poll.* **26** (Ser. A): 39–58.

Turnpenny, A.W.H., Sadler, K., Aston, R.J., Milner, A.G.P. and Lynam, S. (1987). The fish populations of some streams in Wales and northern England in relation to acidity and other factors. *J. Fish Biol.* **31**: 415–34.

Turnpenny, A.W.H., Dempsey, C.H., Davis, M.H. and Fleming, J.M. (1988). Factors limiting fish populations in the Loch Fleet system, an acidic drainage system in south-west Scotland. *J. Fish Biol.* **32**: 101–18.

PART II

SITE CONDITIONS AND CHARACTERISTICS

CHAPTER 4

Selection of a Site for Restoration

G. Howells
University of Cambridge, Cambridge, UK

T.R.K. Dalziel
PowerGen plc, Nottingham, UK

4.1 INTRODUCTION

In this book it is our aim to review a variety of practical methods and procedures aimed at restoring acid waters, and to assess the liming project at Loch Fleet in the light of other selected examples. Some of these examples have already been identified in Chapter 1 (Table 1.1) and other examples, where there is relevant information, will be used throughout this text to develop the principles and practice of restoration. In reality, most of the methods that have been adopted have been based on applications of "lime" (as calcium oxide, calcium hydroxide, or calcium carbonate), or sometimes other materials such as magnesian limestones or calcium silicate slag. Some mention will be made here of the more limited experience reported for other materials and methods.

4.2 PRELIMINARY CONSIDERATIONS

The task of mitigating catchment acid conditions and, in turn, the resulting surface water acidity, can be formulated by reference to simple models of acidification. For example, Sverdrup and Warfvinge (1990) express this in terms of the balance between sources of acidity (deposition, biomass acidity, alkalinity output, and nitrification acidity) and alkalinity generated from weathering and exchange processes. It follows that a steady-state positive balance of alkalinity within a catchment is needed to mitigate acid surface water conditions. The application of alkaline materials such as "lime" has been the common approach, although other techniques to reduce the acid input side of the balance (such as vegetation management or burning) have been tested, some at Loch Fleet (see Chapter 9).

Most projects to ameliorate acid conditions in lakes and rivers seem to have been initiated primarily by concern for lost or declining fisheries. In all the cases quoted here the primary concern has been the health of fisheries – their improvement appears to have been the main aim for any action, but more extensive biological improvements are also anticipated (Olem, 1990). Aside from the declines reported in fisheries (e.g. Chapter 1), what other objectives of lake/river restoration have been identified? In many cases water quality improvement is identified as the initial, related task; broader objectives are to reduce extreme physical and chemical fluctuations that are associated with chemical episodes, to reduce the burden of silt or humic materials during spates, to moderate an excessive flush of acidic bog or mine drainage, to reduce the acidity from atmospheric or soil/mine sources, and to reduce the input or toxicity of potentially toxic metals.

Upland waters classed of low "hydrological order" are naturally subject to extremes of physical and chemical fluctuations, and also have a variable content of particulate and dissolved organic materials. They are manageable through a wide variety of land-use and water manipulations for changing water chemistry – hydrological management plays a role through dilution, retention, and diversion schemes, but the input term via drainage/runoff also is potentially amenable through emission control, land use, soil management, and by direct chemical interventions.

Clearly the action selected must first be determined by the objectives – usually identified as the improvement or restoration of a damaged fishery. Recent physiological studies have specified the water quality required to promote the restoration or improvement of fisheries, as well as habitat requirements, e.g. for spawning; food chain or wider ecosystem requirements seem less manageable. It follows that hydrological management and chemical manipulations are additionally most important measures to consider, although genetic selection of viable and accommodating strains is another.

These considerations lead us to select for treatment and management those sites that are likely to be amenable to restoration, using current and past conditions and liming experience elsewhere as a guide.

4.3 SITE SELECTION

4.3.1 General Principles

Given that the primary purpose of restoration is the restoration of a fishery, the most useful sites for study are those that have a past record of good or valued fisheries. Any treatment should also not seriously impair other legitimate and desired uses of the site.

It follows that candidate sites are those reasonably accessible for treatment and further monitoring of the consequences of the treatments employed as well as for those where the fishery potential can later be exploited.

The recorded fishery decline should be attributable to acidification at least as the primary cause, although secondary influences may play a part; thus correcting this major cause is the object of the restorative treatment.

Since many other activities in the area may influence fishery decline, any site for investigation should be as little changed as possible during the course of restoration, bearing in mind that climatic changes or progressive biological changes (extreme weather events, maturation of forest, climax communities) cannot be controlled.

Every site for investigation is bound to have a "unique" quality and there must be a question of the generality of observations derived from it. For that reason, any investigation must seek out the basic scientific principles and processes involved so that the conclusions can be extrapolated with some confidence to other sites. The development of models, which can be validated with time at the original site and confirmed at similar sites, but then applied to quite different sites, is to be encouraged.

The primary characteristics identified from a wide variety of locations reported in the literature (Olem, 1990) include:

- lakes or streams that have lost fish populations or species, or where population is significantly reduced or anomalous;
- lake size greater than 2 ha, and with reasonable depth, to provide suitable fish habitat, but not so large that sampling is difficult or impossible, or the lake hydrology complex;
- lake retention time greater than 0·5 yr, receiving drainage, not seepage waters;
- streams with year-round flow greater than $0{\cdot}2\,m^3\,s^{-1}$ and less than

$25\,m^3\,s^{-1}$ at peak;

- lakes or streams providing suitable spawning conditions (depending on species of concern), and access for natural recolonisation unless restocking is planned;
- water quality (other than acidity, low calcium, high aluminium) suitable for fish well-being, i.e. sufficient oxygen, low ammonium, low trace metals;
- relatively little accumulation of organic sediments in lakes;
- sites for treatment reasonably accessible (but not to a degree where public interference might jeopardise the outcome)
- stable land use for a reasonable period before and after treatment;
- single or limited ownership, and restricted fishing or angling activity during restoration.

Many of the conditions listed above are recognised in a typical "decision tree" approach (Fig. 4.1; Brown, 1990).

Secondary conditions considered limiting are:

- the presence of non-target species (possibly acidophilic) of conservation interest, or considered rare;
- slow or unwilling permit procedures;
- local or regional adverse public feelings against planned management procedures, e.g. liming or drainage schemes;
- absent or distant support facilities, especially for sample handling and chemical analysis, to maintain field equipment.

4.3.2 Practical Aspects

The selection of a site for restoration in practice depends on a combination of factors, and it should be recognised that not all acid waters are suitable or even amenable to the treatments at present available.

It is essential that any candidate sites for liming or other treatment are first investigated to ensure that conditions, and the likelihood of success, are reasonable, particularly with regard to the primary characteristics listed above. Some recorded data, not only for past fishery status but also for pre-treatment water chemistry, hydrological and soil conditions, and other terrestrial and aquatic fauna and flora, are also essential to assess the relative success (or failure) of the treatments applied.

From a practical point of view, it is important to find support, and active collaboration if possible, from local owners or users of the site. Independent (although legitimate) activity, or thoughtless vandalism, can threaten the success of any project and the sanity of committed researchers!

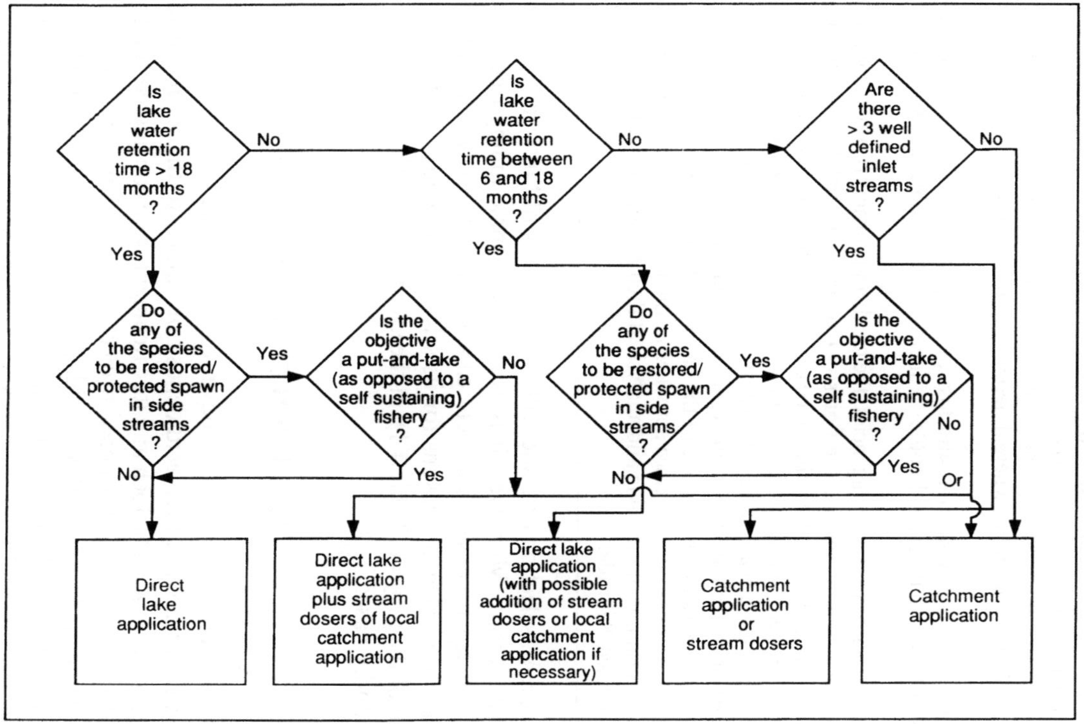

FIG. 4.1 A 'decision tree' to select treatment of lake or catchment by limestone application to restore or protect fish species (after Brown, 1990).

TABLE 4.1
Choice of site

The question of why Loch Fleet was chosen has often been raised. The following conditions were considered favourable: • location in western Scotland, with high acid deposition, similar to that of north-western Europe (e.g. Norway); • similar, related studies were proposed elsewhere in the UK (e.g. western Scotland, west Wales, Cumbria), where different conditions prevailed and treatments differed; • evidence of recent acidification and loss of fish population in the last decade, with expectation that reversal could be met in a reasonable and similar time-scale; • the site was controlled by two agencies, the local land-owners (Cally Estate) who retained the fishing rights, and the Forestry Commission; neither had proposals for change of land use within the time-scale of the proposed project; • access was limited, and the risk of vandalism was thought to be slight; • the site presented a reasonable uniformity of soil and vegetation conditions; a small part of the catchment (about 10%) had been afforested in 1963; • the site location was such that one of the partners (South of Scotland Electricity Board) was well placed to oversee activities in the field, and would undertake a routine programme of sampling and analysis at their Research and Development Centre at East Kilbride.

In the event, selection of a site will involve a mix of "in principle" and pragmatic reasons. Moreover, it is evident that some liming operations may have failed because insufficient care was made in site selection. At Loch Fleet a number of reasons for its selection were identified prior to initiation of the project (Table 4.1); in the event, most of these were found to be soundly based.

4.4 THE LOCH FLEET SITE

Loch Fleet is situated at 340 m above sea level in Galloway, south-west Scotland; the area has a high acid deposition similar to that of northwest Europe, e.g. Norway. The loch and its catchment were considered suitable as a site to undertake the proposed project for a number of reasons. The loch is small (0·17 km^2) and it was acid (pH 4·0–4·5) when the project was initiated in 1984. The loch had once supported a trout fishery and historical records of this and the regional land use were known (see Chapter 3, Fig. 3.1). The catchment was considered manageable by virtue of its size (1·1 km^2) and because there is a single outlet stream. The site is controlled by two agencies; the historic landowners (Cally Estate) retain the fishing

TABLE 4.2
Loch Fleet: environmental data (Bown and Heslop, 1979)

Altitude and topography	Lake 340 m; max. relief + 130 m
Area	Lake 0·17 km^2; catchment excl. loch 1·11 km^2; turnover approx. 0·5 yr
Rainfall, snow	2153 mm yr^{-1}, 42% in Oct.–Jan.; snow approx. 60 days. 80% of rain is of westerly origin
Winds:	
direction	Predominantly westerly (57%) in summer and autumn, easterly in winter
strength	6–11 m s^{-1} (45%), 11–17 m s^{-1} (25%), gales (5·5%)
Air temperature	Cool to cold (550–1100 day deg)
Evapotranspiration	460 mm yr^{-1}
Bedrock and soils	Coarse grained, grey, muscovite–biotite granite, with stony, sandy loam, peaty surface layers, peaty gleys. Thin drifts in Fleet Valley
Vegetation	10% of catchment planted in 1963 (Sitka spruce and lodgepole pine). Rough moorland with *Molinia*, *Calluna*. Forest yields expected as approx. 720 m^3 gross over 60 yr

rights and the Forestry Commission now own and manage the land. Neither had proposals for any changes within the proposed time-scale of the project. The site presented a reasonable uniformity of soil and vegetation conditions, and since access was rather limited, risk from vandalism was thought to be slight (Howells and Brown, 1987).

The underlying geology of the catchment is a coarse-grained, muscovite–biotite granite low in calcium and magnesium, overlain by organic peaty soils or peaty rankers with some glacial drift material. Vegetation is mostly of a rough moorland character dominated by heather (*Calluna vulgaris*) and flying bent grass (*Molinia caerulea*), but the lower (south-west) part (10%) of the catchment was planted with coniferous trees (spruce, lodgepole pine, and larch) by the Forestry Commission in 1963.

Climatic conditions at this exposed, upland site are quite rigorous. The predominantly westerly airstream has a speed in excess of 11 m s^{-1} for more than 25%, and gales for 5% of the time. Precipitation (mean pH $>4·7$) is 2135 mm yr^{-1} (30-yr mean, 1950–80) and snow may be present for 60 days a year (Bown and Heslop, 1979) (Table 4.2). These characteristics of the site are rather extreme in terms of altitude, rainfall, bedrock geology, and poor soils in relation to other upland areas in Britain. It was believed, however, that techniques of acidity mitigation developed here would be applicable to other typically less acid waters and catchments elsewhere in upland Britain.

The major long-term changes in vegetation, farming practice and management, and wildlife are reported for the nearby Cairnsmore of Fleet National Nature Reserve (NCC, 1983) and are summarised (since 1800) in Table 4.3. Vegetation in the area has changed over past centuries from deciduous woodland to birch scrub, heath, and montane species and, more recently, to the predominant planted coniferous forest. Climatic deterioration in Romano-British times brought gradual changes with vegetation stabilised by a mixed grazing regime and occasional burning of encroaching woodlands. In the seventeenth century, sheep monoculture was introduced with regular moorland burning (muirburn) to maintain a stable heathland vegetation. This more radical change also benefited the grouse whose sporting potential was then being appreciated. In more recent times (since 1945) the old sheep monoculture has been replaced by the widespread planting in the area of dense coniferous forest by the Forestry Commission; this has reduced the diversity of habitats with upland moors becoming restricted in extent. Afforestation has significantly altered the hydrological regimes (Hall, 1987), and probably also drainage water quality, and the restriction or prevention of muirburn has probably had substantial consequences for vegetation and soils. The adverse consequences of extensive afforestation on salmonid fisheries have been highlighted for this and other districts of Scotland by Egglishaw *et al.* (1986).

Minerals, both stone for building, and lead, were exploited in the past in the Cairnsmore and adjacent areas. Lead and other metals were mined in the area in the first half of the nineteenth century but the mines were quickly exhausted; the output after 1856 was quite small. Copper and zinc were also worked, mainly opencast, up to the 1914–18 war. A major change to the area was brought about with the construction of the Portpatrick railway (1861), skirting the southern granite pluton west of Skerrow and Grannoch. Not only did this give some incentive to small local industry (e.g. in Gatehouse of Fleet) but opened the area further for game hunting, bringing a change of attitude to the landowners who saw commercial benefit by changing land management for the encouragement of game (wildfowl and deer).

Objective natural history reports of the area about 150 years ago are to be found in *The New Statistical Account of Scotland* (Anon, 1845). Following the Act of Union, and along with a doubling of population in the early years of the nineteenth century, commerce, industry, and agriculture

TABLE 4.3
Brief summary of climate, land-use or other changes in Galloway area since 1800 (*Source*: NCC Report, 1983)

Date	*Climate*	*Land-use changes*	*Other changes*
1800–20	Wet and cold winters, summer droughts in 1815 and 1819	Subsistence agriculture replaced by sheep monoculture with losses in 1810 and 1814. Regular muir-burning.	Upland shepherding, some dyking, bog drainage. Loss of deciduous woods, increase of heather.
1821–40	Severe winters 1823, 1836, and 1838; drought in summer 1826	Feral goats become established. Sheep stocking increased.	Sheep rot in wet, cool summers.
1841–60	Severe winters 1841, 1842, 1845, 1848, and 1855; drought in 1844	Stock disease 1845; potato blight 1846.	Mineral mining for about 20 yr, lead and stone. Upland draining schemes.
1861–80	Severe winters 1861 and 1879; droughts in summers 1864–70; wet years 1872–74	Poor harvests 1868–70.	Wet years 1872–74.
1881–1900	Severe winters 1886, 1891, and 1895; summer droughts 1885, 1893, 1895, and 1896–99		Severe storms 1883, 1891, and 1894.
1901–20	Severe winters 1908, 1917, and 1918; summer droughts 1911; severe snow, rain, and frost 1913	Increase in bracken, *Molinia*, and *Nardus*, more frequent muirburn for sheep and grouse.	Deer herds established; some areas managed for game.
1921–40	Severe winter 1940; cool summers except 1932 and 1933	Extensive coniferous afforestation.	Reduced avian predators needing open range conditions.
1941–60	Many cold winters; summer 1955 and 1959 hot and dry, others cool/warm	Continued forest planting.	
1961–80	Cold winters 1965, 1966, 1969, and 1970; summers 1973 and 1975 very dry	Additional forest planting.	

changed. There is evidence of improved draining of farmlands, application of lime and marl (a limy clay was often used as a manure) to improve barren heathlands, and rich crops of hay were harvested to feed the famous Galloway cattle. Roads and coastal transport were developed for the more effective export of agricultural products and mineral ores from local sources. As a consequence, the small nearby town of Gatehouse of Fleet became a flourishing port with its own local industry, including two water-powered mills and a brick works producing 60 000 bricks a year (Stell, 1986). At the time, it was reported that "The salmon fishings of the Solway and its rivers were . . . of little use having been ruined by stake nets at their mouths". However, Loch Grannoch had pike and perch in considerable numbers, although trout were "few" (in 1844). Similar conditions were reported for four other lochs in the area. For the River Dee (now Loch Ken), trout were said ". . . not to abound, but salmon, sea trout, pike and perch were present". Catches of salmon in the Dee were sometimes large, in spite of the net fishermen in the Solway, with 1426 fish (total weight 800 lb) caught at one pool in a single night in July 1840 (Anon, 1845).

Hydrological and geological conditions were also reported in the last century. In *Fresh Water Lochs of Scotland* (Murray and Pullar, 1910) Loch Fleet is listed and described, and bathymetric soundings based on the survey of Murray in 1903 are given. The geology of the catchment was described by Horne (1870).

More recent information about the management and sporting activities in Loch Fleet and the surrounding area are to be found in the Cally Estate records. Of most relevance to the project and its objectives is the history of brown trout fishery (see Chapter 3).

4.5 DEVELOPMENT OF THE LOCH FLEET PROJECT

4.5.1 Sectors

The project was planned around a sequence of phased activities (Table 4.4). At the start the catchment was divided into a number of discrete "sectors" (Fig. 4.2) on the basis of existing large-scale maps and direct observation of runoff pathways. This provided a set of nine candidate areas for a variety of planned treatments, as well as additional "buffer" zones. These sectors (numbered I to IX) included areas of afforestation and moorland ranging from less than 3 ha to 30 ha. Only sector VII carries a permanent stream (the Altiwhat), although smaller streams in sector IV

TABLE 4.4
Loch Fleet project: schedule of activities

Project activities	1984	1985	1986	1987	1988	1989–94
Installation and instrumentation	Design and install embayments 4 and 8	Test integrity and function				Remove embayments but retain other installations
		Install embayments 5 and 9	Test integrity and function	Use embayment for biology test sites		
	Install flowboxes and water quality probes at IV, VI, VII, and loch outlet		Assessment and interpretation of hydrological record			Install automatic samplers
	Install raingauges at V and IX	Install weather station at loch outlet				
	Establish field laboratory					
Loch bathymetry and sediments	Resurvey loch Sample loch sediments	Test response of sediments to liming	Monitor response of sediments to liming			
Rain and water quality	Select rain and water quality sample sites Set up analytical protocol					
	Instigate sampling programme Set up data handling	Continue sampling programme				Monitor rain and water quality on less intensive programme
		Instigate special investigations: Ground water contribution	Aluminium fractionation, continue post-liming			

(Continued)

TABLE 4.4–*Contd.*

Fishery investigations and other biology	Select sampling sites and methods					Monitor fishery and invertebrate community on an annual basis
	Instigate plankton, invertebrate, and fish sampling	Continue sampling programme …………				
		Survey fish spawning areas				
		Sample macrophyte community	Sample to establish productivity pre- and post-liming …………		Resample macrophytes	Monitor other aquatic and terrestrial components
	Bioassay for eggs, fry and fingerlings (*in situ*) …………		Repeat bioassays …………	Repeat bioassays and test fish in tanks		
				Restock with 1 + fish	Further restocking	
Intervention programme	Select options to be tested		Apply 3 lime treatments on sectors IV, VI, and VII	Burn on Sector VII		
	Set up laboratory tests	Field trials of selected treatments		Monitor soil, vegetation and runoff …………		(Hydrological fine studies and) predictions of treatment effectiveness and duration
	Establish soil conditions			Other lime treatments		
	Survey vegetation					

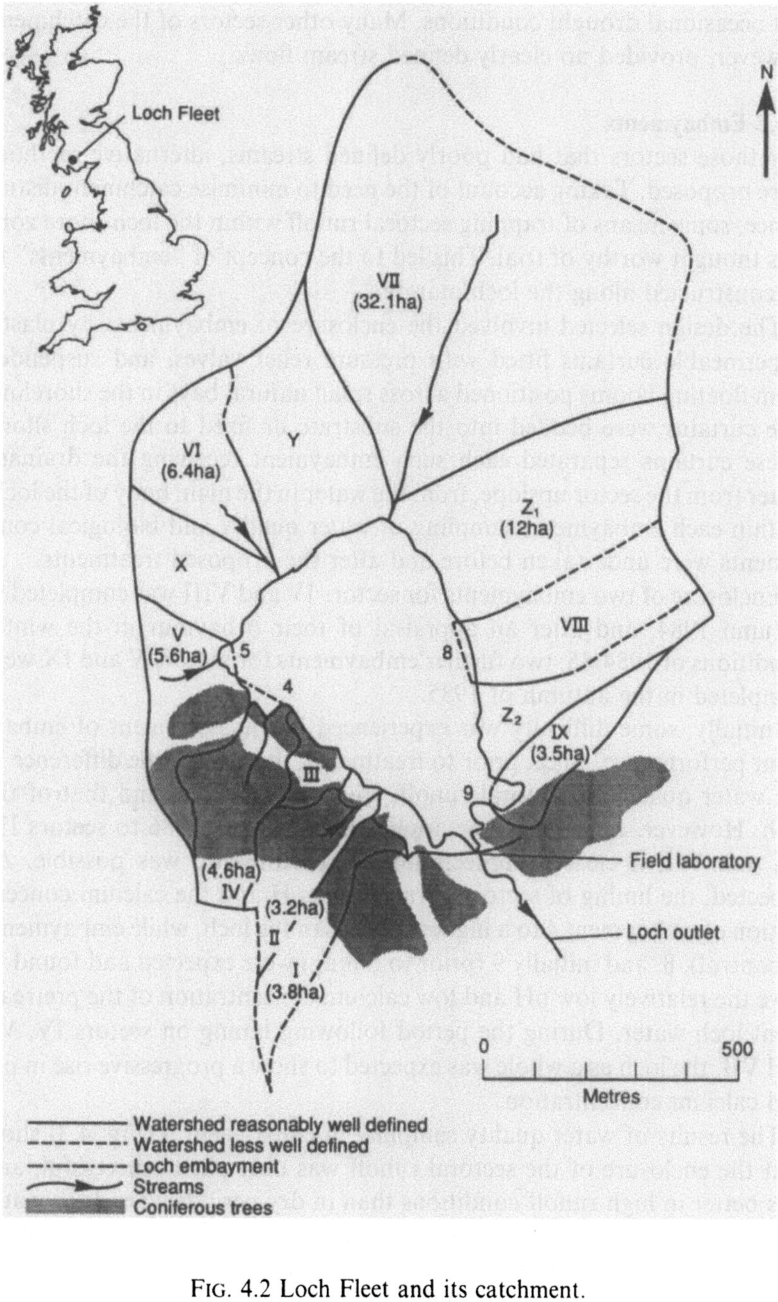

FIG. 4.2 Loch Fleet and its catchment.

(forest) and sector VI (moorland) provided a measurable flow through all but occasional drought conditions. Many other sectors of the catchment, however, provided no clearly defined stream flows.

4.5.2 Embayments

For those sectors that had poorly defined streams, alternative methods were proposed. Taking account of the need to minimise catchment disturbance, some means of trapping sectoral runoff within the loch shore zone was thought worthy of trial. This led to the concept of "embayments" to be constructed along the loch margin.

The design selected involved the enclosure of embayments by plastic impermeable curtains fitted with pressure relief valves, and suspended from floating booms positioned across small natural bays in the shoreline. The curtains were bedded into the substrate or fixed to the loch shore. These curtains separated each such embayment receiving the drainage water from the sector upslope, from the water in the main body of the loch. Within each embayment, sampling of water quality and biological components were undertaken before and after the proposed treatments.

Enclosure of two embayments for sectors IV and VIII was completed in autumn 1984, and after an appraisal of their behaviour in the winter conditions of 1984/85, two further embayments for sectors V and IX were completed in the autumn of 1985.

Initially, some difficulty was experienced in the assessment of embayment performance, since, prior to treatments, there was little difference in the water quality of sectoral runoff, embayment water, and that of the loch. However, after limestone applications in April 1986 to sectors IV, VI, and VII, a clearer appreciation of performance was possible. As expected, the liming of sector IV raised the pH and the calcium concentration of embayment 4 to a higher value than the loch, while embayments 5 (control), 8, and initially 9 (prior to liming) were expected and found to have the relatively low pH and low calcium concentration of the pretreatment loch water. During the period following liming on sectors IV, VI, and VII, the loch as a whole was expected to show a progressive rise in pH and calcium concentration.

The results of water quality sampling in embayment 5 (Fig. 4.3) show that the enclosure of the sectoral runoff was only partly successful, and was better in high runoff conditions than in dry periods when loch water could mix with embayment water through a variety of routes – by seepage below or around the curtain, through the curtain valves, or by wind driven spray and wave action. The individual embayments differed, reflecting

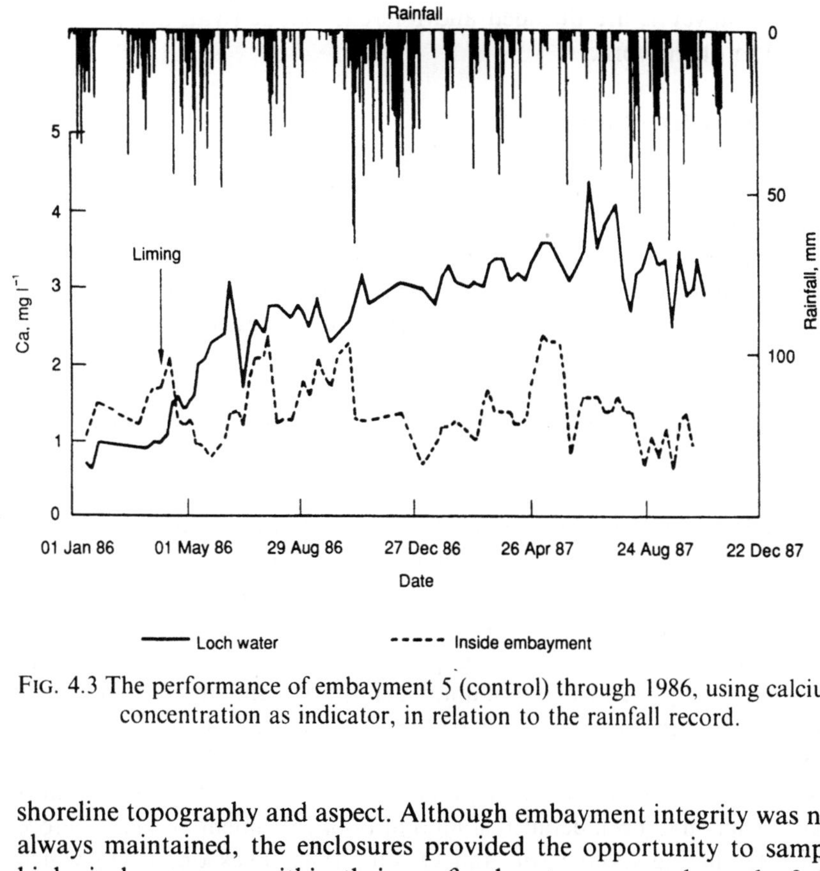

FIG. 4.3 The performance of embayment 5 (control) through 1986, using calcium concentration as indicator, in relation to the rainfall record.

shoreline topography and aspect. Although embayment integrity was not always maintained, the enclosures provided the opportunity to sample biological responses within their confined waters up to the end of the initial project term. In particular, embayment 5 provided some reasonable correspondence to the conditions of the pre-limed loch.

4.5.3 Sampling Programme

The planned schedules of water (loch, embayment, and stream) sampling and biological sampling are shown in Table 4.4, and will be evident in the following chapters in which the results are presented.

In addition to the routine work undertaken by the field staff, and pursued through laboratories of the sponsoring agencies (Howells and Dalziel, 1989) many scientific components of the investigation were commissioned with independent scientists in universities, colleges, and research institutes. Inevitably, it has not been possible within this book to invite all these scientists to participate directly. Some of these investigations, however, are included and reference made (with acknowledgements) to other work.

4.5.4 Longer-term Studies

With the completion of the initial 5 years of the project it was acknowledged that a further 5 years of chemical and biological monitoring would greatly enhance the value of the findings. This will allow the development of more precise predictions (and validation of current ones) of the duration of the various treatments applied, as well as to monitor effects on non-target species within both the aquatic and terrestrial ecosystems.

4.6 COMPARISONS WITH LIMING AT OTHER SITES

4.6.1 Scandinavia

In Sweden, the Lake Gårdsjön liming project was set up to quantify surface water acidification by studying interactions between deposition, buffering processes of the soil and the acidifying contribution of the soil to the aquatic system, and to follow the chemical and biological consequences of lake liming. It was thought important to base such a study in a region of high acid and pollutant deposition, and to have sufficient record of water chemistry and biology for the period prior to liming (Dickson, 1988). The Gardsjon lake (0·31 km^2) and catchment (2·11 km^2) is about 80 km north of Goteburg and about 14 km from the Swedish west coast. There had been detailed studies of both catchment and lake before and after liming (Andersson and Olsson, 1985; Dickson, 1988) which provided a wealth of information on climate, geology, hydrology, soils, runoff water quality, and biological components, including lost fish populations. The 30-year mean annual rainfall is about 1100 mm yr^{-1}, runoff from the thin soils is rapid, and lake retention time is about 1·4 years. The application of finely ground limestone was undertaken in April 1982, calculated from the preliming lake water quality to satisfy criteria of water quality maintenance of pH about 7·0, calcium 350 μeq $litre^{-1}$ and alkalinity 200 μeq $litre^{-1}$ over a 3-year period. Liming was concentrated (80%) on the lake littoral, with 20% delivered to the deeper areas of the lake. In the event, heavy rain events mobilised an excessive part of the dose within the first year, and reacidification had already begun within the planned 3-year term.

At Lakes Lysevatten and Bredvatten, nearby, similar climatic and environmental conditions prevail (Alenäs *et al.*, 1991). Lysevatten is

0·41 km^2 in area and retention time is 2·5 years; Bredvatten has a retention time of about 2 years. Lysevatten has been limed regularly since 1974, aiming to maintain water of circumneutral quality. Initially the liming dose was estimated from the preliming water quality and treated with di-Ca-silicate slag (60 and 40 t, respectively). This material was delivered to shallow areas of the lakes. Later treatments employed 76 and 51 t, respectively, of finely ground (90% $< 0{\cdot}5$ mm) agricultural limestone. At Lysevatten this was undertaken by mixing the lime to a slurry and discharging from a raft in open water areas. The Lysevatten water quality has been maintained through to 1984, with repeated treatments about every two years.

In Norway, Lake Hovvatn was selected to estimate to what extent neutralisation (i.e. liming) is a feasible, technical, and economic means of reducing acidification damage. The site provided the opportunity to compare two liming techniques (Wright, 1985). This site (alt. 500 m) has granitic bedrock and thin soils, with open forest interspersed with peaty areas. Hovvatn (1·1 km^2) has a retention time of 0·8 year and was treated with 200 t of powdered limestone (90% $< 0{\cdot}5$ mm) in 1981, with a further 40 t to Pollen, a small lake (0·05 km^2) draining to Hovvatn. The lime was distributed on snow around the lake littoral, some mixed with snow and delivered by a "blower". Doses were 0·27 and 1·3 t ha^{-1}, respectively, to the two catchments, with an aim to maintain a pH of 6–7 and calcium 50–100 μeq $litre^{-1}$ in the lake water. Brown trout were present until 1948 but could not be found in 1980 in spite of repeated stocking.

In south-central Norway, the two small Tjonnstrond lakes (linked by a stream) drain a 0·25 km^2 catchment; lake retention time is only 2 months. The whole catchment was limed with powdered limestone by helicopter in 1983 using 75 t (3 t ha^{-1}), aiming to maintain acceptable water quality for 30 years. This dose was equivalent to 50 times that to Hovvatn. Before liming, the two linked Tjonnstrond lakes were barren of fish, and stocking over the previous 20 years had failed. The acidity of the lakes was pH 4·4–4·8, with calcium about 50 μeq $litre^{-1}$, and high levels of aluminium were recorded (Rosseland and Hindar, 1988).

In Finland and Sweden several national projects have studied the effects on soil of liming of forest and meadows. Their focus has been on soil changes and the effects on vegetation (see Chapter 11), rather than on the changed quality of runoff and surface waters.

4.6.2 United Kingdom

At Llyn Brianne, south-west Wales, an extensive (17) suite of sites was chosen to investigate the relative contributions of atmospheric and land-use factors in accounting for surface water acidity, with the aim of developing a detailed understanding of acidification processes and of practical management techniques to mitigate acidification. It was recognised that waters in upland Wales had suffered a drop of about 0·5 pH units in the past 100 years, and this had been linked with a decline, and in some cases a loss, of brown trout in some waters of the region (Welsh Water Authority, 1987; Edwards *et al.*, 1990). Afforestation and a decline in agricultural lime applications since 1976 were associated with these changes.

Among the variety of treatments applied to the Llyn Brianne catchments (Hornung *et al.*, 1990) standard agricultural improvement (liming, ploughing, and NPK fertiliser) provided magnesian limestone (agricultural grade) at $10\,t\,ha^{-1}$ on 24% of the catchment. Despite the total of 200 t applied, the runoff stream remains acidic with episodes of pH < 5·0. In an acidic moorland area finely powdered lime (50% < 10 μm) was applied at $9\,t\,ha^{-1}$, resulting in rapid stream water quality improvement that has been sustained over 3 years, although runoff from stagnopodsols and brown podsolic soils show little benefit. In a forest area, stream littoral areas were cleared and magnesian limestone (agricultural grade) applied at $10\,t\,ha^{-1}$; initial response was rapid, attributed to lime falling directly into the stream, and subsequently stream water quality returned to that prior to treatment, although some marginal effects are recorded. A further unplanted area (4 ha) within the forest was treated at $30\,t\,ha^{-1}$ with finely powdered limestone over two periods (October and December 1987) with improved pH and calcium concentrations being maintained through to spring 1988 when pH declined in dry periods, possibly when acidic groundwater or drainage from soil horizons unaffected by lime application was dominant in the stream flow.

Alternative management techniques at Llyn Brianne included forestry contour ploughing and forest felling on bankside verges, with rather little benefit on stream water quality.

Another project in Britain, at Loch Dee in Galloway (Burns *et al.*, 1984), reported a decline in the fishery of the lake and in other waters nearby, which was associated with an increase of intensive afforestation in 1960s. Loch Dee was thought to have become gradually more acidic (a change from pH 6·2 to 5·6) over the past 120 years. Objectives were to assess the effects of acid precipitation and afforestation on the lake ecology, and to develop a realistic management programme to improve the fishery potential. Before liming treatments were undertaken, an intensive monitoring programme was initiated covering water/rain chemistry and biological studies (Welsh and Burns, 1987). The large Dee catchment (150 km^2) drains to the large shallow Loch Dee (1 km^2) with a retention time of only 40 days. Climatic and soil conditions, overall, are similar to those at Loch Fleet. Three subcatchments have been studied: of these, the White Laggan catchment (57 km^2), with a mixture of moorland, peaty areas, and forest, has received a variety of lime applications, mostly by truck delivery alongside the lower stream reach. These treatments were:

spring	1980	– shells in the lower stream reaches:	5 t
autumn	1980	– shells in the lower stream reaches:	5 t
spring	1981	– powdered limestone, direct to stream:	10 t
winter	1981–82	– powdered lime to lower stream reaches:	80 t
		– lime chips on 0·3 ha, 1 km from loch:	80 t
spring	1982	– powdered lime, direct delivery to lower stream reaches:	75 t

The 1981–82 treatments had little effect on lake water quality, while the 1982–83 treatments had only a transitory (1 yr) effect, possibly because of lake sediment capture or by rapid flushing of the lake water. The fishery for brown trout, assessed as critical in 1980, has improved, notwithstanding no significant improvement in water quality (Tervet and Harriman, 1988).

The Rivers Esk and Duddon (Cumbria, UK) have also been used in a catchment liming study (Diamond *et al.*, 1992). Until withdrawal of a liming subsidy in 1976, it was the practice to lime (2000 t yr^{-1}) about 10% of their catchments (about 70 km^2 and 78 km^2, respectively). In June 1980 and September 1983, migrating adult salmon were killed in the lower reaches, just above the tidal limit, and mortality then was linked with low pH ($< 5{\cdot}5$), low concentrations of calcium (< 50–100 μeq $litre^{-1}$) and high aluminium (total) (> 250 μg $litre^{-1}$). There was also evidence of a decline in numbers of returning salmon. Land adjacent to the River Esk (6·4 km^2) was limed at an overall rate of 5 t ha^{-1}, in total 3200 t, in March 1987, and similar applications were made in September 1987 and June 1988. These treatments were followed by a programme of stream water sampling (triggered only at high flows), and biological sampling for invertebrates and fish. Powdered agricultural magnesian limestone (95% < 5 mm, < 10% MgO) was applied to riparian agricultural areas that had routinely received lime prior to 1976, using agricultural methods. While some improvement in stream water pH was detected, calcium concentrations did not increase. An application of 22 t of lime chips to two tributary streams of the Esk and 50 t to the Duddon in 1987–88 also failed to bring any discernible effect on water quality (Diamond *et al.*, 1992).

4.7 SUMMARY

The design and implementation of the Loch Fleet project has provided a detailed investigation of conditions prior to liming and the responses to the variety of treatments applied.

A number of similar investigations elsewhere in Britain and in Scandinavia have many elements in common, particularly their focus on changes in water chemistry targets, changes in aquatic invertebrate communities, and improvement or restoration of fish populations, consistent with their overall common objectives.

Not all the projects reported have had the expected outcome in terms of maintained water chemistry over a reasonable period, or of an improved fishery. Some deficiencies in unsuccessful projects may be attributed to lack of sufficient data and poor understanding, for example of hydrological regimes and pathways, or of soils, or of the fish community, and thus of the quality and quantity of liming agent required.

REFERENCES

Alenäs, I., Andersson, I.B., Hultberg, H. and Romarin, A. (1990). Liming and reacidification reactions of a forest lake ecosystem, Lake Lysevatten, in SW Sweden. *Water Air Soil Poll.* **59**: 55–77.

Anon (1845). *The New Statistical Account of Scotland*, Volume 4. Blackwood, Edinburgh.

Andersson, F. and Olsson, B. (editors) (1985). The Lake Gardsjon – a lake and its catchment, *Ecol. Bull.* **37**: 1–336.

Bown, C.J. and Heslop, R.E.F. (1979). *The Soils of the Country Round Stranraer and Wigtown*. Memoirs of the Soil Survey of Great Britain, Scotland. Macaulay Institute, Aberdeen, pp. 452.

Brown, D.J.A. (1990). Fishery restoration to acid lakes: what do we need to do?, poster at NAPAP 1990 International Conference on Acidic Deposition, 11–16 Feb. 1990, Hilton Head Is., U.S.A.

Burns, J.C., Coy, T.S., Tervet, D.J., Harriman, R., Morrison, B.R.S. and Quine, C.P. (1984). The Loch Dee project. *Fish Manag.* **15**: 145–68.

Diamond, M., Hirst, D., Winder, L., Crawshaw, D.H. and Prigg, R.F. (1992). The effect of liming agricultural land on the water quality of the River Esk, Cumbria. International Conference Acidic Deposition: its Nature and Impacts, Glasgow, UK. *Env. Poll.*, in press.

Dickson, W. (1988). *Liming Lake Gardsjon*. Swed. Nat. Env. Board, 327 pp.
Edwards, R.W., Gee, A.S. and Stoner, J.H. (1990). *Acid Waters in Wales*. Kluwer, Dordrecht, Netherlands, 337 pp.
Egglishaw, H.J., Gardiner, R. and Foster, J. (1986). Salmon catch decline and forestry in Scotland. *Scottish Geogr. Mag.* **102**: 57–61.
Hall, R.L. (1987). Processes of evaporation from vegetation of the uplands of Scotland. *Trans. Roy. Soc. Edin. Earth Sci.* **78**: 327–34.
Horne, J. (1870). Geological Survey of Scotland.
Hornung, M., Brown, S.J. and Ranson, A. (1990). Amelioration of surface water acidity by catchment management, pp. 311–28 in *Acid Waters in Wales* (eds. R.W. Edwards, A.S. Gee and J.H. Stoner). Dordrecht, Netherlands, 337 pp.
Howells, G. and Brown, D.J.A. (1987). The Loch Fleet Project, SW Scotland. *Trans. Roy. Soc. Edin. Earth Sci.* **78**: 241–48.
Howells, G. and Dalziel, T.R.K. (1989). *The Loch Fleet Project: A Report to the Sponsors, 1984–1989*. CEGB, SSEB, & NSHEB, Brit. Coal., 64 pp + annexes.
Murray, J. and Pullar, L. (1910). *Bathymetrical Survey of the Scottish Freshwater Lochs, 1897 to 1909*. Challenger Office, Edinburgh, Vol II, p. 113, Plate XLVI.
NCC (1983). *Land-use History of Cairnsmore of Fleet NNB*. Nature Conservancy Council, Southwest (Scotland) Region, 51 pp.
Olem, H. (1990). *Liming of acidic surface waters*. Lewis Publishers, Michigan, USA, 331 pp.
Rosseland, B.O. and Hindar, A. (1988). Liming of lakes, rivers and catchments in Norway. *Water Air Soil Poll.* **41**: 165–88.
Stell, G. (1986). *Exploring Scotland's Heritage: Dumfries and Galloway, Roy. Comm. Ancient and Historical Monuments of Scotland*. HMSO, Edinburgh, 184 pp.
Sverdrup, H. and Warfvinge, P. (1990). The role of weathering and forestry in determining the acidity of lakes in Sweden. *Water Air Soil Poll.* **52**: 71–78.
Tervet, D.J. and Harriman, R. (1988). Changes in the pH and calcium after selective liming in the catchment of Loch Dee, a sensitive and rapid turnover loch in south-west Scotland. *Aqua. Fish. Manag.* **19**: 191–203.
Welsh, W.T. and Burns, J.C. (1987). The Loch Dee project: runoff and surface water quality in an area subject to acid precipitation and afforestation in SW Scotland. *Trans. Roy. Soc. Edin. Earth Sci.* **78**: 249–60.
Welsh Water Authority (1987). *Llyn Brianne Acid Waters Project, First Tech. Summary Report*. Dept. of Env. & Welsh Office, 91 pp.
Wright, R.F. (1985). Chemistry of Lake Hovvatn, Norway, following liming and reacidification. *Can. J. Fish Aquat. Sci.* **42**: 1103–13.

CHAPTER 5

Deposition Input Considerations

B.R. Stewart and K. Paterson
Scottish Power plc Technology Centre, Glasgow, UK

T.R.K. Dalziel
PowerGen plc, Nottingham, UK

M.V. Proctor
National Power plc, Leatherhead, Surrey, UK

5.1 INTRODUCTION

In any catchment acidity mitigation programme it is essential to understand and quantify the hydrological and biogeochemical processes of the catchment. Quantitative and qualitative measurements of deposition or atmospheric inputs to a catchment are thus important. Catchment hydrological and hydrochemical models (see Chapter 6) require deposition input data and, in association with a complementary programme of stream water or runoff measurement and ionic analysis, measurement of deposition inputs provides data for the calculation of hydrochemical input/output budgets or fluxes. The importance of these are explained and demonstrated in Chapters 12 and 16.

5.2 CATCHMENT INPUT TERMS

Two broad categories describe deposition inputs: wet deposition and dry deposition (Barrett *et al.*, 1987). The delineation of inputs into one or other of these categories is often imprecise and sampling difficult and impractical. Wet deposition refers to atmospheric constituents deposited in solution and in ionic form, whereas dry deposition involves the direct transfer of gases and very small particulate matter. Dry deposited material will eventually be washed from vegetation and ground surfaces in subsequent rain events, contributing to the chemistry of catchment runoff. Dry deposition between rainfall events also contaminates open rain samplers; the consequent sample is termed "bulk deposition". A third form of deposition, known as occult, or interception deposition, involves the trapping of small droplets of condensation (from clouds and fog) in the foliage of vegetation. It has been shown that the concentration of major dissolved ions can be up to two to three times higher in deposition directly impinged from clouds than in typical rain events (Barrett *et al.*, 1987; Irwin *et al.*, 1990). In practice bulk deposition is the usual sample taken for analysis and reported in catchment studies; it is acknowledged to be an underestimate of total deposition.

In addition, accurate measurement of both quantity and the chemical composition of deposition inputs can be affected by other factors. Principally these are to do with the design and siting of the sampling apparatus. Altitude, ground topography, and the proximity of collector or gauge aperture to the ground surface are all factors which may affect quantitative measurements. The design of collectors for chemical analysis of deposition so as to prevent contamination by foreign material from, for example ground splash, and solute concentration by evaporation of water is important. Snowfall measurement has additional problems because of its bulk; in the period from 1985 to 1990 snowfall at Loch Fleet was slight compared with deposition as rain.

5.3 SAMPLING AND ANALYSIS

At Loch Fleet a number of collectors or gauges were used to measure the quantity and quality of deposition inputs (Fig. 5.1). Comparisons were made also between different designs and siting locations. The collectors and gauges used are summarised below.

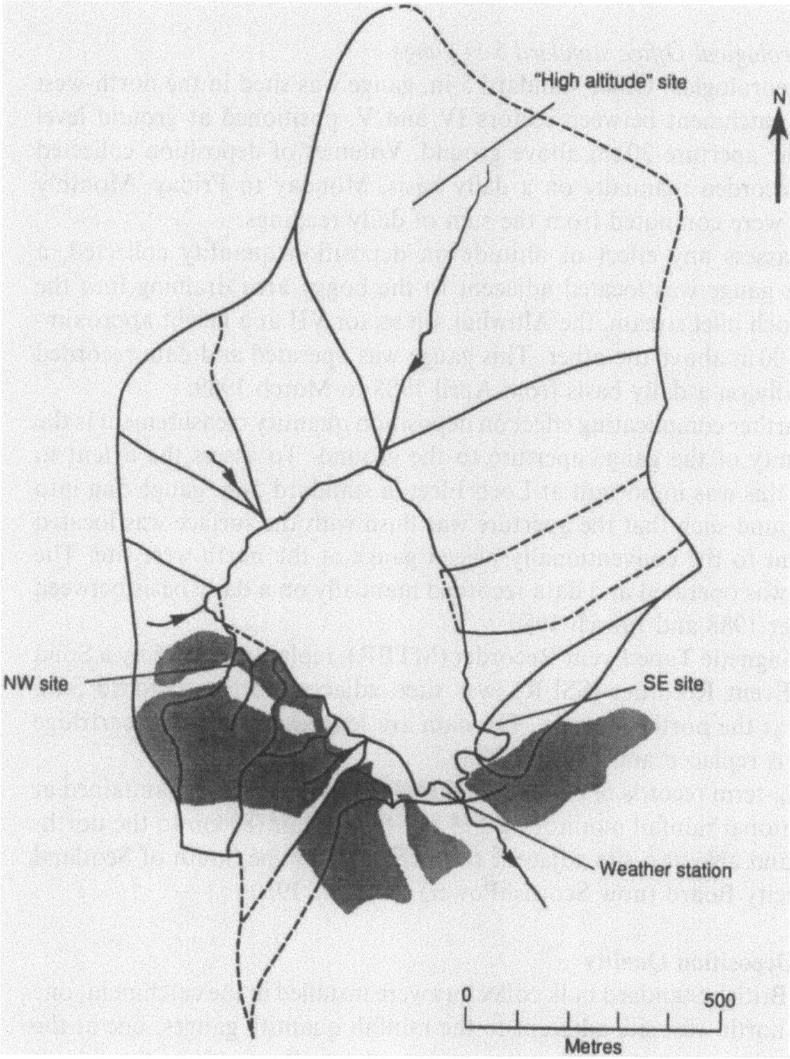

FIG. 5.1 Loch Fleet and its catchment: rain sampling sites.

5.3.1 Deposition Quantity

Meteorological Office standard 5-in gauge

A Meteorological Office standard 5-in. gauge was sited in the north-west of the catchment between sectors IV and V, positioned at ground level with the aperture 30 cm above ground. Volumes of deposition collected were recorded manually on a daily basis, Monday to Friday. Monthly values were computed from the sum of daily readings.

To assess any effect of altitude on deposition quantity collected, a similar gauge was located adjacent to the boggy area draining into the main loch inlet stream, the Altiwhat, on sector VII at a height approximately 100 m above the other. This gauge was operated and data recorded manually on a daily basis from April 1988 to March 1989.

A further complicating effect on deposition quantity measurement is the proximity of the gauge aperture to the ground. To assess the extent to which this was important at Loch Fleet, a standard 5-in. gauge dug into the ground such that the aperture was flush with the surface was located adjacent to the conventionally placed gauge at the north-west site. The gauge was operated and data recorded manually on a daily basis between October 1988 and March 1989.

A Magnetic Tape Event Recorder (MTER), replaced in 1986 by a Solid State Event Recorder (SSER), was sited adjacent to the standard 5-in. gauge at the north-west site. The data are logged on to a tape cartridge which is replaced and read monthly.

Long-term records of deposition quantity have also been maintained at the national rainfall monitoring site at Eskdalemuir (80 km to the north-east), and also at a site adjacent to Loch Fleet by the South of Scotland Electricity Board (now ScottishPower) since the 1930s.

5.3.2 Deposition Quality

Three British Standard bulk collectors were installed in the catchment, one at the north-west site adjacent to the rainfall quantity gauges, one at the weather station at the loch outlet, and another in the south-east corner of the catchment. All were fixed at a height of 1 m. To establish whether the height of the collector had a significant effect on the bulk deposition collected, a further collector was installed in March 1986 at the north-west site. This collector was fixed at the same height as the opening of the 5-in. gauge at the same location (30 cm); the collection bottle was buried in the ground. This collector was operated and samples for chemical analysis taken over the period March to December 1986.

Chemical analysis of deposition collected in the bulk collectors was carried out only when there was a sufficient sample volume (in practice $\geqslant 300$ ml). Thus, during dry periods the analysis results will correspond to a collection time spanning several days; in usual conditions at the site sufficient was collected to allow daily deposition quality analysis. Samples were transferred to the SSEB (now ScottishPower) chemistry laboratory at East Kilbride where analyses were made for conductivity, pH, and the cation, sodium, potassium, ammonium, calcium, and magnesium and the anion, chloride, sulphate, and nitrate. Details relating to the methods of analysis are given in Table 5.1.

5.4 DEPOSITION INPUTS AT LOCH FLEET

5.4.1 Deposition Quantity

Data collected from the "high altitude" gauge and the "ground level" gauge showed that there were no significant differences between the amounts of deposition collected by either gauge compared with those collected by the conventionally positioned standard 5-in. gauges. This was in contrast to findings elsewhere in north-west Britain (Irwin *et al.*, 1990). Nor were there any significant differences between the amounts collected by the conventional 5-in. gauges and the SSER. Consequently, data from the 5-in. gauge at the north-west site were used as daily deposition volumes and used to produce monthly values of deposition input (shown in Fig. 5.2). Since flux calculations and liming treatment effectiveness predictions are based on yearly data from April of each year, the April to March period was taken as the "hydrological year" at Loch Fleet. The annual rainfall quantities recorded over this period for sequential years are shown in Table 5.2. While, in general, more rain falls in the period from October to March than in April to September, high rainfall events occur throughout the year as shown in Table 5.3 for 1988.

5.4.2 Deposition Quality

No significant differences were found in the deposition quality recorded by the three bulk collectors. In general, chemical analysis results of samples from the bulk collector at the north-west site were used for calculating deposition inputs. Occasionally it was necessary to use data from one of the other collectors if, for any reason, the sample from the north-west collector was lost or inadequate. The annual means of deposition acidity (calculated from H^+ concentrations) through the period from 1985 to

TABLE 5.1
Details of analytical methods used to measure major ions in bulk deposition samples

Determinand	*Method*	*Instrument*	*Detection limit*[a] *(mg litre*$^{-1}$*)*	*Notes*
Conductivity	Electrical conductivity	Electronic Instruments, model MC-1, mark V		
pH	Glass and calomel electrodes	Orion Ionanalyser 901		Samples refrigerated at 4°C. Measurement carried out at 20°C after stirring and stabilisation
Chloride, sulphate, nitrate, sodium, ammonium, potassium	Ion chromatography	Dionex model 2120i with Dionex columns	0·2	Samples unfiltered
Sulphur	Inductively coupled plasma emission spectrophotometry	Instrumentation Laboratory Plasma Emission Spectrophotometer	0·032	Samples unfiltered
Sodium	Inductively coupled plasma emission spectrophotometry	Instrumentation Laboratory Plasma Emission Spectrophotometer	0·155	Samples unfiltered
Potassium	Inductively coupled plasma emission spectrophotometry	Instrumentation Laboratory Plasma Emission Spectrophotometer	0·126	Samples unfiltered
Magnesium	Inductively coupled plasma emission spectrophotometry	Instrumentation Laboratory Plasma Emission Spectrophotometer	0·009	Samples unfiltered
Calcium	Inductively coupled plasma emission spectrophotometry	Instrumentation Laboratory Plasma Emission Spectrophotometer	0·086	Samples unfiltered

[a]Detection limit is defined here as 3 times the standard deviation of n blank samples. For this table, $n = 10$.

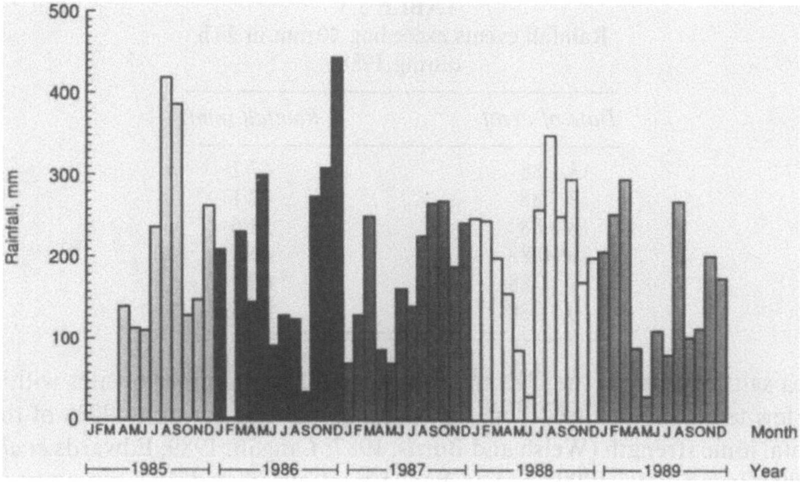

FIG. 5.2 Monthly rainfall at Loch Fleet, 1985–89, NW site.

1990 fall within a range of pH 4·70–4·84. These and cumulated frequencies of rainfall pH values are shown in Fig. 5.3. Less than 4% of rain events have pH less than 4 and more than 50% exceed pH 5.

Table 5.4 shows seasonal and annual weighted mean concentrations for the main ions analysed in bulk deposition samples. It can be seen that there is considerable maritime influence on the deposition inputs at Loch Fleet from the high sodium and chloride concentrations recorded, particularly between October and March each year. This can be explained by this period of the year often being dominated by quite severe westerly winds associated with high precipitation (Fig. 5.2). The importance of high

TABLE 5.2
Annual rainfall quantities recorded at Loch Fleet from 1985 to 1990

Year	*Rainfall*
April 1985–March 1986	2496
April 1986–March 1987	2326
April 1987–March 1988	2474
April 1988–March 1989	2536
April 1989–March 1990	2000
Mean over 5 years, 1985–90	2366

TABLE 5.3
Rainfall events exceeding 40 mm in 24 h, during 1988

Date of event	*Rainfall (mm)*
13.1.88	47·2
13.7.88	47·1
18.8.88	53·6
26.10.88	40·7
9.11.88	43·1
10.11.88	47·5

sea salt loadings in the UK are documented for a number of sites within a few tens of kilometres from the coast, accounting for up to 90% of the total ionic strength (Welsh and Burns, 1987; Langan, 1989; Edwards *et al.*, 1990; Irwin *et al.*, 1990). Ion exchange reactions in acid soils can occur in these conditions leading to a proportion of sea salt cation input (principally sodium) being exchanged for H^+ ions on soil exchange sites as a result. Some of the sodium is retained in the soil and the associated sea salt anion (chloride) is not retained and passes into drainage or stream water. The electrical balance is provided partly by H^+ ions, producing an acid pulse or "episode" in surface water even though the incident rain is of relatively high pH. The incidences of such acid stream water episodes generated by high sea salt loadings and their possible implications for a site such as Loch Fleet are explored in more detail in Chapter 12.

Sulphate, nitrate, and ammonium ions in deposition are often considered to be the major components determining deposition acidity. A proportion of the sulphate deposited is derived from sea salts entrained in the rain; this can be estimated from the magnesium : sulphate ratio in sea water (Erikssen, 1955). Over the years 1985–90 the contribution of sea salt sulphate was 13·2 μeq litre^{-1}, approximately 27% of the total sulphate, 48·2 μeq litre^{-1}, but this varies between years and reflects the influence of wind speed and turbulent air masses across the Atlantic (Irwin *et al.*, 1990). The marine contribution to sulphate is highest each year between October and March. There is no evidence of a significant increase or decrease in either total sulphate or non-marine sulphate over the period sampled. The mean concentrations of nitrate and ammonium over the period sampled were 15·8 and 22·2 μeq litre^{-1}, respectively. The nitrate and ammonium concentrations together are approximately equivalent to the mean non-sea salt sulphate over the same period. In contrast to sulphate, nitrate and ammonium show no consistent seasonal pattern.

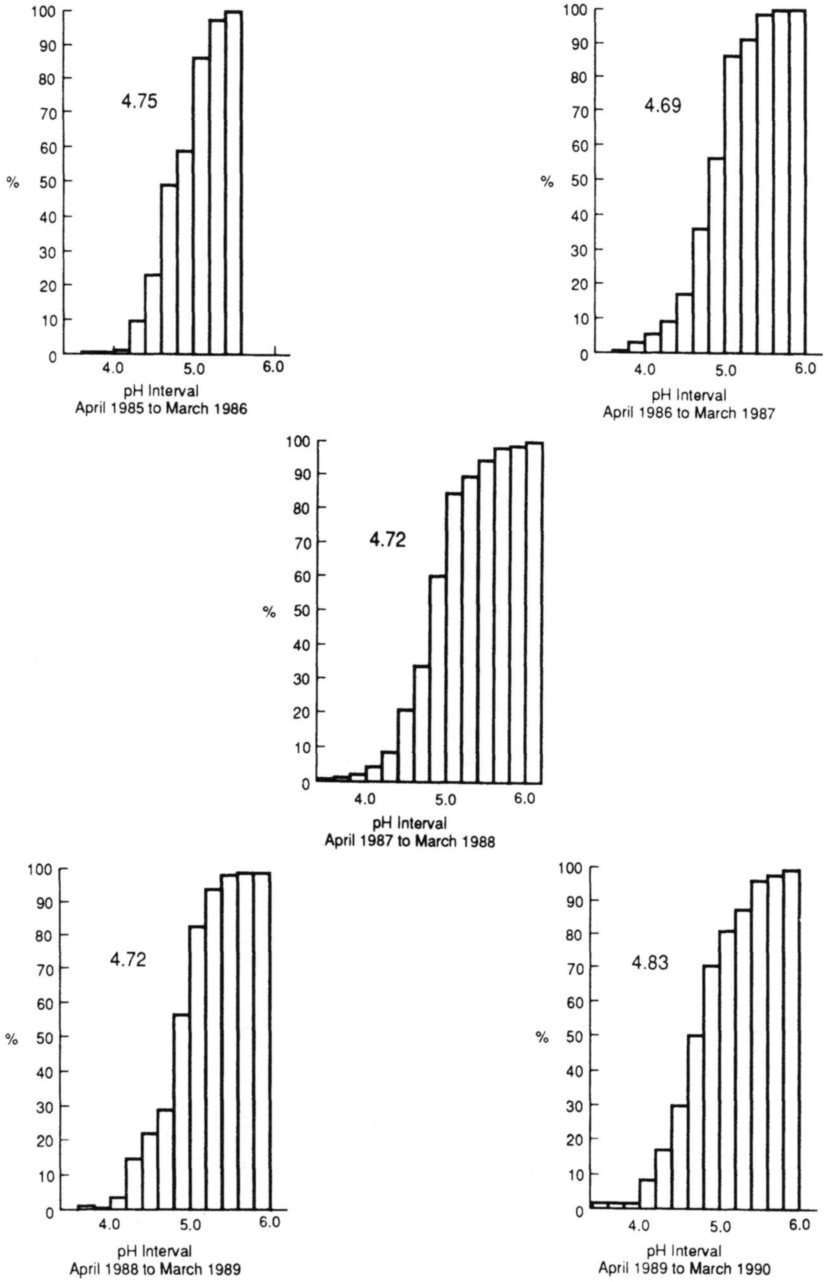

FIG. 5.3 Cumulated pH distribution of rain at Loch Fleet, April 1985 to March 1990. Annual mean (volume weighted) pH values are shown.

TABLE 5.4

Seasonal and annual weighted mean concentrations (μeq litre^{-1}) of major ions in rain

n^a	Sample period	Rain (mm)	pH	H^+	Na^+	NH_4^+	K^+	Ca^{2+}	Mg^{2+}	SO_4^{2-}	NO_3^-	Cl^-
	1985–86											
19	April–June 1985	415	4·58	26	51	28	1	11	16	61	16	56
35	July–Sept. 1985	1078	4·82	16	32	17	2	3	7	39	9	36
25	Oct.–Dec. 1985	473	4·83	15	89	14	< 1	9	21	42	14	105
14	Jan.–March 1986	531	4·80	16	261	26	5	14	57	62	17	342
93	April 1985–March 1986	2497	4·73	17	95	20	2	8	22	48	12	117
	1986–87											
24	April–June 1986	509	4·78	17	58	31	< 1	15	12	41	23	66
21	July–Sept. 1986	292	4·53	30	50	27	1	14	8	48	18	54
44	Oct.–Dec. 1986	1029	4·90	13	152	11	2	10	30	40	8	176
20	Jan.–March 1987	496	4·51	31	78	49	< 1	8	14	60	31	93
109	April 1986–March 1987	2326	4·70	20	103	25	1	11	20	46	18	119
	1987–88											
22	April–June 1987	342	4·82	16	48	38	2	7	8	44	16	50
33	July–Sept. 1987	696	4·60	22	18	23	1	5	2	30	18	20
36	Oct.–Dec. 1987	728	4·65	23	77	20	2	11	13	35	18	83
43	Jan.–March 1988	611	4·79	16	213	15	6	17	44	46	10	267
134	April 1987–March 1988	2359	4·70	20	91	22	3	10	17	38	16	108

	1988–89											
23	April–June 1988	267	4·43	37	23	48	2	20	3	70	38	19
41	July–Sept. 1988	852	4·96	11	37	7	1	14	6	22	6	36
33	Oct.–Dec. 1988	662	4·53	30	137	35	3	12	28	61	26	173
41	Jan.–March 1989	734	4·94	12	199	21	4	13	36	53	13	229
138	April 1988–March 1989	2515	4·73	19	109	23	3	14	20	46	17	127
	1989–90											
28	April–June 1989	232	4·58	29	53	51	2	10	7	77	36	54
21	July–Sept. 1989	452	5·05	8	69	15	2	4	10	30	10	77
4	Oct.–Dec. 1989	489	4·73	19	146	13	2	16	25	53	19	158
6	Jan.–March 1990	818	4·95	11	259	20	2	18	44	86	11	294
59	April 1989–March 1990	2000	4·84	15	164	21	2	14	28	64	16	183

[a]n = number of samples.

TABLE 5.5
Other UK sites: H^+ and SO_4^{2-}-S deposition in the period 1981–86 (from Barrett *et al.*, 1987; Battarbee *et al.*, 1988)

Location	$[H^+]$ $(g\,m^{-2}\,yr^{-1})$	$[SO_4^{2-}-S]$ $(g\,S\,m^{-2}\,yr^{-1})$	*Sample date*
Mid-Wales	0·02–0·03	0·6–1·0	1983–85
North Wales	0·02–0·05	0·8–> 1·0	1986
SW Scotland:			
Galloway (Dee)	0·03–0·04	1·0–1·25	1984–85
Annan	0·00–0·03	1·0–1·25	1981–85
Central Scotland:			
Rannoch	0·03–0·05	1·0–> 1·25	1981–85
Lochnagar	0·03–0·04	0·5–1·0	1981–85
NW England:			
Cumbria	> 0·05	> 1·25	1981–85
Pennines	> 0·05	> 1·0	1986

The 5-yr mean weighted acidity in bulk deposition was 18·3 μeq H^+ litre^{-1} (or pH 4·74). There was no significant trend of increasing or decreasing H^+ concentrations over the sample period. This is consistent with the overall pattern for UK rain monitoring sites, although western sites may be influenced (higher rainfall and higher marine contributions) by variations in westerly weather patterns (Irwin *et al.*, 1990). While concentrations of non-marine components of acidity are lower in the west and north of Britain, wet deposition of acidity, nitrogen, and sulphur is greatest in high rainfall areas. In these areas input deposition of acidity is as high as in southern Scandinavia and Germany.

Analyses of trace elements in bulk deposition samples have shown that these are negligible at Loch Fleet. A reasonable cation : anion balance is obtained routinely, suggesting that most ionic material is accounted for.

5.5 COMPARISON WITH OTHER SITES

For sites in the UK, Barrett *et al.* (1987) list data on bulk rainfall from which some comparison with Loch Fleet bulk deposition chemistry may be made (Table 5.5). Data for Loch Fleet show similarities with those from Loch Dee, a nearby catchment at lower altitude (230 m compared with 340 m at Fleet). At Loch Dee there is, however, a greater sea salt influence (Welsh and Burns, 1987). What is most striking is the range of rainfall quantity among the UK sites (507–2570 mm yr^{-1}) as well as the range of

H^+ concentration, 7–66 μeq litre^{-1}. The sulphate, nitrate, and ammonium concentrations at Loch Fleet fall within the ranges recorded at other remote sites in the UK.

In Wales deposition reflects distance from the coast – in the west, this results in a mean pH of 4·7–5·0. Sea salts are less important at more eastern sites, but pH is little different since acidity is neutralised by ammonium and calcium. In upland areas, acid deposition is greater, pH 4·4–4·6. Although rain in Wales is not as acidic as that reported for eastern UK (pH 4·2) or central Europe (pH 4·0) (Donald and Stoner, 1989), its relatively high rainfall, often exceeding 2000 mm in upland areas, results in a high level of acid deposition (Bird *et al.*, 1990). Annual acid deposition is 0·6 kg H^+ ha^{-1} in upland north Wales, and over 0·4 kg ha^{-1} in southern uplands (Donald *et al.*, 1990).

At Llyn Brianne in Wales, the long-term rainfall mean is about 1800 mm yr^{-1}, and its frequency is high; some daily falls exceed 50 mm. The rainfall varies with site and topography, with higher values at higher altitudes. Mean pH is 4·5, with episodes as low as pH 3·1 (Welsh Water Authority, 1987).

In the north-eastern USA, rain pH is 4·5–5·0 and rainfall 1000–1400 mm (Adirondacks); acid deposition is 0·08 g m^{-2} yr^{-1} (Goldstein *et al.*, 1984).

In Europe, the highest deposition values are reported in Norway (Skreadalen) (H^+ 0·07 g m^{-2}, SO_4^{2-}-S = 1·3 g S m^{-2}), in Sweden (Rorvik) (H^+ = 0·04 g m^{-2}, SO_4^{2-}-S = 1·0 g S m^{-2}), and in Germany (E. Germany, SO_4^{2-}-S = 1·4 g S m^{-2}; but W. Germany, H^+ = 0·05 g m^{-2}) (Irwin *et al.*, 1990). "Acid loading" is generally higher in sites with high rainfall. Acidity in rain shows little change in the past decade, although sulphur emissions have decreased over the same period.

5.6 SUMMARY

At Loch Fleet the mean annual rainfall (2324 mm) had an average pH of 4·74 over the period from 1985 to 1990. This provided an annual "acid loading" of 0·06 g m^{-2}. Sulphate-S (total) deposited was calculated to be 1·6 g m^{-2}.

Most (> 50%) rain events are pH > 5, while less than 4% have a pH below 4·0. Heavy rain events occur in the period from October to March, often with high concentrations of sodium and chloride sufficient to generate acid pulses in runoff. A proportion of sulphate deposition (> 25%) is of marine origin.

No significant trends in acid or sulphur deposition could be detected through the sampling period.

By comparison with other UK sites, acid and sulphur deposition attributable to high rainfall is similar, but appears to be significantly higher than in many European and north American sites.

REFERENCES

Barrett, C.F., Atkins, D.H.F., Cape, J.N., Fowler, D., Irwin, J.G., Kallend, A.S., Martin, A., Pitman, J.I., Scriven, R.A. and Tuck, A.F. (1983). *Acid Deposition in the United Kingdom*. First Report of the UK Acid Rain Review Group. Dept. of Environment, London, 72 pp.

Barrett, C.F., Atkins, D.H.F., Cape, J.N., Crabtree, J., Davies, T.D., Derwent, R.G., Fisher, B.E.A., Fowler, D., Kallend, A.S., Scriven, R.A. and Irwin, J.G. (1987). *Acid Deposition in the United Kingdom*. Second report of the UK Acid Rain Review Group. Dept. of Environment, London, 104 pp.

Battarbee, R.W., Anderson, N.J., Appleby, P.G., Flower, R.J., Fritz, S.C., Haworth, E.Y., Higgitt, S., Jones, V.J., Kreiser, A., Munro, M.A.R., Natkansi, J., Oldfield, F., Patrick, S.T. Richardson, N.G., Rippey, B. and Stevenson, A.C. (1988). *Lake Acidification in the U.K., 1800–1986: Evidence from Analysis of Lake Sediments*. Dept of Environment, Ensis, London, 68 pp.

Bird, S.C., Walsh, R.P.D. and Littlewood, I.G. (1990). Catchment characteristics and basin hydrology: their effects on streamwater acidity, pp. 203–221 in *Acid Waters in Wales* (eds. R.W. Edwards, A.S. Gee and J.H. Stoner). Kluwer, Dordrecht, Netherlands, 337 pp.

Donald, A.P. and Stoner, J.H. (1989). The quality of atmospheric deposition in Wales. *Arch. Env. Contam. Toxicol.* **18**: 109–19.

Donald, A.P., Stoner, J.H., Reynolds, B., Oldfield, F., Rippey, B. and Natkanski, J. (1990). Atmospheric deposition, pp. 39–53 in *Acid Waters in Wales* (eds. R.W. Edwards, A.S. Gee and J.H. Stoner). Kluwer, Dordrecht, Netherlands, 337 pp.

Edwards, R.W., Stoner, J.H. and Gee, A.S. (1990). *Acid Waters in Wales*. Kluwer, Dordrecht, Netherlands, 337 pp.

Erikssen, E. (1955). Airborne salts and the chemical composition of river waters. *Tellus* **7**: 243–50.

Goldstein, R.A., Gherini, S.A., Chen, C.W., Mole, L. and Hudson, R.J.M. (1984). Integrated acidification study (ILWAS): a mechanistic ecosystem analysis. *Phil. Trans. Roy. Soc. Lond. B* **305**: 409–25.

Irwin, J.G., Campbell, G.W., Cape, J.N., Clark, P.A., Davies, T.D., Derwent, R.G., Fisher, B.E.A., Fowler, D., Kallend, A.S., Longhurst, J.W.S., Martin, A., Smith, F.B. and Warrilow, D.A. (1990). *Acid Deposition in the United Kingdom 1986–1988*. Dept of Environment, Warren Spring Laboratory, 124 pp.

Langan, S.R. (1989). Sea salt induced stream water acidification. *Hydrol. Proc.* **3**: 25–41.

Welsh Water Authority (1987). *Llyn Brianne Acid Waters Project*, First Tech. Summary Report. Dept of Environment & Welsh Office, 91 pp.

Welsh, W.T. and Burns, J.C. (1987). The Loch Dee Project: runoff and surface water quality in an area subject to acid precipitation and afforestation in south-west Scotland. *Trans. Roy. Soc. Edin. Earth Sci.* **78**: 249–60.

CHAPTER 6

Hydrological Considerations

A. Dickson
National Power plc, London, UK

6.1 INTRODUCTION

Hydrology is important in acidification and liming studies since it is evident from many investigations that changes (both short and long term) in surface water quality are related to stream flow or lake turnover. This reflects the pathways by which rain reaches surface waters, contributions of groundwater and intercepted fog and mist, and the dynamics and kinds of chemical and biological reactions that occur during this transfer. Since groundwater usually has a long residence time within underlying mineral materials in a catchment, it has higher conductivity and alkalinity. Thus, when atmospheric input is low (drought) the hydrological base flow reflects groundwater composition, rather than rain composition. Conversely, following heavy rains or snow melt, the base flow contributes little to the total, and acidity deposited from the atmosphere, or mobilised during heavy runoff from upper soil layers, will predominate.

Hydrological studies, therefore, are crucial for catchment liming studies.

This chapter provides an account of hydrological investigations at Loch Fleet and reaches some conclusions about the minimum levels of investigation required. Various experimental and modelling projects have been undertaken and the results of these are summarised here and in following chapters. Hydrological studies are required to quantify the transport of water and thus chemical constituents before and after liming. As in any catchment-based study, analysis of the pathways of water transfer and residence time within different components of the catchment are important, allowing the effective duration of treatments to be predicted.

Loch Fleet experiences many of the frontal weather systems that cross northern and western Britain. It has an elevation of 340 m and its proximity to the Irish Sea means that it experiences an additional orographic effect as frontal systems are forced to rise over the land. As a result Loch Fleet has an average annual rainfall in excess of 2000 mm (see Chapter 5). Sustained rainfall events (over 24 h) are common (see Chapter 5) but throughout the project the amount of snowfall has been relatively low, but not atypical for the region.

Several approaches have been taken at Loch Fleet to assess both the general hydrology and its more detailed aspects. Regular measurements of rainfall have been made (see Chapter 5), and streamflow measured, both automatically and manually. These measurements form the principal tool for assessing the overall site hydrology. Further studies of flow pathways have been made using two techniques at Loch Fleet. First, several pits were excavated on the moorland to the east of the loch and in the forested western sectors (see Chapter 8) and measurements made of the flow passing across the different horizons. Secondly, isotope (^{18}O) techniques were used to establish the proportions of stream water arising from rainfall and soil storage during and after a storm (see Chapter 7). Some hydrological modelling was also undertaken with the aim of describing the hydrology of the catchment (Section 6.4, below).

6.2. INSTALLATIONS FOR THE MEASUREMENT OF STREAM FLOW

The principal method of flow measurement at Loch Fleet has been automatic and "continuous" recording at four locations on a 15-min schedule. The whole catchment was considered as a number of separate subcatchments, or sectors, as already described in Chapter 4 (Fig. 4.1). Some of these sectors have well-defined drainage streams. In particular sector VII, the largest in the catchment, contains the main inlet stream, the Altiwhat. This principal inflow stream and the loch outlet stream have been measured with trapezoidal flumes connected to stilling wells fitted with electronic depth recording equipment. Streams from sectors IV and VI have been measured similarly but using V-notch weirs because of the lower expected flow levels that would be more accurately registered than through a trapezoidal flume.

The automatic measurements have been verified and calibrated by manual and electronic methods based on measurements taken manually on an approximately daily basis. This allowed rapid notification of problems with the automatic flow measuring equipment such as the pressure transducers or power supply failures.

The automatic recording equipment has produced by far the greatest bulk of the data concerning hydrology. The usefulness of the data has been enhanced by various studies undertaken first on the accuracy of the information, and secondly on flow duration patterns. The accuracy of the electronic 15-min recordings have been checked by several means. To check for drift in the pressure transducers, the stilling wells themselves have been calibrated every 6 months as well as checked by routine manual measurement of the recorded depth. This was necessary because wood was used in their construction to prevent risk of chemical contamination: the wood could have been subject to warping after periods of wetting and drying.

The resolution and hence the accuracy of the pressure transducers is within plus or minus 4 mm. As would be expected when converting these depths to flows with an equation of the form (flow $= K \times \text{depth}^{3/2}$) the accuracy of the flow calculation relative to the total flow will vary: the accuracy is better at higher flows because these resolution errors become less significant. Informal studies have concluded that the accuracy of the flows is more than adequate for calculating chemical fluxes since low flows tend to carry lower total fluxes of ions.

The accuracy of the flow measurement equipment at sectors VI and VII was also checked using a conventional salt dilution technique. The passage of the salt tracer down the Altiwhat was recorded for several locations and these tests confirmed that the equipment functioned satisfactorily.

The veracity of the data is also checked routinely against the measured rainfall, the expected evapotranspiration, and the deduced sector areas where a simple check between the expected flows and the recorded flows could be made. Normally the correlation between the 15-min flow records from different sectors around the loch proved to be very high and this allowed missing data to be infilled, where necessary, with confidence.

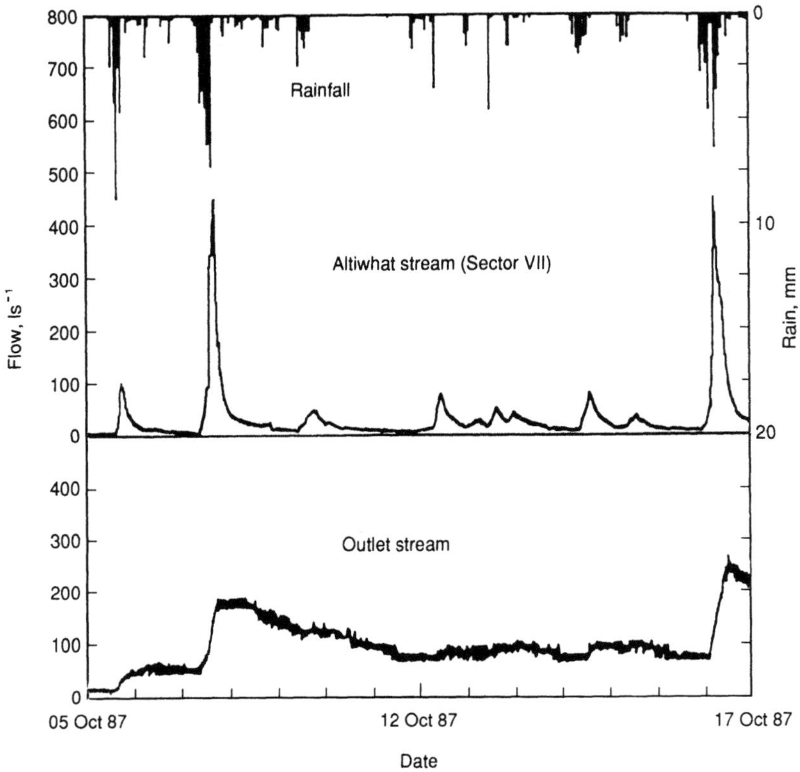

FIG. 6.1 Response of stream water flow (main Altiwhat tributary and loch outlet) in relation to rain events during a limited period in October 1987.

6.3 RESULTS FROM DIRECT FLOW MEASUREMENT

6.3.1 Flow-duration Analysis

Flow-duration analysis has confirmed that the conditions at Loch Fleet are very "flashy" or episodic. Figure 6.1 makes evident the different response of the main inlet stream and that of the outlet, reflecting retention within the loch itself.

Intuitively, this would be expected at Loch Fleet because the impermeable soils of the catchment are usually saturated. Significant amounts of ponding occur around the catchment, particularly at the base of slopes and in the wetland at the head of the Altiwhat (sector VII). Figure 6.2 shows the result of a flow-duration analysis for sector VII. The "higher flows" occur for less than 0·5% of the time. Even at the loch outlet the

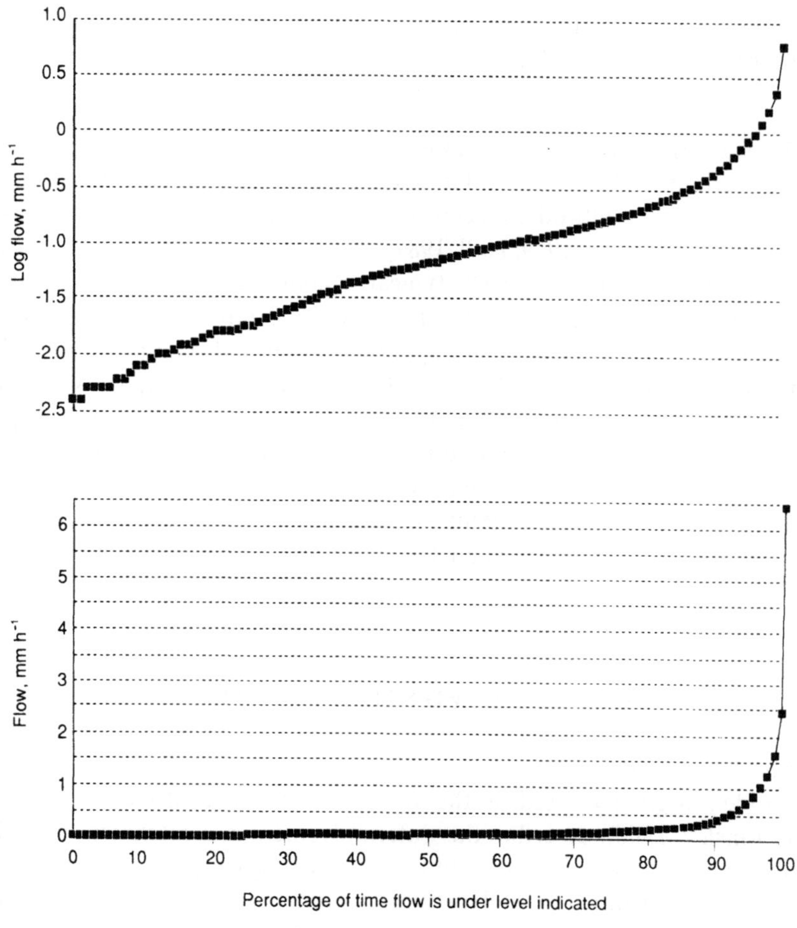

FIG. 6.2 Flow duration analysis for the Altiwhat stream, March to December 1986.

results show a quite rapid response despite temporal attenuation of direct tributary flows through the loch. Individual sectors display different unit runoff characteristics dependent principally on their predominant vegetation types and soil thicknesses. Sector IV, for example, with more than 80% afforestation, has the lowest measured unit runoff of $1{\cdot}65\,m^3\,ha^{-1}\,yr^{-1}$ compared with moorland sector VII with an average of $1{\cdot}88\,m^3\,ha^{-1}\,yr^{-1}$. The unit runoff for the catchment as a whole is between these two at $1{\cdot}80\,m^3\,ha^{-1}\,yr^{-1}$. These differences can be attributed to three factors: higher evapotranspiration for the forested sectors, generally thicker soil in the forested sectors, and potentially different rainfall patterns across the catchment.

6.3.2 Rainfall–Runoff Relationships

The relationship between rainfall and runoff reflects annual climate differences as well as altitude, vegetation, and degree of catchment saturation (Calder and Newson, 1979; Hall, 1987). At Loch Fleet, the ratio of outflow to rain input is about 75%–80%, typical of similar upland sites in Scotland (i.e. 500 mm evapotranspiration and 2000–2500 mm rainfall). Some differences between the sectors are apparent; moorland sectors VI and VII have similar values, while sector IV (80% afforested) has typically a ratio of less than 70%. This is attributable to an expected greater evapotranspiration from mature coniferous forest compared with rough heather moorland. Sectoral data also show that the runoff: rainfall ratio is greater (80–94%) in the first quarter of the year, falling in the summer months, as expected, by higher evapotranspiration as air temperature and foliage activity rises.

6.4 MORE DETAILED STUDIES IN THE FIELD

River Basin Model

A model, originally based on the Stanford Watershed Model (Crawford and Linsley, 1966), describes the water movement in a catchment empirically using a series of fairly traditional hydrological premises. At Loch Fleet the model aimed to infill missing flow data, as well as to help describe the different pathways of flow at Loch Fleet. The model was calibrated on monthly flow data and at this level it accounted for the water balance satisfactorily. The best calibration of the model on the individual sectors indicated, over the water year 1985–86, that sector IV had 68% and sectors VI and VII had 80% as fast runoff (surface or near surface). These values appear quite consistent with the ratios of streamflow to rainfall (Section 6.3.1 above).

Two alternative approaches to field studies have been undertaken at Loch Fleet. The initial studies attempted to establish the predominant throughflow patterns by monitoring runoff and drainage through forest (sector IV) and moorland (sector VIII). This study was carried out prior to liming in 1986 (see Chapter 8). The second approach undertaken at a later stage was to use isotope tracers of water flow (see Chapter 7).

6.5 GENERAL COMMENTS AND CONCLUSIONS

The experience at Loch Fleet has highlighted the importance of hydrological data, essential in the calculation of hydrochemical budgets and hence in the prediction of treatment effectiveness and duration. Large-scale models seem to have limited useful application to the conditions and aims of such a project. Ideally, field data are needed on rainfall and flows on a long-term continuous and frequent basis (perhaps over a decade), as well as detailed information on conditions such as soil water capacity and porosity. In practice, this is difficult and is seldom achieved. At Loch Fleet, however, some independent studies were undertaken to meet some of these needs (see Chapters 7, 8, and 11).

In general, experience at Loch Fleet has indicated that two key hydrological conditions are best recognised prior to liming. First, the possibility that heavy rain could rapidly displace undissolved material placed over the ground to runoff: lime applications should be planned for dry periods if possible. Secondly, qualitative and quantitative information is required on flow pathways: this is vital in estimating the amount of limestone required to strike a balance between a rapid increase of alkalinity in runoff waters and the longer term capability of the treatments to sustain the elevated alkalinity.

Mixing of limestone particles into lower soil horizons, as observed at Loch Fleet (see Chapter 11), will possibly be beneficial in contributing to the long-term robustness of treatments. Prior to treatment the lower horizon waters had higher alkalinity and conductivity than the upper horizon waters: these conditions were reflected in streams during very low (base) flows. Since the peaks of flow appear to be dominated by rainwater and upper horizon soil water and soil chemistry, it was not appropriate to target treatment to conditions in lower soil horizons. Field studies at Loch Fleet demonstrated that it was just as effective to place limestone on the surface rather than incorporate it into soil or in trenches (see Chapter 9). The soil studies have demonstrated, post-liming, that considerable amounts of limestone have percolated vertically into a soil system dominated by lateral drainage. It will be interesting to see the effects of depletion of the lower horizon store on future stream water quality: we may continue to learn more about the hydrological pathways at Loch Fleet in the future.

REFERENCES

Calder, I.R. and Newson, M.D. (1979). Land use and water resources in Britain – a strategic look. *Water Res. Bull.* **15**: 1628–39.

Crawford, N.H. and Linsley, R. (1966). *Digital Simulation in Hydrology, Stanford Watershed Model IV*, Tech. Report 39.

Hall, R.L. (1987). Processes of evaporation from vegetation of the uplands of Scotland. *Trans. Roy. Soc. Edin. Earth Sci.* **78**: 327–34.

CHAPTER 7

Shallow Flow Pathways or Pre-event Water?

K.H. Bishop
Swedish University of Agriculture Science, Vindeln, Sweden

K.S. Richards
University of Cambridge, Cambridge, UK

7.1 INTRODUCTION

A sound description of the flow pathways and residence times of water is widely recognised to be a prerequisite for understanding and modelling the mechanisms of surface water acidification (Hornung *et al.*, 1986). Similarly, a quantitative description of how catchment liming will reverse acidification, in particular the duration and intensity of reversal in different situations, presupposes an adequate knowledge of catchment hydrology.

Direct observations of flow pathways with plot studies that intercept throughflow have often found stormflow pathways to be highly localised. The rapid flow of water through these pathways, either at the surface or through systems of macropores or "pipes", implies a rapid flux of rainwater which bypasses much of the soil.

Another line of investigation over the last two decades has employed stable isotopes to determine how much rainwater reaches a stream during a storm event (Fig. 7.1). Invariably, more than half, and often more than three-quarters, of the storm runoff is shown by isotope methods to be "old" water displaced from the catchment by rain, rather than the rainwater falling during an event. This is consistent with the notion of piston,

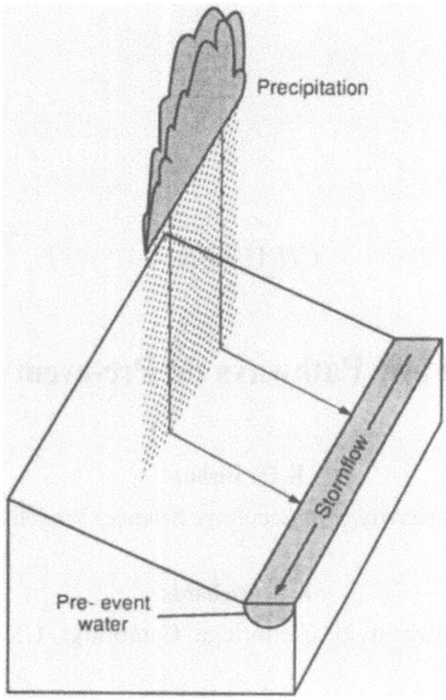

FIG. 7.1 Two component mixing model of stormflow generation.

or displacement, flow developed by forest hydrologists (Hewlett and Hibbert, 1967).

Given the importance of hydrology in the Loch Fleet project, it was decided to employ both plot studies and isotope hydrology to arrive at an understanding of the catchment hydrology. As one of the few sites where the two approaches have been combined, Loch Fleet provides an excellent opportunity to explore the apparent contradiction between the localised stormflow pathways indicated by direct observations and the large amount of displaced pre-event water indicated by isotope studies.

The contradiction is exemplified in the empirical research at Loch Fleet, where plot studies (see Chapter 8) suggest that 87%–99% of storm runoff travels within a few centimetres of the surface of peat soils, whereas isotope hydrograph separations show the majority of quickflow in storm hydrographs to be pre-event water displaced by new rainwater.

Some measure of this apparent contradiction may stem from the difficulty in making a direct comparison between the results of the isotope studies and the plot studies. The former specifies the origins of runoff in terms of a simple dichotomy between the pre-event water and the incoming rain, while the latter points to the contributions to quickflow runoff from specific soil horizons. This chapter, in addition to summarising the results of the isotope study, will quantify the implications of the isotope study in terms that facilitate comparison with the plot studies.

7.2 THE ISOTOPE HYDROGRAPH SEPARATION METHOD

A classic hydrological technique has involved separation of the origins of runoff generated by a storm into baseflow and quickflow components with a graphical analysis of the shape of the falling limb of the hydrograph. Graphical analysis, however, has no means of physically identifying the components of the hydrograph that were ostensibly being separated, and no additional information about the hydrological processes is provided.

A measure of physical significance was lent to hydrograph separation when electrical conductivity (Pinder and Jones, 1969; Calles, 1985) or silica content (Hooper and Shoemaker, 1986) were used as tracers to "label" the different components in the hydrograph. Unfortunately, the use of these tracers has been hampered by their interactions with the catchment. A technique better suited to hydrograph separation is based on the stable isotopic composition of the atoms in water itself, specifically the ratio between oxygen-18 (^{18}O) and oxygen-16 (^{16}O). This ratio is measured by mass spectrometry and reported in the literature as the deviation of the sample isotope ratio from that of Standard Mean Ocean Water (SMOW). This deviation is denoted as $\delta^{18}O$, a unitless ratio which is typically very small and written as per mille (parts per thousand). Water derived from meteoric sources in higher latitudes of the northern hemisphere is commonly depleted in ^{18}O compared with SMOW, yielding negative $\delta^{18}O$ values.

As a highly conservative tracer within temperate catchments, the $\delta^{18}O$ of water is well suited for use in hydrograph separation when the $\delta^{18}O$ of precipitation in a storm event is different from the $\delta^{18}O$ of the pre-event water. Pre-event water refers to all water in the catchment prior to the onset of a storm event. During a storm, the stream is a mixture of pre-event water and rainwater, each of which has a characteristic $\delta^{18}O$ signature (Fig. 7.2). The $\delta^{18}O$ of the stream during a storm can be used to determine the proportion of each of these two components in the stream.

Using the two-component conceptualisation of runoff generation, two

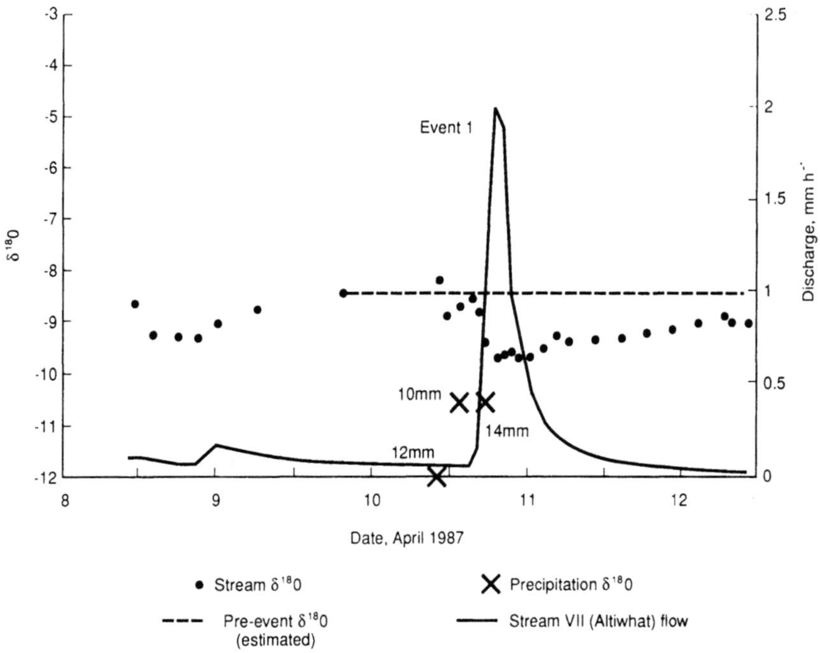

FIG. 7.2 Components of rain and pre-event water in Altiwhat stream as indicated by the oxygen-18 ratio during a high flow event (1), April 1987.

mass balance equations (equations (1) and (2)) can be postulated:

$$O_p * V_p + O_r * V_r = O_s * V_s \tag{1}$$

$$V_p + V_r = V_s \tag{2}$$

where O_p, O_r and O_s are the $\delta^{18}O$ values of the pre-event water, rainfall, and streamflow, respectively, while the volumes of these waters are V_p, V_r, and V_s, respectively. These two equations can be solved for the fraction of pre-event water in the stream at a given time:

$$X = V_p / V_s = (O_s - O_r)/(O_p - O_r) \tag{3}$$

In these equations the $\delta^{18}O$ of stream water is measured directly from stream samples. The $\delta^{18}O$ of the rainfall is the volume weighted average of the rain in a storm event up to the time when X is calculated. The concentration of the pre-event component was assumed to be equal to the $\delta^{18}O$ of stream water at low flow prior to the onset of the storm hydrograph.

7.3 SAMPLING PROGRAMME

The streams draining Loch Fleet subcatchments (sectors) VI and VII (Fig. 4.1) were chosen for the isotope hydrograph separation (IHS) investigations. The largest single stream feeding Loch Fleet drains sector VII which comprises 32 ha, or about one-third of the loch's catchment area. The subcatchment is distinguished by the large rift in which the stream flows to an alluvial fan. A smaller stream arises from sector VI (6·4 ha). The even, peat-covered slopes of that sector were considered more representative of the Loch Fleet catchment as a whole, and particularly the slopes on sector Z_2, where many of the detailed plot studies were conducted (Chapter 8).

Hourly stream samples were collected from sector VII between 8 and 19 April 1987, with interruptions over the weekends. Stream VI was sampled in a similar manner between 3 and 19 July 1987. Rainfall samples were collected during these periods in a rain gauge near the outlet of the loch. The rain gauge was emptied every weekday morning, and later in the day as well if significant amounts of rain had fallen. This sampling regimen yielded three events with a difference between rainfall $\delta^{18}O$ and stream water $\delta^{18}O$ large enough to allow for IHS.

7.4 PRINCIPAL FINDINGS

Event 1 exhibited the typical pattern of $\delta^{18}O$ dynamics in rainfall and stream water for the storm hydrographs from sectors VI and VII (Fig. 7.3). That pattern reveals two qualitatively significant features. The first is that when storm rainfall has a $\delta^{18}O$ that differs from that of the stream prior to the storm, the stream water $\delta^{18}O$ shifts in the direction of that in the rain during the storm. However, the shift in stream $\delta^{18}O$ is extremely damped. This damping, which is quantified in subsequent calculations, is the basis of the contention that most of the water in storm runoff is not event water that fell on the catchment during the storm. Note also that as the flow ebbs after a storm, the stream $\delta^{18}O$ moves back towards its original value, but not completely. The incomplete "recovery" in low flow $\delta^{18}O$ of water is due to a shift in the $\delta^{18}O$ of water stored in the catchment. That shift occurs because the rainfall which does not run off to the stream is added to the pre-event store, thereby changing its $\delta^{18}O$.

The characteristics of the three storm events and their hydrograph separations are summarized in Table 7.1. The uncertainty in these IHS

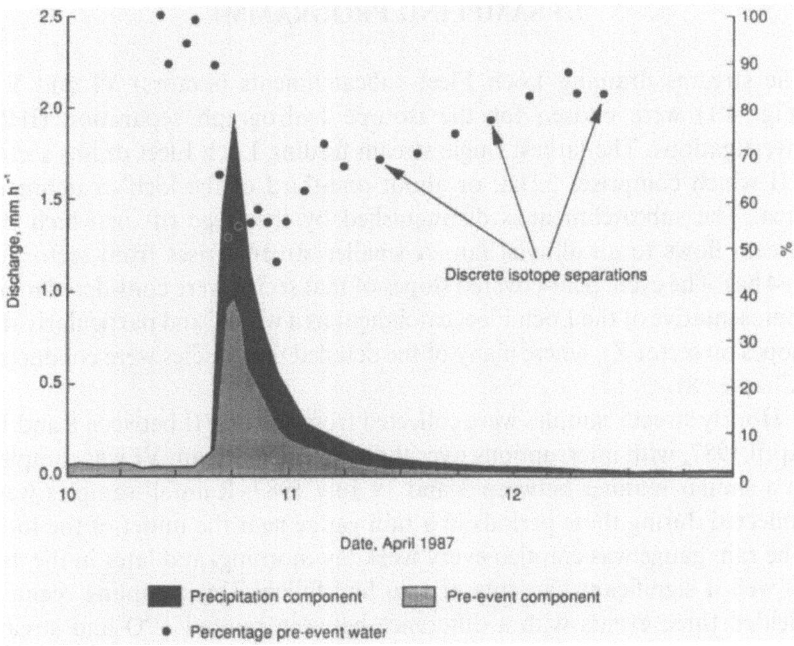

FIG. 7.3 Isotope hydrograph separation and the percentage of pre-event water in the Altiwhat stream; event (1).

TABLE 7.1
Summary of hydrograph separations for three events at Loch Fleet

Event	*1*	*2*	*3*
Sector	VII	VI	VI
Duration (h)	36	17	43
Precipitation (mm)	36·2	19·0	24·2
Total discharge (mm)	12·5	10·4	24·9
Pre-event component (%)	60	86	89
Minimum pre-event component (%)	52	84	86

Note: Runoff coefficients were not calculated because the receding limb of the hydrograph was often truncated by the arrival of another storm event.

TABLE 7.2
Isotope hydrograph separations

Principal author	*Date*	*Rainfall (mm)*	*Catchment area (ha)*	*Average pre-event fraction (%)*	*Minimum pre-event fraction (%)*
Obradovic	1987	–	6 000	60	50
Rodhe	1987	14–49	4–660	68–99	53
Herman	1980	45	1 870	77	–
Sklash	1979	6–35	120	–	65–80
Sklash	1979	17–38	100	–	60–80
Sklash	1976	21–28	100	70	–
Cherry	1975	50	–	50–90	50–60
Fritz	1974	–	–	50–70	–
Mook	1974	8–16	650	"minor"	10
Crouzet	1970	6–31	570–9 100	54–99	–

From Bernier (1985).

investigations has been conservatively estimated as less than 15% (Bishop and Richards, 1988). The IHS investigations indicate that the major part of water in the stream throughout a storm event is pre-event water. The minimum fraction of pre-event water in the stream coincided roughly with each event's peak flow. The fraction of pre-event water then increased over the next few hours to between 85% and 90% (Fig. 7.3).

The minimum fraction of pre-event water from the event 1 IHS (sector VII) was 52%, compared with minimum pre-event fractions of more than 80% from sector VI during events 2 and 3. While there are not enough hydrograph separations to establish the statistical significance of this difference in the results from the two sectors, the smaller fraction of pre-event water from event 1 on sector VII may arise from differences in catchment physiography, storm intensity, and/or antecedent moisture conditions.

For the purposes of this contribution, the most significant result of these IHS investigations is the consistently large component of pre-event water in all of the storm hydrographs. These results are similar to the findings from a variety of other documented catchments (Table 7.2), yet such a large component of pre-event water appears to contradict the results of the plot study on sector Z_2 (Leech and Nisbet, 1987; see also Chapter 8). In that study a series of trenches was dug through the saturated peat soils to bedrock. One trench was located at the top of the hillslope near the watershed (see Fig. 8.1). Another trench was dug at the base of the hillslope near the loch, and the remaining trenches were spaced along the slope between the upper and lower trenches. Throughflow in the soil profile at different levels on the 1 m wide upslope face of the trenches was collected in plastic troughs. A large proportion, 87%–99%, of the stormflow was collected in the uppermost trough in each trench. This trough collected water running through the loosely matted organic detritus overlying the peat soils of the hillslope.

The slopes and shallow soils (less than 1 m deep) in the vicinity of the plot studies on sector VIII were broadly similar to those of sector VI. In the latter sector, the storm runoff sampled in the stream was consistently more than 75% pre-event water according to the IHS. If the interpretation of both the plot studies and the IHS are correct, then most of the water travelling superficially downslope is pre-event water displaced by the storm rainfall. It is, however, difficult to see how the superficial flow, which fits Horton's (1942) classification of "concealed overland flow", can include so much pre-event water.

It would appear that a choice must be made between the palpable evidence of so much downslope flow being collected in near-surface troughs and the inference from the damped response of the isotope signal in stream water that the greater part of the storm hydrograph is composed of pre-event water. Any choice between these two alternatives is likely to have significant implications for the understanding of how a catchment responds to liming. Before one considers how to balance the evidence from the plot studies and the IHS, it may, however, be useful to clarify the magnitude of the perceived contradiction. This is especially true as the evidence from the plot studies concerning the route of water flow (within the upper 5 cm of the soil) is not directly comparable with the evidence from the IHS about the origins of the pre-event water (from within the hillslope).

That comparison would be facilitated by determining how large a store of pre-event water, measured as a depth over the sector, is needed to produce the damped stream $\delta^{18}O$ response to rainfall. Knowing that volume, and the moisture content of the soil profile, an estimate can be made of the minimum depth of soil needed to hold the reservoir of pre-event water inferred from the IHS.

7.5 THE EFFECTIVE PRE-EVENT RESERVOIR

The IHS is based on a division of streamflow into event water that enters a catchment during a storm and pre-event water already in the catchment prior to a storm. The rainwater that enters a catchment during an event and does not run off or return to the atmosphere during that same event is added to the store of pre-event water. This should change the resultant $\delta^{18}O$ of the store. That change is evidenced at Loch Fleet by the fact that while the rate of streamflow after a storm may return to pre-storm levels, the $\delta^{18}O$ of low flow waters does not return to the pre-storm value. The "ante-storm", pre-event $\delta^{18}O$ is shifted towards the $\delta^{18}O$ of the rainfall to yield a "post-storm" $\delta^{18}O$. The magnitude of that shift, and mass balance considerations (equations (1) and (2)), can be used to estimate the effective size of the pre-event reservoir prior to the storm (Fig. 7.4).

In this mass balance calculation, evapotranspiration during the short period of the event is ignored. All rainwater that did not run off, according to the IHS, is assumed to have entered the pre-event reservoir. Since the two-component mixing model of runoff generation assumes that the pre-event component is a uniform and well-mixed reservoir, which is almost certainly not the case, the qualifying adjective "effective" is used when describing the calculated pre-event reservoir.

The average depth of the effective pre-event reservoir in sector VI was 135 mm prior to event 2 and 165 mm prior to event 3. On sector VII, the effective depth was 178 mm prior to event 1. For sector VI, assuming a 75% volumetric moisture content in the upper soil horizons prior to the onset of the storm events (Environmental Advisory Unit, 1986) and an average pre-event reservoir depth of 150 mm, that reservoir could be contained in the upper 20 cm of soil. While that depth is inconsistent with the inference that almost all downslope flow travels through the upper 5 cm of soil, it is not as large a discrepancy as one might have expected when contrasting concealed overland flow with a pre-event component of more than 85% in stormflow from sector VI. The mean age of water in the effective pre-event reservoir is estimated to be roughly 1 month for the 150 mm depth of the effective soil reservoir and a mean annual runoff of 1702 mm.

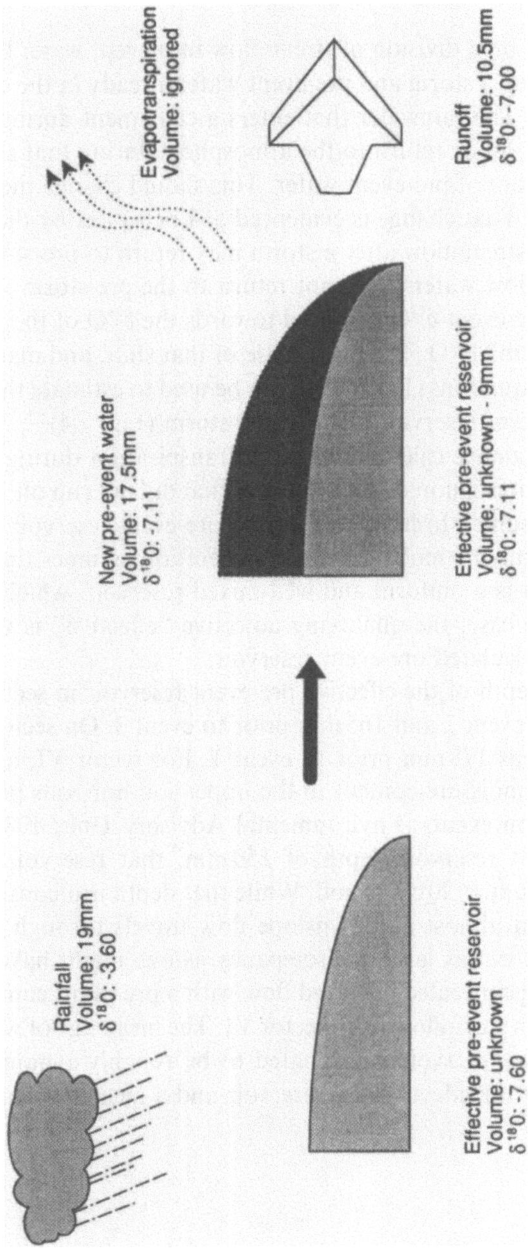

FIG. 7.4 Calculation of the pre-event reservoir contribution to runoff, sector VI, event 3.

7.6 CONCLUSIONS

Two storm hydrographs from sector VI were separated. In both storms about 20 mm of rain yielded hydrographs in which an average of 85%–90% of the storm flow was pre-event water. The hydrograph from a 36 mm storm event in Loch Fleet's main tributary draining sector VII was also separated to show that 60% of the event's runoff was pre-event water.

Despite its significance, such information from IHS is not usually incorporated into hydrochemical models (Hooper *et al.*, 1988). This results, in part, from an intuitive scepticism about the large proportion of pre-event water in the storm hydrograph indicated by the IHS. While such intuitions often are not strongly grounded in empirical data, they would appear to be more justified at Loch Fleet where most runoff in plot studies has been observed to move very near the surface. To address the scepticism, tests were made by Bishop and Richards (1988) to try to falsify the IHS technique. The $\delta^{18}O$ of water from different locations in the catchment prior to a storm was compared with the pre-event $\delta^{18}O$ estimated from stream baseflow. The possibility of shifts in the $\delta^{18}O$ of pre-event water reaching the stream during storm events was investigated by comparing hydrograph separations generated by consecutive rainstorms with widely different $\delta^{18}O$ values that shifted the stream $\delta^{18}O$ in different directions. The effective depths of the pre-event reservoir required to generate the IHS results were compared with the estimated size of the pre-event reservoir in the soils. None of these tests succeeded in disproving the IHS results.

The fact remains that the IHS results do not show that the greater part of runoff is rainfall running over the top of a saturated, low-hydraulic conductivity peat soil, as suggested by the plot studies. This chapter has used the shift in the $\delta^{18}O$ content of the pre-event reservoir in response to rainfall to estimate the effective size of that reservoir. On sector VI, only a 20 cm depth of soil is needed to contain the effective pre-event reservoir, assuming complete mixing of water within the soil. This probably underestimates the actual depth of the pre-event reservoir participating in runoff generation, because any immobile or otherwise unmixed water in the upper 20 cm will lead to a greater depth of the pre-event reservoir. Nonetheless, the 20 cm depth of the effective pre-event reservoir provides a useful measure for comparison with the plot study, which shows the great preponderance of the runoff to be moving within 5 cm of the ground surface.

Questions can be raised about the results of both plot studies and IHS studies. Soil characteristics are known to be spatially heterogeneous, and when a profile is exposed for throughflow monitoring, it is inevitable that visually apparent horizon boundaries are selected which, on a wider spatial scale, may be anomalous. Plot studies conventionally focus on small (about 1 m wide) profile exposures, but data more directly comparable with the lumped information from IHS studies may require much more laterally extensive sampling (e.g. Dunne and Black, 1970). Furthermore, throughflow pits can disturb the local hydraulic potential field and encourage the distortion of flow pathways away from, or into, a sampling point, thereby under- or over-estimating throughflow volumes (Atkinson, 1978). However, questions can also be asked of the IHS methodology which assumes, without complete verification, that a unique isotopic content of pre-event water in a storm hydrograph can be identified from the $\delta^{18}O$ content of stream baseflow. The age and $\delta^{18}O$ of soil moisture may vary with distance up-slope and depth down-profile. Varying water flow patterns during storm events might vary the sources of the pre-event water reaching the stream. Hence the $\delta^{18}O$ of the pre-event quickflow component of storm runoff might vary, which would lead to errors in the IHS. At the present time these criticisms do not appear strong enough to warrant the dismissal of either the plot study findings or of the IHS at Loch Fleet. The essential plausibility of the two methods notwithstanding, it remains a possibility that either plot studies or the IHS are seriously in error. If, however, the ambiguities in the techniques can span a gap of just some 15 cm, then both the plot studies and the isotope studies would provide a complementary hydrological picture of near surface pathways for stormflow in which water had a mean residence time of the order of 1 month.

If the apparent contradiction between the concentration of flow in superficial pathways can be reconciled with the large component of pre-event flow, then a similar resolution is a possibility in other catchments. The possibility of a reconciliation between plot studies and IHS, or the rejection of one of these methods, would require experimental programmes capable of rigorously testing both approaches.

REFERENCES

Atkinson, T.C. (1978). Techniques for measuring subsurface flow on hillslopes, pp. 73–120 in *Hillslope Hydrology* (ed. M.J. Kirkby). John Wiley & Sons, London.

Bernier, P.Y. (1985). Variable source areas and storm-flow generation: an update of the concept and a simulation effort. *J. Hydrol.* **79**: 195–213.

Bishop, K.H. and Richards, K.S. (1988). A study of water flow pathways and residence times in the catchment of Loch Fleet using Oxygen-18 as a natural tracer. Loch Fleet Project. Consultancy report for the Loch Fleet Project, 50 pp.

Calles, U.M. (1985). Deep groundwater contribution to a small stream. *Nord. Hydrol.* **16**: 45–54.

Dunne, T. and Black, R.D. (1970). An experimental investigation of runoff production in permeable soils. *Water Res. Res.* **6**: 478–90.

Environmental Advisory Unit (1986). *The Loch Fleet Study*. Land Intervention Treatment Report No. 6. University of Liverpool, Department of Botany.

Hewlett, J.D. and Hibbert, A.R. (1967). Factors affecting the response of small watersheds to precipitation in humid areas, pp. 275–90. In *Proc. Internat. Symp. on Forest Hydrology*, Pennsylvania State University, Pergamon.

Hooper, R.P. and Shoemaker, C.A. (1986). A comparison of chemical and isotopic hydrograph separation. *Water Res. Res.* **22**: 1144–54.

Hooper, R.P., Stone, A., Christophersen, N., de Grosbois, E. and Seip, H.M. (1988). Assessing the Birkenes model of stream acidification using a multi-signal calibration methodology. *Water Res. Res.* **24**: 1308–16.

Hornung, M., Adamson, J.K., Reynolds, B. and Stevens, P.A. (1986). Influence of mineral weathering and catchment hydrology on drainage water chemistry in three upland sites in England and Wales. *J. Geol. Soc. Lond.* **143**: 627–34.

Horton, R.E. (1942). Remarks on hydrological terminology. *Trans. Amer. Geophys. Union* **14**: 446–60.

Leech, A.F. and Nisbet, T.R. (1987) *Acid Transfer Through Vegetation and Soils*, Report to UK Dept of Environment. Macaulay Land Use Research Institute, Aberdeen, UK, 233 pp.

Pinder, G.F. and Jones, J.F. (1969). Determination of the groundwater component of peak discharge from the chemistry of total runoff. *Water Res. Res.* **5**: 438–45.

CHAPTER 8

Interactions Between Rain, Vegetation, and Soils

A.F. Nisbet
National Radiological Protection Board, Didcot, Oxon, UK

T.R. Nisbet
Forestry Commission Research Station, Farnham, Surrey, UK

8.1 INTRODUCTION

An understanding of the chemistry of Loch Fleet and its drainage waters necessitates a study of the chemical interactions between rainfall, vegetation, and soil. The vegetation can exert a strong influence on bulk deposition inputs through physical interception and chemical reactions at leaf surfaces. Further modifications resulting from interaction with soil depend on the hydrological pathways taken from soil to stream and lake.

Detailed plot studies were undertaken within the Loch Fleet catchment to study rainfall–vegetation–soil interactions in forested and moorland sectors (Fig. 8.1). Twelve plots were established, six within Sitka spruce (*Picea sitchensis* (Bong.) Carr.) (planted 1963) on deep peat (sector I), and six within lodgepole pine (*Pinus contorta* Douglas ex Loud) (planted 1963) on peaty gley soils (sector IV). One of the main drainage ditches in the area of Sitka spruce was instrumented for the study on forest soil hydrology. Four soil pits were excavated up the moorland hillslope (sector Z_2) to investigate moorland soil hydrology.

Each of the following components was sampled at fortnightly intervals

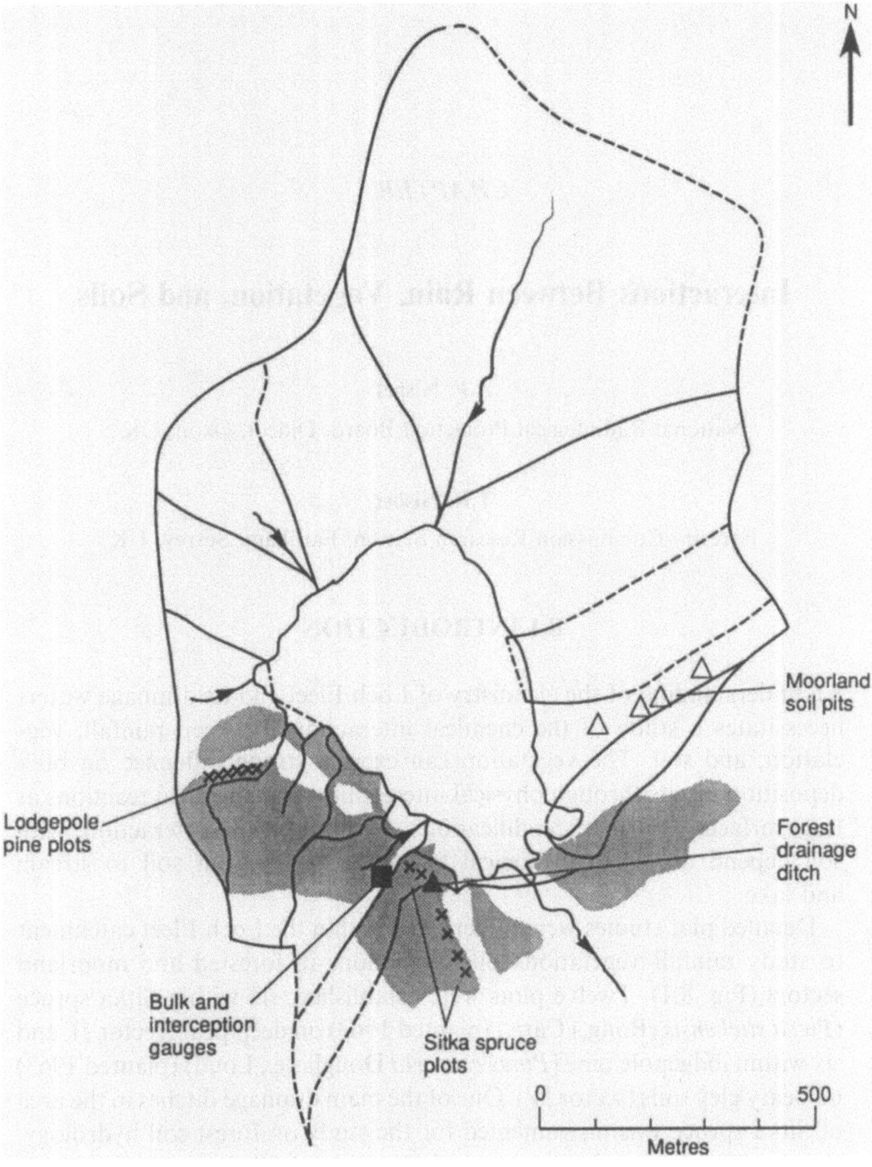

FIG. 8.1 Loch Fleet and its catchment: sample locations for plot studies of interaction between rain, vegetation, and soils.

over the 2-year period 1985–86, and analysed for a wide range of anions and cations:

- bulk precipitation and interception deposition, collected by gauges (site A) installed at canopy height in a forest ride between sectors I and II;
- throughfall and stemflow, collected by a network of randomly sited gauges within each forested plot and an area of moorland;
- soil water throughflow and ditchflow, collected by a system of gutters installed in soil pit faces on forest and moorland slopes.

Bulk precipitation, interception deposition, throughfall, and stemflow were collected fortnightly and bulked monthly for chemical analysis. Soil water throughflow was collected fortnightly for analysis. All samples underwent the following analyses: pH and conductivity; potassium and sodium by atomic absorption spectrophotometry (AAS); calcium and magnesium by inductively coupled plasma emission spectrometry (ICP); sulphate and chloride by Dionex ion exchange chromatography; ammonium, nitrate, aluminium, and silicon by colorimetric techniques.

Because the nutrient status of vegetation exerts a marked influence on rainfall–vegetation interactions, foliage samples were taken from the forested plots during October of each year and subsequently analysed for nitrogen, phosphorus, potassium, and magnesium following a sulphuric acid–hydrogen peroxide digestion. Sulphur was determined by X-ray fluorescence of finely milled foliage pressed into discs.

8.2 HYDROLOGICAL PATHWAYS

8.2.1 Bulk Precipitation, Interception Deposition, Throughfall, and Stemflow

In 1985 and 1986 annual bulk precipitation was 101% and 109%, respectively, of the long-term 30-year average (2153 mm yr^{-1}) for the site. Volumes collected under the filter gauge, representing interception deposition, were more than double those in the standard rain gauge (Table 8.1), indicating a high incidence of mist and fog events in both years. Throughfall volumes, particularly under Sitka spruce stands, formed a very high proportion of bulk precipitation (66%–85%), indicating low interception losses. Stemflow volumes were within the expected range for young Sitka spruce trees (14%–17% of bulk precipitation inputs). Those for lodgepole pine were much higher and extremely variable because of the uneven nature of these stands.

TABLE 8.1
Annual volume weighted mean concentrations and annual fluxes, 1985

		Volumes (mm)	pH/H^+	Na^+	Ca^{2+}	Mg^{2+}	K^+	NH_4^+-N	NO_3^--N	Total $SO_4^{2-}-S$	Cl^-
Bulk precipitation[a]	μeq litre^{-1}	2004	4·53/29·5	73·5	13·0	16·5	3·8	25·0	19·3	56·7	85·4
	kg ha^{-1}		0·60	33·9	5·25	4·04	3·02	6·95	5·36	18·19	60·66
Interception[a]	μeq litre^{-1}	4735	4·43/37·2	353·0	30·9	58·4	8.4	44·3	45·7	101·6	341·4
Sitka throughfall[a]	μeq litre^{-1}	1713	4·21/61·7	159·6	50·4	43·6	33·5	21·4	26·4	147·7	175·2
	kg ha^{-1}		1·06	62·88	17·37	9·14	22·37	5·08	6·74	40·59	106·48
Lodgepole throughfall[a]	μeq litre^{-1}	1298	4·17/67·6	157·0	41·9	37·0	24·6	18·6	31·4	125·2	172·4
	kg ha^{-1}		0·87	46·8	10·85	5·90	12·50	3·33	5·69	26·13	79·48
Sitka stemflow[b]	μeq litre^{-1}	284	4·19/64·6	165·2	37·9	35·4	39·1	14·3	14·3	121·5	180·3
	kg ha^{-1}		0·18	10·82	2·16	1·22	4·36	0·56	0·58	5·55	18·19
Lodgepole stemflow[b]	μeq litre^{-1}	596	4·16/69·2	164·3	24·4	23·9	32·5	16·4	21·4	87·9	175·2
	kg ha^{-1}		0·41	22·55	2·92	1·74	7·58	1·35	1·78	8·40	37·08

[a]February–December 1985.
[b]April–December 1985.

8.2.2 Forest Runoff Pathways

Within the forest, the surface throughflow pathway dominated drainage with the greater part (more than 95%) of runoff during wet periods being routed along plough furrows. This resulted from the deep peat soils remaining close to saturation throughout the year. Rainfall of sufficient intensity or duration to exceed the limited canopy and soil storage capacity quickly generated surface runoff. The intense drainage network afforded by the plough furrows and ditches produced a very flashy runoff response, usually lasting less than 12 h. During dry periods ditch flow was low, consisting of basal seepage from ditch faces.

8.2.3 Moorland Runoff Pathways

Surface throughflow also dominated drainage from the moorland slope with 87%–99% of downslope flow routed through the open network of *Molinia* and *Calluna* roots and stems. Annual volumes were large, ranging from 240×10^3 to 630×10^3 litres per metre length across the slope. The moorland peat and peaty ranker soils remained saturated for much of the year, generating surface throughflow throughout 1985, other than a brief (3-week) period during October. The response to rainfall was almost immediate, resulting in a very spiked hydrograph.

8.3 CHEMICAL INTERACTIONS

8.3.1 Bulk vs Interception Comparisons

An annual volume weighted mean pH of 4·53 and annual H^+ flux of 0·6 kg ha^{-1} for bulk deposition (Table 8.1) are within the range published for other sites in south-west Scotland (Nicholson *et al.*, 1980; Burns *et al.*, 1984; Irwin *et al.*, 1990). On a seasonal basis, greatest rainfall acidity was recorded during the spring of each year coincidentally with maximum concentrations and fluxes of sulphate and nitrate. The inefficient capture of mist and fog by the standard bulk deposition gauge results in ion inputs to the forest being underestimated. The interception gauge, acting as an inert trapping surface for cloud water, considerably increased the concentration of most elements, with sea-derived ions showing fourfold increases and non-marine ions a doubling in concentration when compared with bulk deposition (Table 8.1). The pH values remained similar. Flux data for interception deposition were not calculated because of the difficulty in correcting for the increased surface area of the double cylinder of polyethylene coated wire mesh that comprised the interception gauge.

8.3.2 Forest Vegetation Interactions

The hydrological study has shown that rainfall–vegetation interactions rather than soil–water interactions dominate drainage water chemistry. Generally, the chemical composition of rainwater tends to be markedly changed after its passage through the forest canopy and forest floor. At Loch Fleet, the forest canopy was found to increase deposited rainwater acidity for eleven months of the year, both in terms of pH and H^+ ion fluxes (see Table 8.1). This is in contrast to other coniferous throughfall studies in the UK (Cape and Lightowlers, 1988) where rainfall was effectively neutralised within the canopy, particularly during the summer months (Miller, 1984). At Fetteresso, on the Scottish east coast, similar aged stands of lodgepole pine (planted 1963) neutralised incoming rainfall for more than six months of the year despite greater hydrogen ion inputs there (Leech and Nisbet, 1987). The inability of the forest canopy to neutralise rain-derived acidity at Loch Fleet was thought to reflect, in part, the poor nutrient status of the trees. Results from foliar analysis (Table 8.2) highlighted a deficiency of nitrogen in both Sitka spruce and lodgepole pine. To compensate for this, the Sitka spruce and lodgepole pine canopies removed nitrogen, principally as the ammonium ion, from bulk precipitation in exchange for hydrogen, throughout summer and autumn 1985. This NH_4^+–H^+ exchange in the tree canopy resulted in a marked increase in throughfall acidity in autumn 1985, which could be reversed in the case of Sitka spruce by the application of urea-N to the forest floor. Miller *et al.* (1976) also recorded a marked decrease in throughfall acidity following the application of nitrogen fertiliser to nitrogen-deficient stands of Corsican pine in north Scotland. The uptake of nitrate-N by the tree canopy at Loch Fleet was only recorded for Sitka spruce and this was restricted to a short period in summer 1985. At this time the net effect of canopy uptake of ammonium and nitrate was a very small overall reduction in throughfall H^+ flux. However, the pH remained slightly lower in throughfall than bulk precipitation (i.e. 4·38 compared with 4·42).

Concentrations and fluxes of sea salts in throughfall and stemflow showed the same seasonal trends as rainfall, but were considerably enriched owing to interception and wash-off of sea salts from the forest canopy.

Other factors also contributed to the increased acidity of bulk deposition beneath the forest cover, particularly early in the year. During winter and spring the main acidifying processes were considered to be canopy interception of mist and fog, and subsequent wash-out of this

TABLE 8.2
Foliage summary data

	N[a]	P[a]	K[a]	Mg[a]
Sitka spruce (n = 18)				
Optimum[b]	> 1·5	> 0·18	> 0·7	> 0·07
Observed, 1984	1·39 ± 0·024	0·28 ± 0·009	0·8 ± 0·033	0·12 ± 0·006
Lodgepole pine (n = 18)				
Optimum[b]	> 1·4	> 0·14	> 0·5	> 0·05
Observed, 1984	1·13 ± 0·028	0·10 ± 0·003	0·54 ± 0·020	0·07 ± 0·004

[a]Expressed on a percent basis.
[b]From Binns *et al.* (1980).

TABLE 8.3
Foliar sulphur : nitrogen ratios

		S : N[a]
Sitka spruce	1985	0·045
	1986	0·054
Lodgepole pine	1985	0·027
	1986	0·031
Typical conifer[b]		0·030

[a]On a gram atom basis.
[b]Kelly and Lambert (1972).

deposited acidity in association with sulphate and nitrate. Sulphate, being very mobile, can contribute to overall acidification by increasing the leaching of base cations from soils. Annual fluxes of sulphate beneath the canopy of Sitka spruce at Loch Fleet were far in excess of bulk precipitation inputs (Table 8.1). This gain in sulphate may be attributed either to enhanced canopy interception or to increased canopy leaching of sulphate. There is much controversy over the relative importance of these two processes. Some studies (e.g. Lindberg and Garten, 1988) indicate that dry deposition provides more than 85% of the sulphate enrichment in throughfall, while others (e.g. Cape *et al.*, 1987) suggest that leaching is the major component. Foliage data on sulphur : nitrogen ratios (Table 8.3) for Sitka spruce suggest that canopy leaching may make a significant contribution to sulphate deposition in throughfall, particularly during the metabolically active summer months. Normally sulphur : nitrogen ratios approach a typical 0·03 in conifer foliage, irrespective of species (Kelly and Lambert, 1972). Ratios for Sitka spruce were much higher at around 0·05, implying the presence of a substantial foliar pool of inorganic sulphate, potentially available for leaching. Support for the operation of a within-tree sulphate cycle comes from other studies in nitrogen-deficient, sulphur-rich ecosystems (Hesse, 1957; Turner *et al.*, 1980).

It is important to note that large gains in potassium beneath the forest canopy during summer and autumn occurred without correspondingly large reductions in hydrogen flux. This failure to neutralise incoming rainfall acidity through H^+-base cation exchange at leaf surfaces suggests neutral salt leaching (i.e. K_2SO_4) as the principal mechanism for potassium enrichment in throughfall. This predominantly leaching role for potassium is also supported by Ulrich (1983) who considers calcium and magnesium to be the major ions involved in H^+ exchange reactions. Because throughfall enrichment of calcium and magnesium was much less than for potassium, base cation exchange was not considered to be an important rainwater–canopy interaction in either Sitka spruce or lodgepole pine at Loch Fleet.

Stemflow was consistently more acidic than bulk precipitation during both years, its chemistry tending to resemble that of throughfall. Although stemflow constituted a smaller hydrological input than throughfall, it did make a significant contribution to the total ion flux reaching the forest floor.

8.3.3 Moorland Vegetation Interactions

As in the forest, incoming rainfall was modified as it passed through *Calluna* and *Molinia* vegetation canopies on the moorland slope (Table 8.4). Throughfall remained slightly more acidic during spring and summer, and slightly less acidic during autumn and winter. The latter coincided with the period of senescence and minimal growth of the vegetation when concentrations of calcium, magnesium, and potassium peaked as a result of foliar leaching. These base cations together with sulphate remained significantly enhanced in throughfall throughout the year. The moorland foliage, like that in the forest, was found to remove ammonium ions from bulk precipitation for most of the year, presumably through an NH_4^+–H^+ exchange mechanism.

8.3.4 Water–Surface Soil Interactions

Throughfall and stemflow were further acidified on passing through the litter and surface soil layers of the forest (Table 8.5). Acidification of this surface throughflow probably resulted from ion exchange reactions; the upper soil having a very high exchangeable acidity. Concentrations of potassium, ammonium, nitrate, and sulphate were all much lower in throughflow than throughfall, indicating root uptake and/or soil microbial incorporation. This tight nutrient cycle would be expected of trees growing on nutrient-poor soils. Aluminium, which was undetectable in throughfall and stemflow, reached high concentrations in surface throughflow. These high aluminium concentrations increased further in ditchflow, whilst H^+ decreased, thereby reflecting the contribution of seepage waters from the mineral material at the base of the peat.

Surface throughflow over moorland remained more acidic than throughfall throughout the year and showed enrichment of most ions except ammonium, potassium, and nitrate, which were removed by root

TABLE 8.4
Volume weighted mean concentrations (March to November 1986) for moorland slope soil throughflow at four soil pits

	Pit 1		*Pit 2*			*Pit 3*			*Pit 4*		
	A	*B*	*A*	*B*	*C*	*A*	*B*	*C*	*A*	*B*	*C*
H^+ (as pH)	4·25	4·24	4·23	4·22	4·48	4·19	4·19	4·20	4·08	4·12	4·15
Ca^{2+}	39·4	38·9	46·4	47·9	45·9	40·4	45·4	44·9	26·4	26·4	27·9
Mg^{2+}	54·3	54·3	56·8	60·9	54·3	55·1	60·9	62·6	47·7	45·3	45·3
Na^+	229·6	227·4	225·7	239·6	233·5	225·7	232·2	240·9	225·7	213·9	220·4
K^+	14·6	14·6	14·1	13·3	10·2	12·5	13·8	14·1	12·3	10·0	11·5
Al^{3+}	28·9	30·0	31·1	33·3	45·6	36·7	37·8	40·0	54·4	54·4	48·9
Si	0·84	0·84	1·00	0·99	1·09	1·06	1·06	1·10	1·11	0·87	0·76
NH_4^+–N	3·6	4·3	3·6	2·9	2·1	3·6	2·9	5·7	2·9	2·1	6·4
NO_3^-–N	12·1	16·4	19·3	22·9	7·9	22·9	31·4	36·4	32·9	25·7	25·0
Cl^-	240·8	236·6	240·0	249·0	237·5	240·3	253·8	263·4	240·8	224·2	236·6
SO_4^{2-}–S	102·8	111·5	109·7	120·9	117·8	107·2	117·1	119·6	108·4	110·3	107·8

Notes: A: surface flow through the root mat, 0–5 cm depth; B: flow through the upper organic horizon, 5–20 cm depth; C: flow through the lower organic horizon above granite bedrock, variable depth 20–50 cm. Concentrations in μeq per litre, except for Si, in mg per litre.

TABLE 8.5

Seasonal volume weighted mean concentrations (summer 1985 to spring 1986) for forest slope soil throughflow

		A	*B*	*C*	*TF*	*DF*
H^+ (as pH)	Su[a]	3·91			4·38	4·05
	Au	3·91	4·01		4·42	4·00
	Wn	3·65		4·10	3·88	4·00
	Sp	3·93		3·63	4·04	4·02
Ca^{2+}	Su	43·4			40·4	41·4
	Au	18·5	48·4		45·4	23·9
	Wn	66·3		14·5	73·3	27·4
	Sp	17·5		71·3	58·9	31·4
Mg^{2+}	Su	18·9			23·9	41·2
	Au	16·5	28·0		42·0	32·9
	Wn	94·7		32·9	106·2	56·0
	Sp	21·4		134·2	69·1	68·3
Na^+	Su	189·6			80·4	228·7
	Au	123·0	178·3		155·7	158·3
	Wn	552·6		164·8	415·2	223·9
	Sp	239·6		342·6	261·3	263·9
K^+	Su	1·3			41·7	2·0
	Au	1·5	13·3		39·1	2·8
	Wn	9·0		12·8	30·4	8·2
	Sp	2·6		8·2	15·6	6·1
Al^{3+}	Su	33·3			0·0	48·9
	Au	24·4	26·7		0·0	36·7
	Wn	27·8		25·6	0·0	36·7
	Sp	12·2		13·3	0·0	38·9
Si	Su	0·47			0·00	2·23
	Au	0·40	0·59		0·00	1·44
	Wn	0·24		0·50	0·00	1·78
	Sp	0·30		0·95	0·00	1·99
NH_4^+-N	Su	6·4			10·0	2·1
	Au	7·9	13·6		18·6	4·3
	Wn	5·7		13·6	47·1	6·4
	Sp	2·1		55·0	57·9	4·3
NO_3^--N	Su	7·1			15·0	7·9
	Au	5·0	5·0		21·4	10·0
	Wn	2·1		18·6	36·4	19·3
	Sp	0·7		9·3	62·1	14·3
Cl^-	Su	162·3			95·5	191·5
	Au	128·7	192·4		175·5	147·0
	Wn	772·1		191·3	444·2	261·4
	Sp	231·0		442·0	285·1	303·1
SO_4^{2-}-S	Su	79·8			114·6	112·1
	Au	63·6	148·3		125·9	82·2
	Wn	135·8		81·0	181·9	109·7
	Sp	96·6		316·5	201·2	138·3

[a]Su = summer; Au = autumn; Wn = winter; Sp = spring.

Notes: A: surface flow along the plough furrow, 0–5 cm depth; B: flow through the upper organic horizon, 5–60 cm depth; C: flow through the lower organic horizon, 60–80 cm depth; TF: Sitka spruce throughfall; DF: drainage ditch flow. Concentrations in μeq per litre, except Si, in mg per litre.

TABLE 8.6
Seasonal volume weighted mean concentrations (winter 1985 to autumn 1986) for moorland slope soil throughflow

		BP	*TF*	*A*	*B*	*C*
H^+ (as pH)	Wn[a]	4·37	4·71	4·25	4·21	4·46
	Sp	4·31	4·27	4·22	4·21	4·40
	Su	4·42	4·33	4·25	4·26	4·60
	Au	4·58	4·66	4·24	4·21	4·55
Ca^{2+}	Wn	16·5	41·9	50·4	57·9	56·4
	Sp	13·5	27·9	51·9	54·4	50·9
	Su	13·0	24·9	30·9	31·9	35·4
	Au	14·0	28·9	42·4	44·9	44·9
Mg^{2+}	Wn	48·6	93·8	67·5	79·8	75·7
	Sp	29·6	37·0	63·4	70·8	60·9
	Su	14·0	23·9	32·9	32·9	35·4
	Au	31·3	51·9	54·3	59·3	58·4
Na^+	Wn	267·8	617·4	254·3	321·7	284·3
	Sp	148·7	142·2	238·3	262·6	250·9
	Su	73·5	71·7	167·4	165·2	207·0
	Au	171·3	199·1	225·7	239·1	221·3
K^+	Wn	5·9	16·1	17·1	18·7	6·4
	Sp	3·3	8·2	18·4	19·2	12·8
	Su	4·1	17·9	2·8	2·6	9·0
	Au	3·1	26·6	10·0	7·9	5·4
Al^{3+}	Wn	0·0	16·7	34·4	35·6	54·4
	Sp	0·0	7·8	31·1	31·1	46·7
	Su	0·0	4·4	33·3	33·3	42·2
	Au	0·0	3·3	31·1	36·7	45·6
Si	Wn	0·00	0·09	1·30	1·17	1·52
	Sp	0·00	0·05	1·18	1·26	1·36
	Su	0·00	0·12	0·36	0·35	0·77
	Au	0·00	0·11	0·90	0·86	0·82
NH_4^+-N	Wn	35·7	10·7	5·0	4·3	5·0
	Sp	19·3	54·3	5·0	4·3	2·9
	Su	26·4	11·4	2·1	2·1	1·4
	Au	2·1	0·7	0·7	16·4	8·6
NO_3^--N	Wn	13·6	7·1	30·0	28·6	4·3
	Sp	28·6	52·1	32·1	41·4	14·3
	Su	12·9	14·3	0·0	0·0	0·0
	Au	9·3	12·1	1·4	0·0	0·0
Cl^-	Wn	314·4	678·9	316·1	414·1	376·9
	Sp	169·0	161·4	257·5	277·5	274·9
	Su	84·2	78·0	158·0	149·3	167·3
	Au	193·8	244·8	242·3	253·8	224·5
SO_4^{2-}-S	Wn	100·3	100·9	103·4	104·0	94·7
	Sp	64·8	97·8	119·6	135·2	123·4
	Su	49·2	67·3	85·4	84·7	113·4
	Au	43·6	64·2	99·1	115·3	110·3

[a]Wn = winter; Sp = spring; Su = summer; Au = autumn.

Notes: A: surface flow through the root mat, 0–5 cm depth; B: flow through the upper organic horizon, 5–25 cm depth; C: flow through the lower organic horizon above granite bedrock, 25–40 cm depth; BP: bulk precipitation as measured in forest ride; TF: *Calluna/Molinia* throughfall. Concentrations in μeq per litre, except Si, in mg per litre.

uptake. Aluminium, which was present at low concentrations in throughfall, once again increased markedly in surface throughflow. Overall, soil throughflow chemistry varied little with position on the slope, soil depth, or flow during rainfall events (Table 8.6). This latter feature is indicative of a rapid buffering of rainfall within the surface root mat through ion exchange processes (Bache, 1984).

It was not possible to calculate accurate flux data for soil drainage because the upslope contributing area of each soil pit was unknown. Nevertheless total soil element throughputs per metre of contour length were calculated; as expected, the surface throughflow pathway (and therefore rain–vegetation interactions) dominated the total throughput/export of all ions into the loch.

8.4 CONCLUSIONS

This study has highlighted the importance of rain–vegetation interactions and soil hydrology in controlling the transfer of acidity into Loch Fleet. The dominance (87%–99%) of the surface throughflow pathway within both forested and moorland slopes resulted in rain/vegetation interactions exerting a major control over runoff chemistry. Forest and moorland vegetation modified bulk precipitation inputs leading to further acidification of drainage waters. Three main processes were identified as controlling rain/vegetation interactions at Loch Fleet:

(1) canopy capture and washoff of dry deposition and intercepted mists and fog, leading to a potential decrease in throughfall pH;
(2) canopy uptake of essential nutrients in exchange for hydrogen, leading to a potential decrease in throughfall pH; and
(3) canopy leaching of neutral salts and their subsequent recycling causing no change in pH.

As a consequence of these processes and the dominance of the surface throughflow pathway, all slopes within the Loch Fleet catchment generated large volumes of acidic runoff throughout the year in response to heavy rainfall. This was associated with high concentrations of aluminium.

ACKNOWLEDGEMENTS

The authors are indebted to the Department of Environment for funding this work and to Macaulay Land Use Research Institute for assistance during the project. Additional funding and assistance from the Loch Fleet project (CEGB, NSHEB, SSEB, and British Coal) for work at Loch Fleet is also gratefully acknowledged.

REFERENCES

Bache, B.W. (1984). Soil–water interactions. *Phil. Trans. Roy. Soc. Lond.* **B305**: 393–407.

Binns, W.O., Mayhead, G.J. and Mackenzie, J.M. (1980). *Nutrient Deficiencies in British Forests, An Illustrated Guide*. For. Comm. Leaflet No. 76. HMSO, London.

Burns, J.O., Coy, J.S., Tervet, D.J., Harriman, R., Morrison, B.R.S. and Quine, C.P. (1984). The Loch Dee project: a study of the ecological effects of acid precipitation and forest management on an upland catchment in S.W. Scotland, 1 Preliminary investigations. *Fish. Manag.* **15**: 145–67.

Cape, J.N. and Lightowlers, P.J. (1988). *Review of Throughfall and Stemflow Chemistry Data in the United Kingdom*, Final Report to Dept of Environment. ITE, Bush Estate, Penicuik, Midlothian.

Cape, J.N., Fowler, D., Kinnaird, J.W., Nicholson, I.A. and Paterson, J.S. (1987). Modification of rainfall chemistry by a forest canopy, pp. 155–69 in *Pollutant Transport and Fate in Ecosystems* (eds. P.J. Coughtrey, M.H. Martin and M.H. Unsworth). Blackwell, Oxford, 414 pp.

Hesse, P.R. (1957). Sulphur and nitrogen changes in forest soils of East Africa. *Plant Soil* **9**: 86–96.

Irwin, J.G., Campbell, G.W., Cape, J.N., Clark, P.A., Davies, T.D., Derwent, R.G., Fisher, B.E.A., Fowler, D., Kallend, A.S., Longhurst, J.W.S., Martin, A., Smith, F.B. and Warrilow, D.A. (1990). *Acid Deposition in the United Kingdom 1986–1988*, Third Report of UK Review Group on Acid Rain. Dept of Environment, Warren Spring Laboratory, 124 pp.

Kelly, J. and Lambert, M.J. (1972). The relationship between sulphur and nitrogen in the foliage of *Pinus radiata*. *Plant and Soil* **37**: 395–407.

Leech, A.F. and Nisbet, T.R. (1987). *Acid Transfer Through Vegetation and Soils*, Final Report to Dept of Environment. Macaulay Land Use Research Institute, Craigiebuckler, Aberdeen, 233 pp.

Lindberg, S.E. and Garten, C.T. (1988). Sources of sulphur in forest canopy throughfall. *Nature* **336**: 148–51.

Miller, H.G. (1984). Deposition–plant–soil interactions. *Phil. Trans. Roy. Soc. Lond.* **B305**: 339–52.

Miller, H.G., Cooper, J.M. and Miller, J.D. (1976). Effect of nitrogen supply on nutrients in litterfall and crown leaching in a stand of Corsican pine. *J. Appl. Ecol.* **13**: 233–48.

Nicholson, I.A., Cape, N., Fowler, D., Kinnaird, J.W. and Paterson, I.S. (1980). pH and sulphate content of precipitation over northern Britain, pp. 142–43 in *Proc. Internat. Conference Ecological Impact of Acid Precipitation* (eds. D. Drablos and A. Tollan). SNSF Project, Norway, 383 pp.

Turner, J., Johnson, D.W. and Lambert, M.J. (1980). Sulphur cycling in a Douglas fir forest and its modification by nitrogen applications. *Oecologia Plantarum* **1**(15): 27–35.

Ulrich, B. (1983). Interaction of forest canopies with atmospheric constituents: SO_2, alkali and alkali earth cations and chloride, pp. 33–45 in *Effects of Accumulation of Air Pollutants in Forest Ecosystems* (eds. B. Ulrich and J. Pankrath). Reidel, Dordrecht, Netherlands.

PART III

LIMING DESIGN AND PRACTICE

CHAPTER 9

Liming Design and Strategy at Loch Fleet

R.M. Bell
Environmental Advisory Unit, Liverpool, UK

9.1 INTRODUCTION

At Loch Fleet the liming strategy was based on both laboratory and field trials of lime applications to moorland soils, the design of the latter being based on the results of the former. Innovative treatments, for example moorland burning and forest fertilisation, were also attempted, and have been compared with the performance of the liming treatments.

On the basis of these trials, effective methods of treating the soils of an upland catchment have been developed, and have the potential for modifying the water chemistry of the loch sufficiently to permit fish survival and reproduction.

Preliminary work was undertaken to characterise a variety of liming/neutralising agents on the effects of particle size, and on application "dose" or rate. Further trials were made to identify treatments with the greatest potential for modifying soil leachate. It was recognised at the start that for treatments to be acceptable, they should not cause significant and persistent adverse effects on the vegetation or other components of the terrestrial ecosystem.

9.2 ASSESSMENT OF LIMING AGENTS

9.2.1 Materials Tested (Vizard, 1985, 1986)
Samples of fifteen commercially available limestones (six Carboniferous, two Jurassic, three Permian, four Cretaceous) and of two marine shells (scallops) were assessed. The samples were prepared to a uniform particle size distribution and treated with dilute sulphuric acid under controlled conditions. The rate of neutralisation of the sulphuric acid, defined by the rate of change in pH value, was determined for each material in turn.

One limestone sample was studied more extensively to determine the effect of repeated treatments with dilute sulphuric acid and also to determine the relationship between rate of neutralisation and particle size distribution. The particle size distribution of each limestone sample as supplied was also determined.

Each sample was analysed to determine the content of 23 chemical constituents.

9.2.2 Leachate Chemistry and Vegetation Effects
A further laboratory trial comprised a variety of liming agents applied in a range of formulations at different application rates to undisturbed peaty soil cores with their associated vegetation taken from an area adjacent to the Loch Fleet catchment (Hockin and Bell, 1988).

The aim of the trial was to identify practical treatments with the best potential for modifying the soil leachate chemistry. To be successful the liming agent must decrease acidity by increasing the pH, and increase the carbonate ion concentrations, so enhancing the buffering capacity of the system. A decrease in aluminium leached to the soil water should then occur. Persistent adverse effects on moorland vegetation species were also assessed in this trial.

The liming agents investigated included calcium carbonate as limestone and chalk, and calcium hydroxide (hydrated lime). Caustic reagents, such as soda lime, were seen to be of no practical value because of reaction with the peat matrix and degradation of the humic acids causing the soil water percolate to be strongly coloured and possibly of excessively high pH.

The materials were applied in these trials at three application rates, equivalent to $5\,t\,ha^{-1}$, $20\,t\,ha^{-1}$, and $50\,t\,ha^{-1}$. Hydrated lime was applied as a fine powder, while various particle sizes of limestone and chalk were used. Monolith soil columns ($0{\cdot}25\,m \times 0{\cdot}4\,m$) containing peaty soil cores were set up in a field laboratory and watered twice weekly with a simulated rain (pH 4·71) at a rate characteristic of the site

TABLE 9.1
Chemical composition of limestone samples (means)

	Limestones			*Chalk* *Cretaceous* (n = 4)	*Marine shells* (n = 2)
	Carboniferous (n = 6)	*Jurassic* (n = 2)	*Permian* (n = 3)		
CaO^a	54·6	39·4/49·0	32·3	53·5	55·2/55·5
MgO^a	0·7	0·5/0·8	17·6	0·4	0·2/0·3
$Al_2O_3{}^a$	0·1	0·9/3·4	0·8	0·2	< 0·1/0·1
Zn^b	26	27/28	31	18	4/15
Cd^b	< 1	< 1	< 1	< 1	< 1
Mn^b	167	250/295	348	471	40/55
Cu^b	16	7/9	5	4	5/5
Pb^b	55	20/30	27	21	20/25

[a]As % dry weight.
[b]ppm dry weight.

(2000 mm yr^{-1}). Treatments investigated were: limestone (3 mm dust); limestone dust (< 63 μm); washed limestone chips (3 mm); lump chalk; chalk powder; hydrated lime; and soda lime.

Of prime importance in this test was that there was no evidence of preferential leakage paths through the lysimeters during the experimental period. Hydraulic conductivities were measured by falling head methods and produced levels of approximately 2–4 cm day^{-1} (MAFF, 1982). Particle densities varied between 1·32 and 1·43 g cm^{-3}, the bulk density of the peat was 0·15 g cm^{-3} (MAFF, 1982), and the available water capacity varied between 63% and 69% (Stakman *et al.*, 1969) with a total porosity of 88% and 92% (Stakman and van der Harst, 1969).

9.2.3 Results

9.2.3.1 Chemical Tests (*Vizard, 1985*)

The rate of neutralisation of sulphuric acid by the samples was related to their calcium content, the highest rates being for the high calcium Carboniferous and Cretaceous limestones and for scallop shell samples.

Jurassic limestones have lower calcium content and high inert (silica) content. Permian limestones contain the lowest calcium content, but had higher magnesium (Table 9.1); the magnesium component contributed to

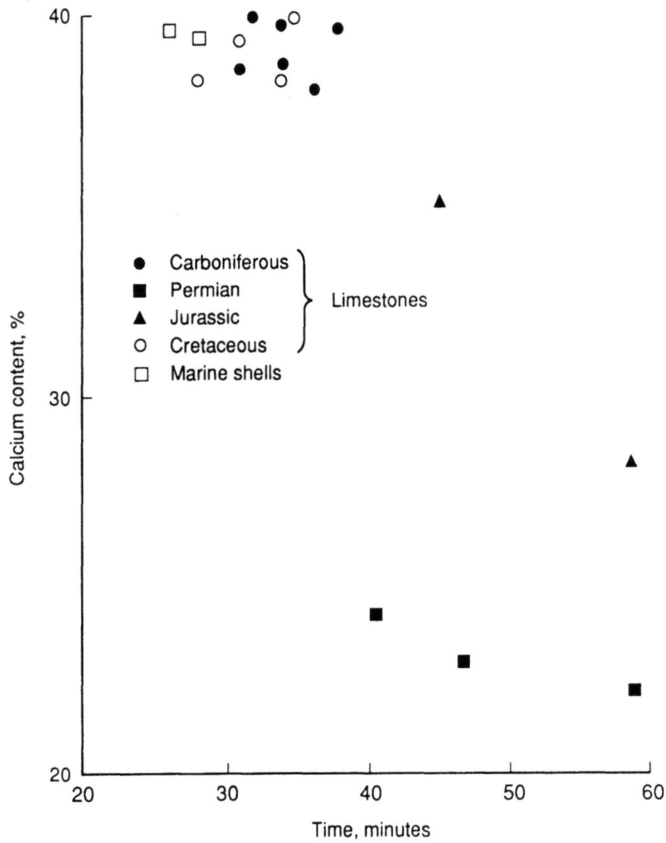

FIG. 9.1 Time to change pH (from 4 to 5) of sulphuric acid using possible liming agents.

the rate of neutralisation of the sulphuric acid but is less effective in improving water quality for fish survival since calcium content per unit weight is less and magnesium provides no physiological benefit for fish.

Calcium concentrations were higher in the leachates from the Carboniferous, Cretaceous, and scallop shell samples than from Permian and Jurassic samples, in approximate proportion to the determined rates of neutralisation. Magnesium concentrations reflect the magnesium component of Permian limestones and contribute to the neutralisation of strong acids.

Overall, the high calcium Carboniferous, Cretaceous, and marine shell samples gave the highest rates of neutralisation of sulphuric acid (Fig. 9.1).

The rate of neutralisation of sulphuric acid by Carboniferous limestone

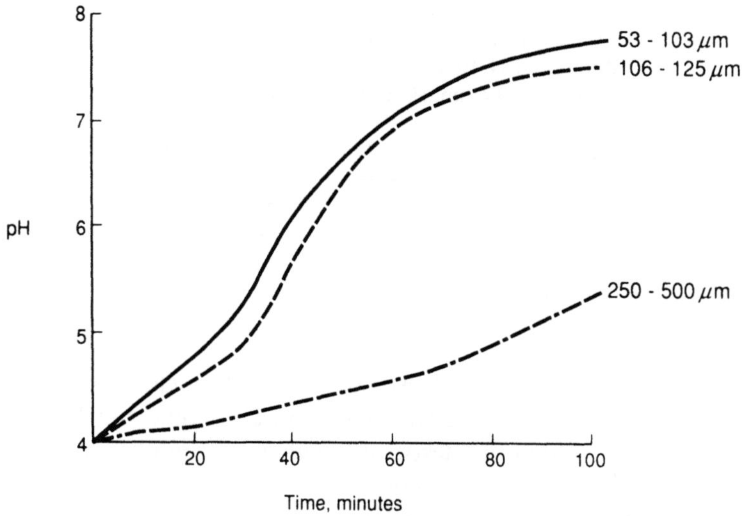

FIG. 9.2 Carboniferous limestone A: effect of particle size.

A increases as the particle size is reduced in approximate proportion to the surface area of the particle (Fig. 9.2); particle size distribution is thus of prime importance. A combination of finest practicable particle size distribution and high calcium content is thus the most effective material (Sverdrup, 1985; Hindar, 1987, 1988) (Table 9.2).

The doses of limestone required to raise the alkalinity of natural waters will vary according to their chemical composition. This requirement has been tested for Swedish waters (Lessmark and Thornelof, 1986; Hultberg

TABLE 9.2
Sedimentation rates and dissolution rates (at pH 4·0) for limestone particles of various sizes (after Hindar, 1985)

Particle diameter (mm)	*Time required for 10 m fall*	*Time required for dissolution*
2	2·4 s	9 yr
0·6	0·5 min	10 months
0·2	4 min	1 month
0·06	1 h	3 days
0·02	7 h	8 h
0·006	3 days	1 h
0·002	1 month	5 min

TABLE 9.3
Doses of limestone required for waters of different pH to raise alkalinity to >0·1 meq litre^{-1} (Lessmark and Thornelof, 1986)

pH	*Dose* ($g\,m^{-3}$) *as* $CaCO_3$
4·0–4·5	30
4·6–5·0	20
5·1–5·5	15
5·6–6·0	10

and Nystrom, 1988) (Table 9.3). At Loch Fleet, the yield of alkalinity in runoff from catchment liming was uncertain and could not have been predictable from laboratory tests, although some indication could be found from field plot studies (Section 9.2.4).

Such neutralising agents contain low levels of other components which need to be taken into consideration depending upon their possible effects on fish. A limited programme of bioassay to identify any possible adverse effects of the selected liming agents on brown trout fry was also undertaken. No adverse effects were reported. Full chemical analysis shows differences in calcium and magnesium, mainly as carbonates; the other major constituent is usually silica. Most products tested contain appreciable concentrations of zinc, manganese, iron, lead, and fluoride.

9.2.3.2 Soil Leachates and Vegetation

The results presented in Fig. 9.3 show the pH of the soil water percolate over the experimental period, while those presented in Fig. 9.4 show the cumulative amount of calcium that is leached. Figure 9.5 shows the cumulative amount of aluminium that is leached. In most treatments an increase in the amount of liming agent applied caused an increase in the soil water pH and calcium content, and a decrease in the aluminium concentration. An exception occurs with hydrated lime, which is thought to have "caked" so as to make most of the material applied ineffective.

Data show the base saturation of the various treatments at the end of the 12-month experimental period. A substantial increase, from approximately 10% base saturation initially, is shown with all of the liming agents at all depths of the soil profile.

The results indicate that any of the liming agents tested (limestone, chalk, and hydrated lime) can modify the water chemistry of moorland

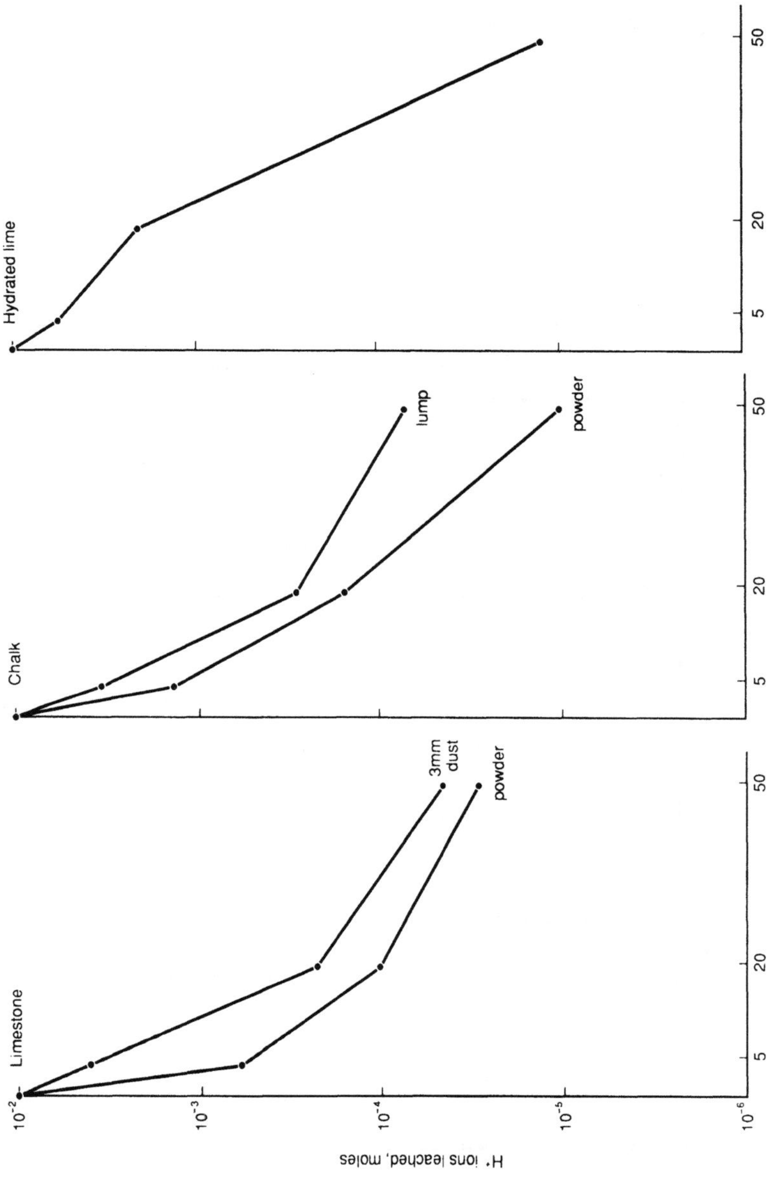

FIG. 9.3 Hydrogen ions leached in one year through peat lysimeters treated with various liming agents applied at different rates.

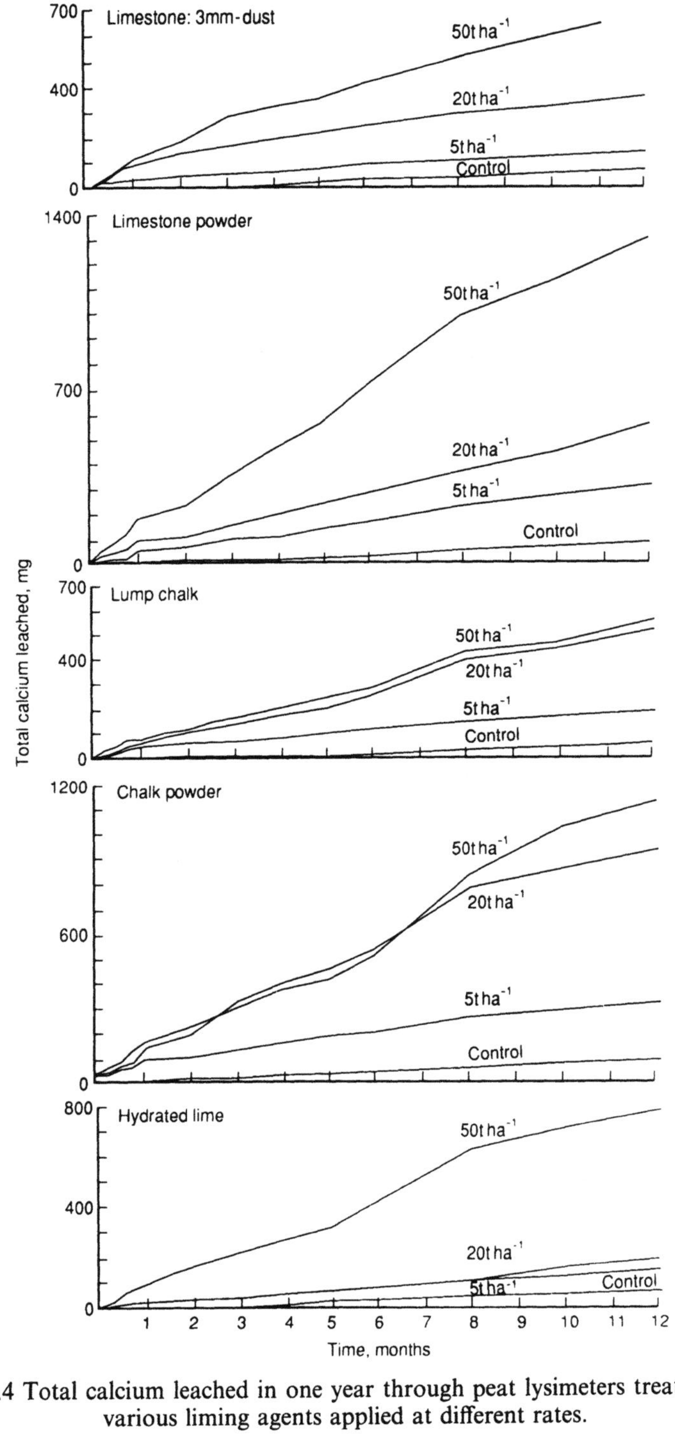

FIG. 9.4 Total calcium leached in one year through peat lysimeters treated with various liming agents applied at different rates.

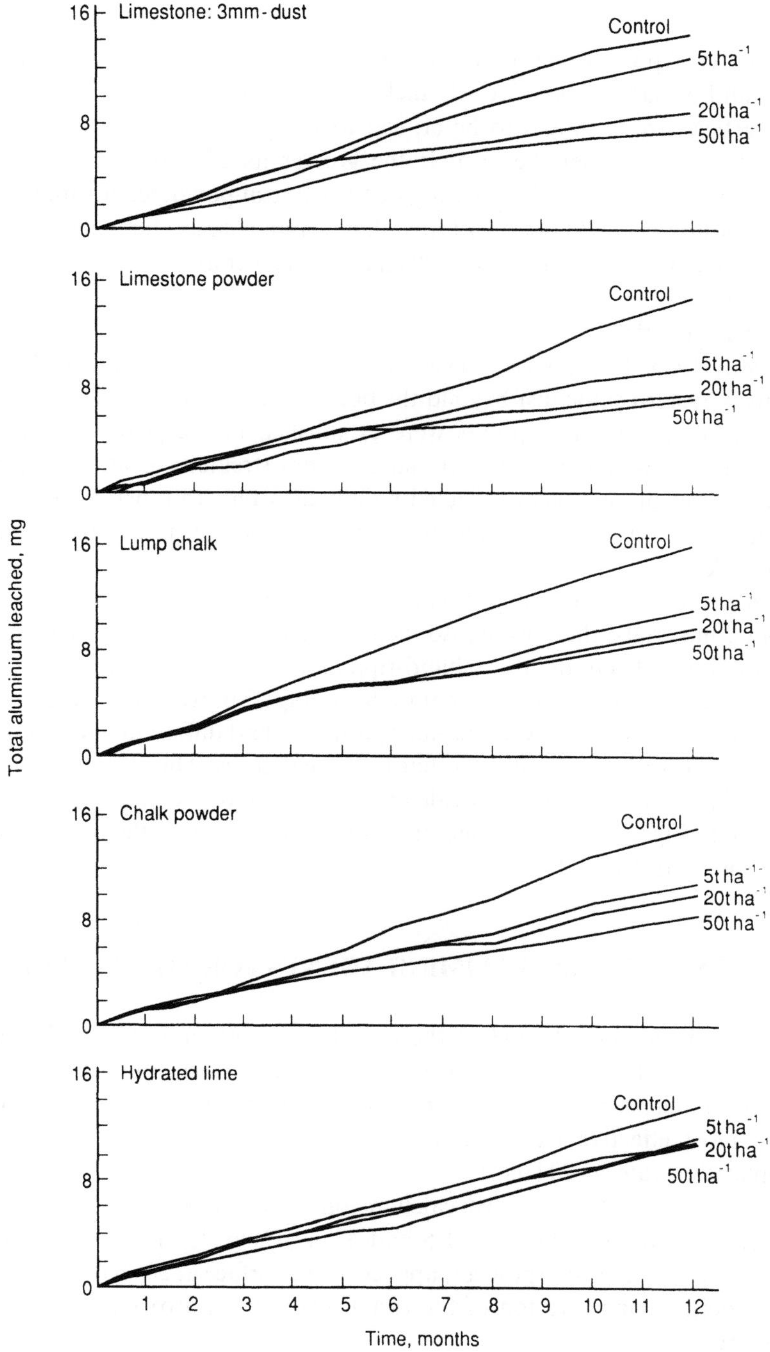

FIG. 9.5 Total aluminium leached in one year through peat lysimeters treated with various liming agents applied at different rates.

soil cores. Application rates of 50 t ha^{-1} are shown to be needlessly excessive, and 5 t ha^{-1} to be relatively ineffective, as shown in Fig. 9.6.

Particle size is shown to be critical to the performance of the liming agent, with the finer the particle the greater its efficiency. As a consequence, it was proposed to investigate the value of powdered formulations of limestone, chalk, and hydrated lime, and coarser lump chalk and limestone 3 mm dust at 10 and 20 t ha^{-1} in field trials.

9.2.4 Field Trials

In May 1986, 16 plots each measuring 7 m × 7 m were established on a convex eastern slope just beyond the boundaries of the Loch Fleet catchment. Water sampling points were established in the peat to intercept surface flow at uphill (control) and downhill (treated) locations. Triplicated cores of peat were collected within each plot to a depth of 20 cm, which was approximately the full depth of peat overlying the granite bedrock.

In addition to the straightforward surface application of the plot with liming materials, these were also incorporated into the peat matrix at an approximate 15 cm depth by hand-ripping across the contours.

The results confirmed the laboratory findings, but were inherently more variable because of uneven drainage pathways and differential leaching of the peat *in situ* compared with that from peat cores. The effects upon the runoff pH and the dissolved calcium and aluminium concentrations are shown in Figs. 9.3–9.5. Changes in soil base saturation in the upper 10 cm are shown in Fig. 9.6.

9.3 SELECTION OF LIMING AGENTS FOR LOCH FLEET

In these field trials the pH of the runoff from untreated (control) areas varied between 3·9 and 7·1 over the 1-yr period. Surface applications of all of the liming agents were effective in increasing the pH of the runoff water. The most effective treatment is identified as limestone powder at an application rate of 20 t ha^{-1}.

Variation in the dissolved calcium ion concentration of runoff from untreated plots was between 1·5 and 45 μeq Ca yr^{-1}. The most effective liming agent (limestone powder) applied at the surface at 20 t ha^{-1} increased the calcium concentration of the runoff water to approximately 100 μeq Ca $litre^{-1}$.

As a result of these tests a single material, carboniferous limestone in a

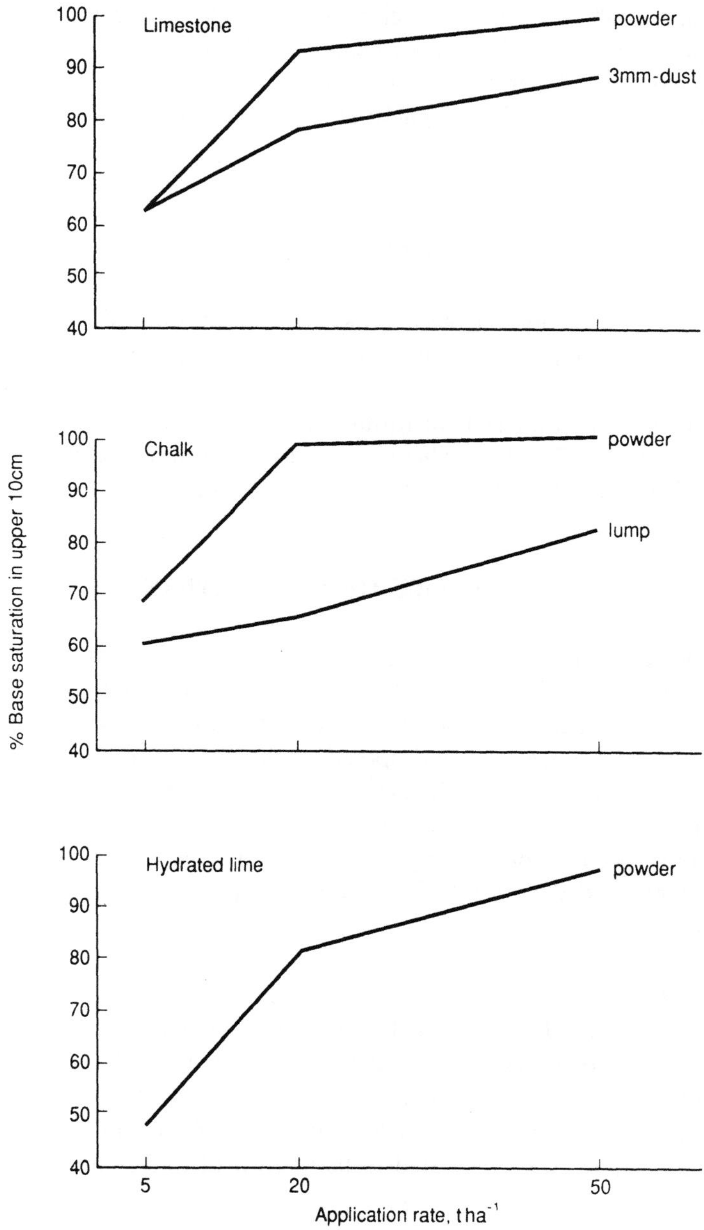

FIG. 9.6 Base saturation (%) over one year in the upper 10 cm of peat lysimeters treated with various liming agents applied at different rates.

size range, 100% < 63 μm, was selected as suitable for use at Loch Fleet. This was supplied by Imperial Chemical Industries (ICI) Buxton, and designated "ICI Filter Dust". Analysis of this material gave 98·53% calcium carbonate, 0·52% magnesium carbonate, and 0·05% aluminium oxide.

9.4 FORMULATIONS

Practical considerations suggested that the application of this material in powder form might prove difficult particularly in handling the material and spreading uniformly on rough moorland and under forest canopy. Consequently, two alternative formulations were used which overcame some of these difficulties. These were limestone slurry and limestone pellets. Further details are given in the following section.

9.5 METHODS OF APPLICATION

9.5.1 Dry Powder

Where this was used at Loch Fleet, it was lifted in 750-kg loads by helicopter from a central depot nearby to each of the sectors to be treated, where the bags were split and spread by hand. This ensured that the limestone was restricted to the delineated sectors, which could not have been achieved by aerial dispersion by helicopter. The fine dry powder could have been supplied using a conventional type hopper delivery by helicopter provided with suitable vibration equipment in a practical liming exercise.

9.5.2 Slurry

This was produced by mixing the limestone powder with loch water pumped to a mixing tank on the hillside above the forest (1 kg to about 2 litres of water). The slurry was then fed from a hose pipe by gravity through the brashed tree stands in a reasonably uniform manner. The total water application was calculated to be equivalent to 3·5 mm of rainfall.

9.5.3 Pellets

The fine limestone provided by ICI was also used in the form of pellets that could be dropped from a helicopter (Plate I). The process involved mixing the limestone with 2% brebent bentonite, 0·5% pre-gelatinised maize starch, and approximately 12%–13% water. The pellets, approximately 11 mm × 20 mm, dry off and harden and are then suitable for transport. On contact with excess water after application in the field, the pellets break up, releasing fine particles of limestone. Toxicity tests of the pellet leachate using brown trout fry have shown it to be innocuous.

9.6 OTHER TRIALS

9.6.1 Fertilisers

Two small forested sectors (I and II) were used to test a hypothesis, supported by limited observations within tree stands, that in this nitrogen-deficient location foliar uptake of ammonia from rain could be inhibited by the application of urea-N fertiliser (see Chapter 8). The foliar uptake was thought to result in more acidic conditions in the rooting zone. In these two sectors a small, dry limestone application (5 t ha^{-1}) was made by hand in spring 1987, with an additional (NPK) fertiliser (total nitrogen 14·5%, total phosphorus as P_2O_5 13·5%, total potassium as K_2O 13·0%) application to sector II during early summer tree growth.

9.6.2 Moorland Burning

In an effort to manipulate the mineral cycling in accumulated litter, as well as that of the highly organic peaty soils at Loch Fleet, it was proposed to burn an extensive area in sector VIII. It was expected that a transient flush of mineral ash would improve the water acidity in the short term, and that subsequently the exchangeable acidity in the soil would be reduced. In the prevailing climatic conditions at Loch Fleet, however, only a superficial burning of the moorland vegetation was achieved.

9.6.3 In-stream Liming

At an adjacent site, with water quality closely approximating that of Fleet, trials were made of in-stream liming using limestone materials in the stream bed, and a bankside lime doser to treat the stream flow. Details of these and other alternatives to catchment liming are given in Chapter 10. The results of these trials are reported in Chapter 13.

9.7 TREATMENTS APPLIED AT LOCH FLEET

The first series of treatments (sectors IV, VI, and VII) at Loch Fleet were undertaken in April 1986, and a second series (sectors I, II, Z_1, VIII, and

TABLE 9.4
Loch Fleet catchment treatments, 1986 and 1987

Sector	*Treatment*	*Nominal application rate*	*Sector totals*
I	Fertiliser	1 t ha^{-1} below tree canopy	Lime: 0 t Fert: 4 t
II	Lime + fertiliser	5 t ha^{-1} } below tree 1 t ha^{-1} } canopy	Lime: 18 t Fert: 4 t
IV	Limestone slurry	20 t ha^{-1} below forest canopy	Lime: 110 t
V	None	None/"control" sector	—
VI	Limestone dust	30 t ha^{-1} spread uniformly over moorland	Lime: 135 t
VII	Limestone dust	10 t ha^{-1} in and around wetland area	Lime: 106 t
VIII	Muirburn	—	—
IX	Limestone pellets	7·5 t ha^{-1} over moorland	Lime: 15 t
Z_1	Limestone dust	5 t ha^{-1} uniformly over moorland	Lime: 60 t
		Catchment total:	Lime: 445 t Fert: 8 t

IX) in 1987. Table 9.4 summarises these treatments, including nominal application rates and the total applied.

The water chemistry changes consequent to these treatments are reported in Chapter 12.

9.8 OTHER RELEVANT STUDIES

9.8.1 Introduction

The Loch Fleet project followed a sequence of investigations into the characteristics of liming materials, biological investigations of fish responses, and pilot and trial applications prior to the formulation of a variety of applications and procedures to match conditions at the site.

On the basis of published reports, other projects have not always been

PLATE I (a) Limestone pellet spreading using helicopter. (b) Recently spread limestone pellets on the ground.

able to follow such a sequence of experiments and trials, but notably some studies have been made, both to develop the more fundamental science of liming and to match the planned applications to the characteristics of the selected sites. Some selected examples have been introduced in Chapter 1, Table 1.1. Recent reviews of liming operations, particularly in North America, have recently been published (Fraser and Britt, 1982; Gloss *et al.*, 1989; Olem, 1990). The majority of liming programmes have used direct application to lakes and rivers. Only a selection of examples relevant to catchment liming is given here. In some cases the lime dose has been estimated by titrating the lake or stream water, or the acid catchment runoff, by the selected neutralising agent to calculate the quantity needed for neutralisation over a specified time. This has not always provided a realistic estimate, partly because of "in lake" processes, which consume or generate alkalinity, and partly because short-term climatic events flush out the applied limestone at a rate in excess of that estimated, and before it has dissolved. In the case of catchment liming, while the reserves of exchangeable acidity or alkalinity within the soil can be calculated, processes may not be sufficiently well quantified and mobilisation is dependent on flow pathways. Topics of investigation or assessment have included: suitability of sites, alternative methods of application, alternative neutralising agents, costs, cost-effectiveness, as well as potential problems.

9.8.2 Swedish National Programme

While the Swedish national programme is comprehensive and substantial, rather few preliminary studies of a general nature were undertaken prior to treatment. Of the 5000 lakes treated, only a few have been subject to intensive and reported investigation. Several thousand kilometres of rivers and streams have also been treated. Most of the treated waters received powdered limestone directly from boats, helicopters, or bankside dosers (Anon, 1985).

The most comprehensive study was undertaken at Lake Gårdsjön, where powdered limestone was added direct to the lake as a slurry by boat. The dose applied was designed to raise the acid water quality to conditions suitable for fish, taking account of the morphological and hydrological conditions of the lake. The dose was calculated from the components of acidity contributed by deposition and runoff from adjacent littoral areas to the lake, and for an effective period of 5 years, as well as the expected dissolution rate of the limestone used (Hultberg and Nystrom, 1988). Of the 110 t applied, 80% was dissolved in 3·3 years, 74% of it in the first year following liming. Floods and high discharge flows during the first year caused greater than expected loss of calcium.

The outcome of the extensive Swedish liming programme has been used to validate the principles and formulation of a liming model (Sverdrup *et al.*, 1984, Sverdrup, 1985). The model includes consideration of lime particle size, dissolution characteristics and kinetics, lake morphometry and behaviour, sedimentation and sequestration, and initial lake and runoff chemistry. A reasonable match between predicted and observed water quality immediately after liming and during the progressive reacidification process has been achieved. While rather few data are as yet available to apply the model to catchment liming, the same principles apply. The use of data for Loch Fleet has shown that it can be applied there (see Chapter 16).

9.8.3 Norwegian Liming Project

A number of sites for liming trials have been selected in southern Norway, including both lakes and rivers. Experiments were undertaken on a wide variety of neutralising materials, of alternative treatment methods, chemical and biological effects, and responses of stocked fish (Baalsrud, 1985). Lime doses were calculated on the basis of the chemical characteristics of materials used, acidified water quality and the target desired, and hydrological characteristics of the water bodies. At Lake Hovvatn, littoral limestone application was employed, with slow release of the dissolved calcium carbonate, but the retention in littoral soils and sediments seems not to have been investigated. At Tjonnstrand, where the catchment was limed at a rate of 3 t ha^{-1}, there is evidence of an increase in soil base reserve and only 3·5% of lime applied was transferred to the lake over the first year and 11% after 4 years (Johannessen, 1985).

Additional studies underway in Norway include the liming of discharge areas, the systematic testing of alternative neutralising agents, and the optimisation of stream liming dosers (Hindar, 1988). While studies of the effective dissolution of applied lime have confirmed the use of Sverdrup's model for reacidification of larger lakes, some evidence is found of uptake in lake sediments and subsequent ion exchange within the water column. However, studies of catchment or littoral soils seem seldom to have been reported.

9.8.4 United Kingdom

A few liming applications have been undertaken in the United Kingdom, namely at Loch Dee in Galloway, on the Esk catchment in Cumbria, and at several sites in Wales (see Table 1.1 for references). These projects are all experimental, and all involve some combination of stream and catchment liming, consistent with the national conditions of rapid stream flows in acidified waters. In all cases limestone and other applications have been somewhat pragmatic, based on former upland management techniques rather than the calculation of a "lime requirement" for a target water quality. In some instances material of coarser grade has been used, or applications have been rendered ineffective by subsequent heavy runoff, so that only some applications can be judged as successful.

Further results of these projects, including water quality changes, fishery success, and other biological responses are reported in later chapters.

9.8.5 North America

In the USA an intensive study of Woods Lake (Adirondacks) includes investigation of soil characteristics, terrestrial ecology components, as well as lake conditions. At Woods Lake, lime (1000 t of pelleted limestone) was distributed on two subcatchments (100 ha) in 1989 (10 t ha^{-1}). The aim was to increase lake water from pH 5·0 to circumneutral values, increase calcium from 100 μeq litre^{-1} to more than 200 μeq litre^{-1}, and alkalinity to more than 100 μeq litre^{-1}. After the first year some improvement in water quality has been noted, but falls somewhat short of the target. However, direct lake liming had only transient effects and subsequent catchment liming has not yet been reported (Porcella, 1989, and pers. comm.; Olem, 1990).

In Canada several lakes in Ontario and Nova Scotia have been treated directly by limestone addition; most studies seem to have focused on post-liming aquatic changes (Dodge *et al.*, 1988).

9.9 CONCLUSIONS AND SUMMARY

The Loch Fleet project has confirmed a number of other current studies of liming, namely the selection of the most appropriate agents for correcting acidity, generating alkalinity, and increasing calcium concentration in surface waters. The use of the fine particulate grade of limestone (less than 63 μm), and an effective threshold of lime application has also been demonstrated. In contrast to some purely pragmatic liming operations, or on the other hand some that have attempted (e.g. by lake titration models) to calculate an effective application dose, within the Loch Fleet project an experimental approach was employed. This tested the efficacy of alternative liming materials, alternative doses and application regimes, and pilot scale responses on runoff and vegetation.

While a large body of knowledge has accumulated regarding the techniques of direct lake or stream liming, relatively few studies have been directed towards those suitable for lakes with brief retention, or streams with rapid flows. This gap is all the more serious when we consider that these lakes and rivers are those most at risk of acidification, and their fish species (salmonids) the most sensitive to its effects.

A further achievement of the Loch Fleet project has been the close coordination of diverse studies on soil and water quality changes, as well as the biological responses, both of the target species, brown trout, and their supporting aquatic fauna. Additional studies have addressed non-target effects on terrestrial and aquatic vegetation.

While a number of studies elsewhere, certainly amongst direct liming projects, have followed similarly diverse approaches, the Loch Fleet project serves as a model of broad ecological scope which has seldom been matched. Further studies (see Chapters 13 and 16) will provide greater insight into the prediction and time-scale of liming activities, as well as the biological responses in the longer term, when reacidification will occur.

REFERENCES

Andersson, F. and Olsen, B. (editors) (1985). *Lake Gardsjon: An Acid Forest Lake and Its Catchment, Ecological Bulletins*, No. 37. Swedish Research Councils, Stockholm, Sweden, 336 pp.

Anon (1985). *Action Program against Air Pollution and Acidification*. Nat. Swedish Environmental Protection Board, 77 pp.

Baalsrud, K. (editor) (1985). *Lime Treatment of Acid Water*. Dept Environment and Directorate of Nature Management, Trondheim, Norway, 147 pp.

Dodge, D.P., Booth, G.P., Richman, L.A., Keller, W. and Tomassini, F.D. (1988). An overview of lake neutralization experiments in Ontario. *Water Air Soil Poll.* **41**: 75–84.

Fraser, J.E. and Britt, D.L. (1982). *Liming of Acidified Waters: A Review of Methods and Effects on Aquatic Ecosystems*, FWS/OBS-80/40.13. Fish and Wildlife Service, US Dept of Interior, Washington, USA, 189 pp.

Gloss, S.P., Schofield, C.L. and Marcus, M.D. (1989). *Liming and Fisheries Management Guidelines for Acidified Lakes in The Adirondack Region*, Biol. Report 80 (40.27). Fish & Wildlife Services, US Dept of Interior, Washington, USA, 59 pp.

Hindar, A. (1985). Lime treatment as a countermeasure, pp. 23–39 in *Liming of Acid Water*, Final Tech. Report on the Liming Project (ed. K. Baalsrud). Dept of Environment and Directorate of Nature Management, Trondheim, Norway, 147 pp.

Hindar, A. (1988). Liming acidic waters in Norway: national policy and research and development. *Water Air Soil Poll.* **41**: 17–24.

Hockin, D.C. and Bell, R.M. (1988) Development of a practical land intervention programme for areas subject to acidification. Annex III-14, in *Report of the Intervention Phase (2) 1986–1988, Loch Fleet Project*, ed. G. Howells and T.R.K. Dalziel, CEGB, 102 pp + annexes.

Hultberg, H. and Nystrom, U. (1988). The role of hydrology in treatment duration in the limed Lake Gardsjon, pp. 95–134 in *Liming of Lake Gardsjon* (ed. W. Dickson). Swedish Environmental Protection Board, 327 pp.

Johannessen, M. (1985). Description of project executed, pp. 40–56 *Lime Treatment of Acid Water* (ed. K. Baalsrud). Dept of Environment and Directorate of Nature Management, Trondheim, Norway, 147 pp.

Lessmark, O. and Thornelof, E. (1986). Liming in Sweden. *Water, Air Soil Poll.* **31**: 809–15.

MAFF (1982). *Techniques for Measuring Soil Physical Properties.* Ministry of Agriculture Fisheries and Foods, HMSO, London.

Olem, H. (1990). *Liming of acidic surface waters*, Lewis Publishers, Michigan USA, 331 pp.

Porcella, D.B. (1989). Mitigation of acidic conditions in lakes: an overview of an ecosystem perturbation experiment. *Can. J. Fish Aquat. Sci.* **46**: 246–48.

Stakman, W.P. and van der Harst, G.C. (1969). *Determination of Soil Moisture Retention Curves*, II. Institute for Land and Water Management Research, Wageningen, the Netherlands.

Stakman, W.P., Valk, G.A and van der Harst, G.C. (1969). *Determination of Soil Moistuie Retention Curves*, I. Institute for Land and Water Management Research, Wageningen, the Netherlands.

Sverdrup, H.N. (1985). Calcite dissolution kinetics and lake neutralisation. Doctoral dissertation, Univ. of Lund, Sweden. LUTKDH/1002/1-169/1985, 169 pp.

Sverdrup, H.N., Rasmussen, R. and Bjerle, I. (1984). Simple model for the reacidification of limed lakes taking the simultaneous deactivation and dissolution of calcite in sediments into account. *Chim. Scripta* **24**: 53–66.

Vizard, R.N. (1985). *Assessment of the Rate of Neutralisation of Sulphuric Acid by Selected Limestones and Marine Shells.* Loch Fleet Project Interim Report No. 4. British Coal YRL/17385, 3 pp + tables.

Vizard, R.N. (1986). *Investigation of Selected Particles Within Loch Fleet Sediments.* Loch Fleet Project Interim Report No. 6. British Coal YRL/17858, 4 pp + illus.

CHAPTER 10

Alternatives to Catchment Liming

A.W.H. Turnpenny
National Power plc, Southampton, UK

10.1 INTRODUCTION

Direct liming of acid waters to prevent fish mortalities has been practised for several decades in hatcheries and other waters (Jensen and Snekvik, 1972) and is more commonplace than catchment liming. The scientific basis of direct water liming is well established in the literature, whilst relatively little scientific investigation of catchment liming had been reported prior to the start of the Loch Fleet project in 1984. For this reason, a major objective of the Loch Fleet project was to advance the science of catchment liming and only where alternative forms of treatment could complement catchment liming were they considered as part of an integrated strategy. In particular, the possible role of direct lime dosing or in-stream treatment targeted to spawning streams to provide a higher degree of protection for the early life-stages of salmonids was investigated.

In this chapter a brief review of alternatives to catchment liming is presented, including lake surface and bottom liming, littoral liming, and treatment methods for running waters and spawning beds. Only the principles will be described here, and the relative merits and disadvantages over catchment liming. The practical requirements for these alternative methodologies are well covered in the literature (Bengtsson *et al.*, 1984;

Gunnerod and Baalsrud, 1985) and reference will be made to other sources of information as appropriate. More comprehensive reviews of other liming methods and results will be found in Fraser and Britt (1982), Baalsrud (1985), Sverdrup (1985), and Olem (1990).

A detailed account of spawning ground treatments tested within the Loch Fleet project will also be given.

10.2 LAKE LIMING

10.2.1 Surface Liming

The direct application of lime to acidic lakes has been widely practised in Scandinavia (Baalsrud, 1987; Lessmark and Thornelof, 1986), North America (Fraser and Britt, 1982; Brocksen and Emler, 1988; Schreiber, 1988) and to a minor extent in Britain (Underwood *et al.*, 1987; Gee and Stoner, 1989). In Sweden, a national liming programme had treated some 3 000 lakes by 1985, invariably with quick positive effects and few negative effects on the biota (Lessmark and Thornelof, 1986). The main limitations of direct lake treatment are lack of benefit to lake feeder streams, often the preferred spawning and nursery habitat of salmonid fish, and the necessity of frequent retreatment intervals compared with catchment liming.

The deacidifying agent most frequently used for lake liming is limestone, where calcium carbonate ($CaCO_3$) is the active agent, though other materials have been used, including burnt lime (CaO), slaked lime ($CaOH_2$), lye (NaOH), and soda (Na_2CO_3) (Skogheim *et al.*, 1986) and even sea water (Matzow *et al.*, 1985). Nutrient additions have also been investigated in some eutrophic waters (Davison, 1987).

Particle size has a pronounced effect on dissolution rate and hence effectiveness (Hogberg, 1983). Small particles have a large surface area to mass ratio and dissolve rapidly; they also remain in suspension longer (see Table 9.2). Larger particles falling to the lake bed gradually become coated with humic substances and inorganic iron, manganese, and aluminium compounds, where these are present in the water, and this inhibits further dissolution (Bengtsson *et al.*, 1984; Sverdrup *et al.*, 1984). A size fraction of 0–0·2 mm is often used for lake liming, the required dose rate being dependent on the initial pH of the water (see Table 9.3).

Lime is applied to a lake by one of several methods. Application by boat or raft, in the form of a slurry, is the preferred method because limestone dust is unpleasant to handle. Slurrying also separates small particles from large ones, allowing them to dissolve before sinking to the

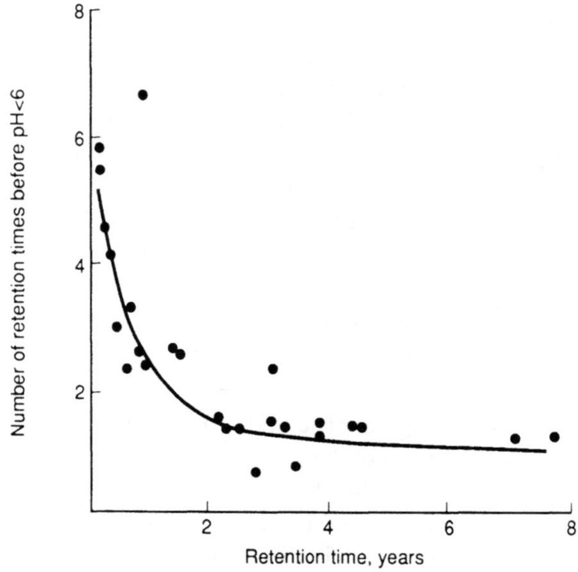

FIG. 10.1 Relationship of lake retention time and treatment effectiveness of lake liming (after Lessmark and Thornelof, 1986). The time taken for a limed lake to reacidify (to pH less than 6) is determined by the number of times the lake volume is replaced.

bed (Warfvinge *et al.*, 1984). Powdered limestone may also be applied to the ice cover of lakes, or lake margins, using agricultural implements (Hindar, 1987). Helicopters have been used at remote locations to spread powdered lime, but this method is not ideal because it results in patchy distribution and drift of dust on to surrounding land. To achieve maximum dissolution, lime should be applied over the whole lake surface with the major portion over deep areas (Hindar, 1987).

Longevity of effectiveness of direct lake liming is largely dependent on lake retention time. Lakes with a long retention period (greater than 2 years) reacidify after fewer retention periods than do short turnover lakes (Fig. 10.1), an effect caused by particle coating of sedimented material. When retention time is greater than 3 years, deacidification for longer than 1·3 times the retention time becomes uneconomic (Sverdrup and Warfvinge, 1984). For retention times shorter than 0·3 years, continuous dosing of the lake inflow may be a more practical solution.

The changes in water chemistry that result from direct liming of waters are somewhat different from those resulting from catchment liming. When lime is applied to soils, the resulting rise in the pH of soil and runoff waters inhibits mobilisation of potentially toxic trace metals (aluminium, heavy metals) from soil reserves into watercourses (see Chapter 2; Loch Fleet project data, see Table 12.3). The effect of lime applied to the water directly is to alter speciation or complex formation within the water body. For aluminium, this generally results in the formation of species or complexes of lower toxicity than before (Johannessen and Skogheim, 1985), although temporary increases in aluminium toxicity may be experienced immediately after liming as a consequence of freshly precipitated hydroxides (Dickson, 1983) (but see Chapter 3).

The behaviour of heavy metals (e.g. lead, zinc, copper, and manganese) following liming is rather variable as some metals have amphoteric properties. In theory, raising the pH could bring precipitates or complexes of some metals from the lake bed back into solution, but practical experience in Norway shows that the heavy metal content of the water column is reduced after liming, by precipitation in the higher pH (Johannessen and Skogheim, 1985) with remobilisation during reacidification (Olem, 1990).

A consequence of not liming the catchment itself is that large volumes of acidic runoff may enter the lake periodically and remain identifiable around the margins of the lake, or on the surface in thermally stratified lakes (Johannessen and Skogheim, 1985). Fish generally avoid these areas, but in doing so may be denied access to important feeding areas.

10.2.2 Lake Bed and Littoral Liming

Surface liming of lakes rarely leads to complete instantaneous dissolution and a proportion of the material settles on to the lake bed where gradual dissolution extends the life of the treatment, until particle coating inactivates the remaining material. This has been used as a deliberate feature in some liming projects, for example at Lake Store Hovvatn (Rosseland and Hindar, 1988) and at Lake Lysevatten (Alenäs *et al.*, 1991).

Alternatively, lime may be spread along the littoral zone where wave action causes particles to abrade, increasing dissolution and reducing the coating effect. It is common practice in Sweden during lake surface liming to spread any surplus lime in the littoral zone (Lessmark and Thornelof, 1986).

Littoral liming of Lake Store Hovvatn in southern Norway has been described by Johannessen (1985) and Rosseland and Hindar (1988). This lake has a retention time of 9 months. In 1981, 200 t of powdered limestone (90% $< 63\,\mu m$) was spread over ice on a 3-km long strip of the shoreline, in a band 10–20 m wide. The lime was first deposited as a 2 m strip by a manure spreader, then further dispersed by tractor and snow-blower. After ice-out, about 14% of the lime had dissolved (compared with 65%–70% for water column liming). By autumn 1987, 55% of the lime had dissolved: about 15% lower usage than would be expected from whole-lake liming. The project was successful in maintaining a brown trout (*Salmo trutta*) population over six years.

Targeted liming of inshore gravels for the benefit of lake-spawning fish has been described by Gunn and Keller (1980) and Booth *et al.* (1986) in Canada. Here, limestone gravel is mixed with the spawning substrate to provide protection against acidic runoff waters upwelling through the gravel.

10.3 TREATMENT OF RUNNING WATERS

10.3.1 Streambed Treatment

A variety of basic materials have been applied directly to the streambed as a means of deacidifying running waters. These include limestone gravel (Burns *et al.*, 1984; Watt *et al.*, 1984; Lacroix, 1991), shell sand (Hindar, 1985), and scallop shells (*Chlamys opercularis*: 37% calcium) (Burns *et al.*, 1984). The effectiveness of these treatments on the stream as a whole is limited. The sizes of materials used are generally large (up to 50 mm) to keep them in place, so avoiding the need for continuous dosing systems. Consequently, dissolution rate is slow and particle coating becomes a problem over time. Watt *et al.* (1984) conducted experiments using streambed gravel treatments over three years but with little success. Treatment effectiveness was inversely proportional to flow rate, rendering treatment less effective in high flows, when most needed. Dissolution was also reduced at low temperatures, so that under winter conditions uneconomic quantities of limestone would have been required to ensure adequate water quality.

A more effective use of streambed liming is to protect intra-gravel water quality on spawning grounds. The principle is the same as used for lake shore spawning ground treatments (Section 10.2.2). Since the water velocity is very much lower within the gravel matrix compared with that of overlying running water, usually by several orders of magnitude (Wickett, 1958), contact time with the limestone surface is much greater. Levels of interstitial pH and calcium can be raised significantly by this means and trace metal levels lowered (Gunn and Keller, 1980; Rosseland and

Skogheim, 1986). This in itself will not benefit recruitment success where the swim-up fry are obliged to emerge into toxic surface water conditions, and supplementary treatment of surface waters using lime wells or dosers may need to be instigated towards the end of the incubation period.

A successful treatment of this type is reported by Lacroix (1992), who applied 200 t of crushed (40 mm grade) dolomitic limestone to the bed of Fifteen Mile Brook, Nova Scotia (mean discharge 0·29 $m^3 s^{-1}$) to protect salmonid spawning beds. The material was applied over eight separate areas in a 300-m long section, to form bars 30–50 cm deep. Mean surface pH in an untreated upstream control section was 4·97 compared with pH 5·3 below the treated section. The neutralising effect was reduced at low water temperature and high stream discharge, but the duration of low pH episodes was sufficiently short to prevent lethality. Intra-gravel pH values were raised from 5·37 in the control section to 7·07 in the treated section. The number of redds cut and the subsequent fry densities were increased significantly in the treated stream section. No reduction in treatment effectiveness was found over a 3-year trial period, indicating that particle coating was not a problem in this case.

10.3.2 Lime Dosers and Lime Wells

Surface water dosing systems are in widespread use in Scandinavia for the treatment of flowing waters (Hindar, 1985; Lessmark and Thornelof, 1986). Systems for small streams (less than 3 $m^3 s^{-1}$) can be water powered. Electrically-powered dosers have been used to dose rivers of up to 100 $m^3 s^{-1}$, but can only be used where a power supply is available.

Material may be dosed as a dry powder or as a slurry. Dry powder dosers use a lime particle size of 0–0·2 mm, giving 50% instantaneous dissolution, the rest sedimenting out and dissolving more slowly. Slurry dosers use fine suspensions of 0–15 μm material giving 90%–100% instantaneous dissolution (Tidestrom and Moberg, 1984). Control of dosing rate can be manual, but then may require several adjustments per day, or automatic. Automatic dosing is achieved electronically in the more sophisticated dosers, or mechanically by water flow in simpler systems. All dosing systems require frequent attention to overcome freezing, blockages, and to recharge lime silos. A 100 $m^3 s^{-1}$ flow doser can use up to 40 t of lime powder per day (Hindar, 1985).

Lime wells offer an inexpensive and effective means of controlled surface dosing into short segments of small streams (typically less than 2 $m^3 s^{-1}$), making them useful for treating spawning streams. A lime well (Fig. 10.2) consists of a tank containing coarse limestone grit or shell sand

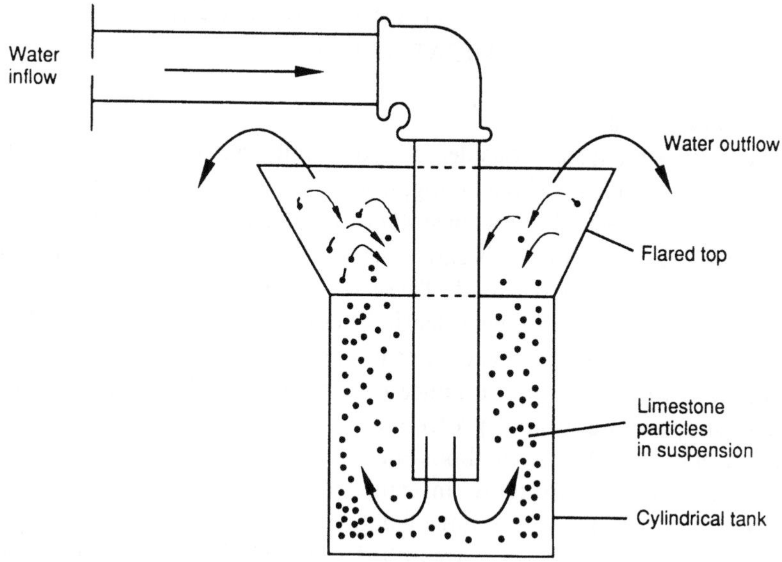

FIG. 10.2 Lime well: principles of operation (after Sverdrup, 1985).

through which water upwells thereby dissolving the material and carrying it into the stream.

Sverdrup (1985) has described the functional principles of the lime well. The aim is to keep the limestone in motion so that particles continually collide and abrade, preventing surface coat formation and aiding dissolution. To achieve this motion, water must be injected with some force at the bottom of the lime well, requiring that the source water is sited a metre or so above the top of the lime well to give an adequate hydraulic head. Once the vertical velocity of the water is above the fluidisation velocity of the particles, they will be brought into suspension. There is then a danger of losing suspended material into the stream without dissolving it. The top of the lime well is widened out to slow the flow in the upper levels to a point below the minimum fluidisation velocity, allowing particles to fall back down. Blockage of supply pipes and freezing are the main threats to the reliability of lime wells.

Dosing proportional to flow can be attained by running several lime wells in parallel, each one fed from a different stream level (Hindar, 1985). As the stream level rises, so more lime wells come into play. Ideally, at least 30% of the stream should pass through the lime well for satisfactory treatment (Tidestrom and Moberg, 1984).

10.4 LOCH FLEET PROJECT SPAWNING-GROUND TREATMENTS

10.4.1 Background

In-stream treatments are considered a potentially useful adjunct to catchment liming, as well as to lake liming, where it is required to protect critical spawning and nursery habitats in streams. By allowing precise targeting of treatments, in-stream methods can be selected to protect surface waters or intra-gravel waters, as required. These forms of treatment may be particularly helpful when ecological sensitivities preclude the use of terrestrial treatments such as source headwater liming.

The need to test the performance of lime wells and streambed treatments for spawning ground protection was identified at the start of the Loch Fleet project. Whilst methods and materials for in-stream treatments have been fairly thoroughly investigated and are well documented (see Section 10.3.2), the amount of data to demonstrate their influence on water quality within the streambed is minimal. Accordingly, a set of experiments was designed with the specific objective of monitoring responses of streambed chemistry to surface dosing and direct streambed applications of limestone, whilst at the same time testing the biological response using implanted salmonid eggs. It was of special interest to examine streambed chemistry and the effectiveness of treatments during acid episodes at high stream discharges, necessitating a high sampling frequency during these (usually short) periods.

The experiments could not be carried out around Loch Fleet itself. The loch is fed by only a single stream of sufficient size to support spawning (the Altiwhat) and this stream had been selected for the source headwater liming trials described earlier (see Chapter 9). A second experimental area was selected, located on the Fell of Fleet, some 2·5 km to the north-east of Loch Fleet. Here, a pair of streams join close to the crossing of a forestry road, allowing access for a mobile analytical laboratory. The resulting single stream discharges eventually into the River Dee, to the east of Loch Fleet. The site was chosen for its proximity to Loch Fleet, and the similarity of its streams to the Altiwhat, having a common underlying granitic geology, and being of similar size (1–2 m width), discharge (95% $< 0{\cdot}4\,m^3\,s^{-1}$ below the stream confluence, compared with $0{\cdot}25\,m^3\,s^{-1}$ for the Altiwhat), and gravel characteristics (coarse, angular granite gravel, less than 50 mm). There being two stream limbs, it was possible to design two independent forms of treatment side by side, with an untreated control area upstream (Fig. 10.3).

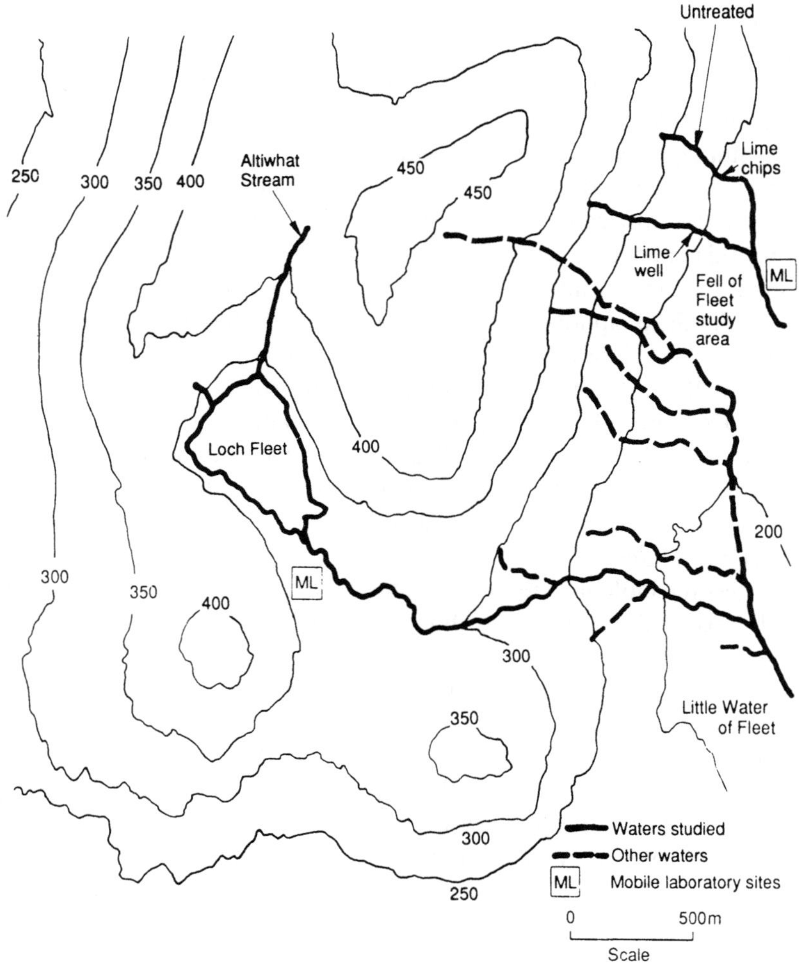

FIG. 10.3 Loch Fleet and adjacent area: fishery investigation sites. Altitude contours (50 m intervals) are shown.

Experiments were conducted over three spawning seasons between 1985 and 1988. Only a small sample of the findings is presented here: a complete description of the work is in preparation (Turnpenny *et al.*, pers. comm.). Additional biological data from the study are given in Chapters 3 and 13.

10.4.2 Treatments Applied and Experimental Methods

The layout of the two treatment zones and untreated control zone is shown in Fig. 10.3. Each zone comprised a stable bed of moderately uniform gravel to a depth of 30–40 cm above bedrock. Gravel permeability was measured by a standpipe method (Turnpenny and Williams, 1982) and was high, greater than 8×10^4 cm h^{-1} in all three beds. In 1985 a streambed treatment was used in one of the lower zones by application of a 25-mm layer of 2–6 mm limestone chips, lightly raked into the gravel surface. For the second treatment, a small (0·25 m^3) lime well was installed a few metres upstream of the test section. This was fed from a header pool about 1·5 m above the surface of the lime well. Under low flow conditions, a small flow (5 litre min^{-1}) percolated through the lime well but, at high stream levels, a siphonic action was initiated, giving flow rates of up to 24 litres s^{-1} (15% of total flow). The well was charged with crushed limestone (50% 0·1–2 mm, 50% 2–6 mm) as required.

In 1986 artificial spawning beds of more uniform granite gravel were constructed. The same lime well dosing arrangement was used but the streambed treatment was modified by mixing the limestone chips uniformly into the top 25 cm of gravel at a rate of 1 part limestone to 7 parts gravel by volume.

To test the biological response, salmonid eggs were implanted in plastic mesh boxes held in streambed housings (Fig. 10.4). These were designed to be hydraulically "transparent", i.e. to have no influence on gravel permeability (Turnpenny *et al.*, pers. comm.), and allowed egg boxes to be removed periodically for inspection and replaced without disturbing the streambed. Porous ceramic sampling blocks (25-mm cubes) were fitted alongside the egg boxes at depths of 7·5, 15·0 and 22·5 cm, respectively, and buried plastic tubes conveyed intra-gravel water samples to the stream bank, from where they were delivered to a mobile analytical laboratory by a peristaltic pump. For each experimental bed there were 16 sample lines (3 depths × 5 replicates + 1 surface sample line). Determinations were made of total aluminium, calcium, and pH using a continuous flow system, sequentially sampling with a repeat sampling interval of 135 min. This frequency was adequate to monitor fluctuations in stream chemistry during high flow episodes, which normally lasted at least 6 h (from base flow to peak flow and back to base flow), and generally 12–24 h or more.

10.4.3 Chemical and Biological Findings

10.4.3.1 Steady Flow Conditions

Flow conditions during periods sampled in the 1985–86 season were comparatively stable and provided the opportunity to test the performance of treatments under steady flow conditions (Fig. 10.5). Under these

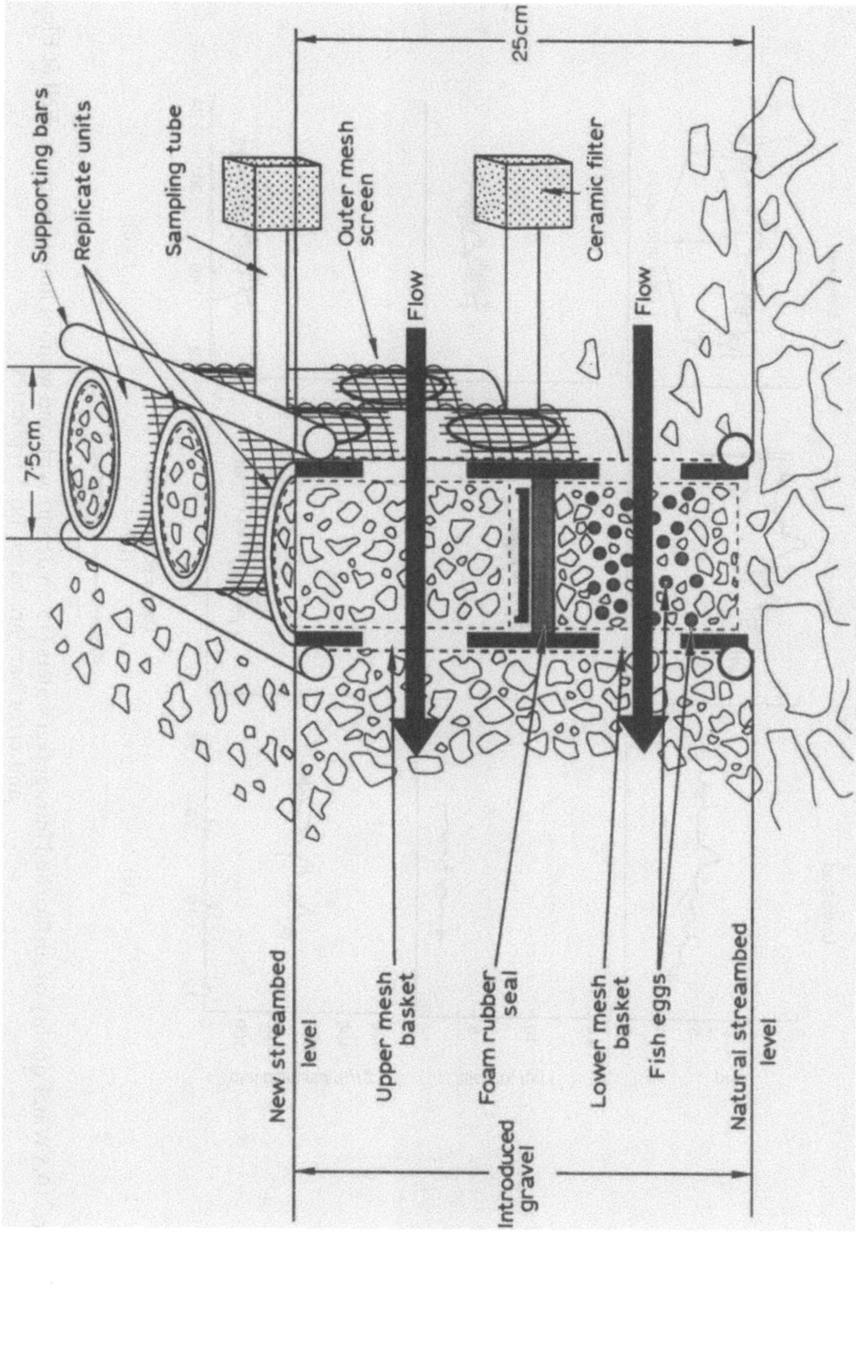

FIG. 10.4 Intragravel egg boxes and sampling units: diagrammatic section.

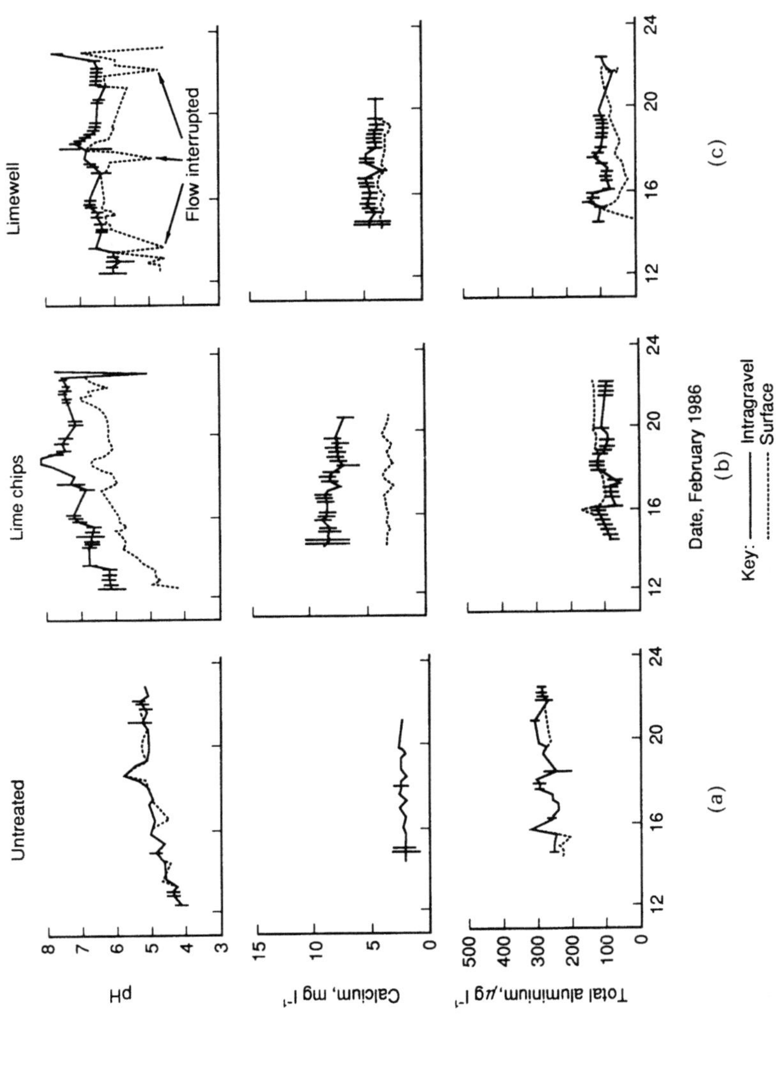

FIG. 10.5 Water quality of surface and intragravel waters (15 cm depth), with and without lime treatments, Fell of Fleet. Means and error bars are based on 5 replicates.

TABLE 10.1
Surface and interstitial water quality for untreated and treated spawning beds: median values for the period November 1986–April 1987, Fell of Fleet experimental site (600 samples)

	pH	*Calcium* (μeq *litre*$^{-1}$)	*Aluminium* (μg *litre*$^{-1}$)
Untreated			
Surface	4·50	77	444
7·5 cm	4·56	83	395
15·0 cm	4·54	94	425
22·5 cm	4·53	83	490
Lime chips			
Surface	4·49	77·5	441
7·5 cm	5·16	148·5	304
15·0 cm	5·24	162	314
22·5 cm	5·19	155·5	323
Lime well			
Surface	4·89	105	300
7·5 cm	5·82	187	272
15·0 cm	5·85	191	260
22·5 cm	5·91	184	164

conditions, interstitial waters of the untreated bed were more or less in equilibrium with the overlying surface waters, whilst those in the lime chip treated bed were about 1 pH unit higher, with higher calcium and lower aluminium levels. The degree of improvement was enhanced with strembed depth (Table 10.1), a finding to be expected since the water exchange rate declines with increasing streambed depth (Wickett, 1958; McNeill, 1962) and hence the contact time increases. The pH and calcium content of surface waters flowing over the treated gravels was only marginally improved over those of the untreated section though, curiously, aluminium concentrations were reduced to about one-third.

The lime well produced significantly improved surface water conditions (Fig. 10.5), and also resulted in improved interstitial water quality, although not to the same extent as the streambed application. The fact that interstitial waters were less acidic than surface waters, even with surface dosing, was due to the presence of undissolved limestone particles flushed from the lime well into the streambed, thereby providing a secondary streambed treatment. Occasional failures of the lime well due to blockage or freezing are indicated by sudden pH drops, which were quickly reversed once flows were reinstated. Interstitial water quality was sustained during these events by the material deposited in the streambed.

The biological response corresponded with the water quality conditions. Hatching rates of brown trout (*Salmo trutta*) eggs were 30%–33% in the untreated section, 40%–82% in the bed treated with lime chips, and 52%–84% in the lime well treated section.

10.4.3.2 Fluctuating Flow Conditions

Flows were much more variable in the 1986–87 season and data were collected for a series of spates which yielded acid episodes. Data for a 4-week period from November to December 1986, starting from the time when eggs were planted, are shown in Fig 10.6. Table 10.1 summarises water quality measurements taken over the entire egg incubation period from November 1986 to April 1987.

The interstitial water chemistry in the untreated bed (Table 10.1) was slightly (0·03–0·06 pH units) but significantly ($p < 0{\cdot}001$) less acidic than that of surface waters, presumably because of the intrusion of more alkaline groundwater into the base flow, a feature demonstrated in the Loch Fleet catchment (see Chapter 12). During spates, interstitial water chemistry closely matched surface water chemistry (Fig. 10.6(a)), signifying that exchange between the two environments is rapid, and that the resistance to flow of the streambed gravel provides little physical buffering against chemical fluctuations of surface waters for streambed biota. Gravels of lower permeability may well be more protective in this respect.

The interstitial waters of the lime chip treated bed, as under steady flow conditions, again were substantially less acidic than surface waters (by 0·74–0·82 pH unit over the whole season: Table 10.1), with higher calcium levels (by up to 80 μeq litre^{-1}) and lower aluminium concentrations (up to 137 μg litre^{-1} lower). More importantly, these differentials were largely maintained during severe episodes when surface water pH values fell by half a pH unit or more (Fig. 10.6(b)).

Results from the lime well in 1986–87 (Fig. 10.6(c)) were disappointing, with the main deacidifying effect being due to dissolution of particles deposited in the streambed. Stream discharges were generally much higher than in the previous year, and it became apparent that the capacity of the lime well was too small to cope with the high flows. A great deal of the lime material was flushed out in an undissolved state, accounting for the streambed treatment effect.

The longevity of direct streambed treatment was shown to be limited in this study, the effectiveness of the treatment diminishing over the course

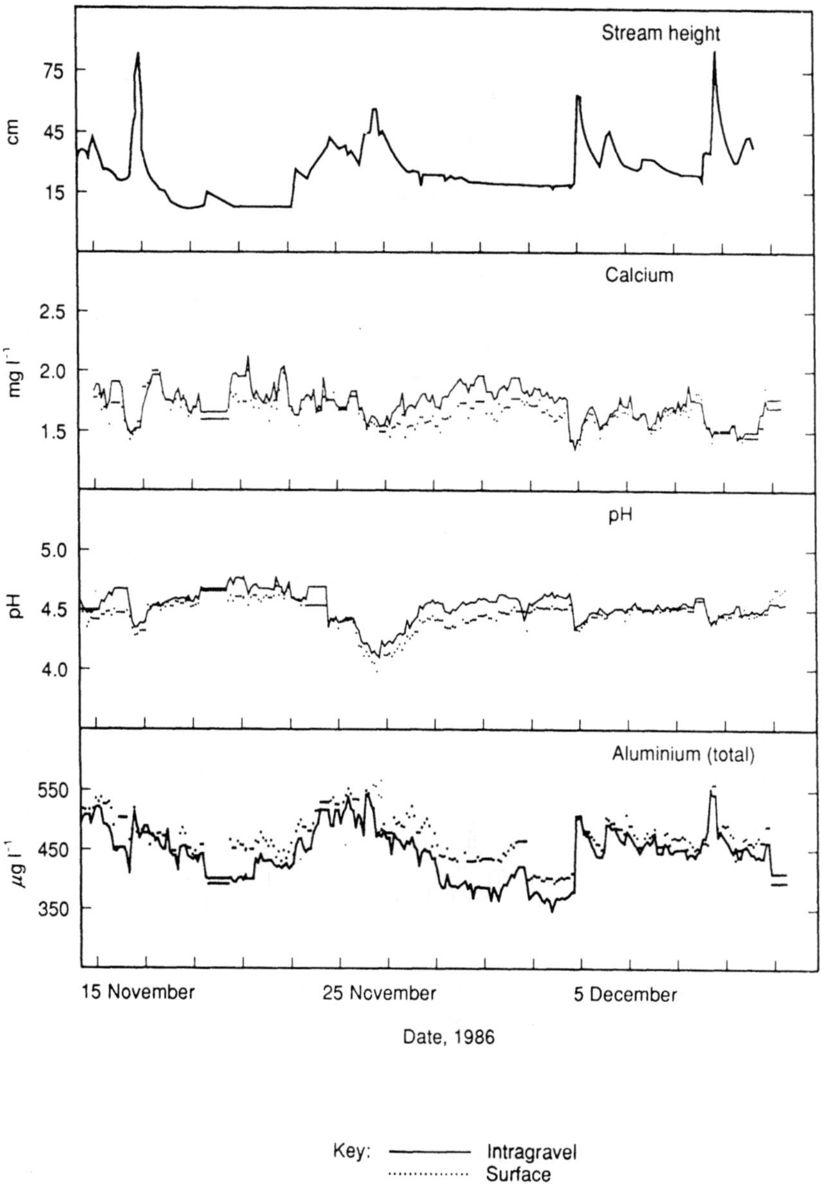

FIG. 10.6 (a) Changes in water quality of surface and intragravel water during flow varations in untreated reach, Fell of Fleet.

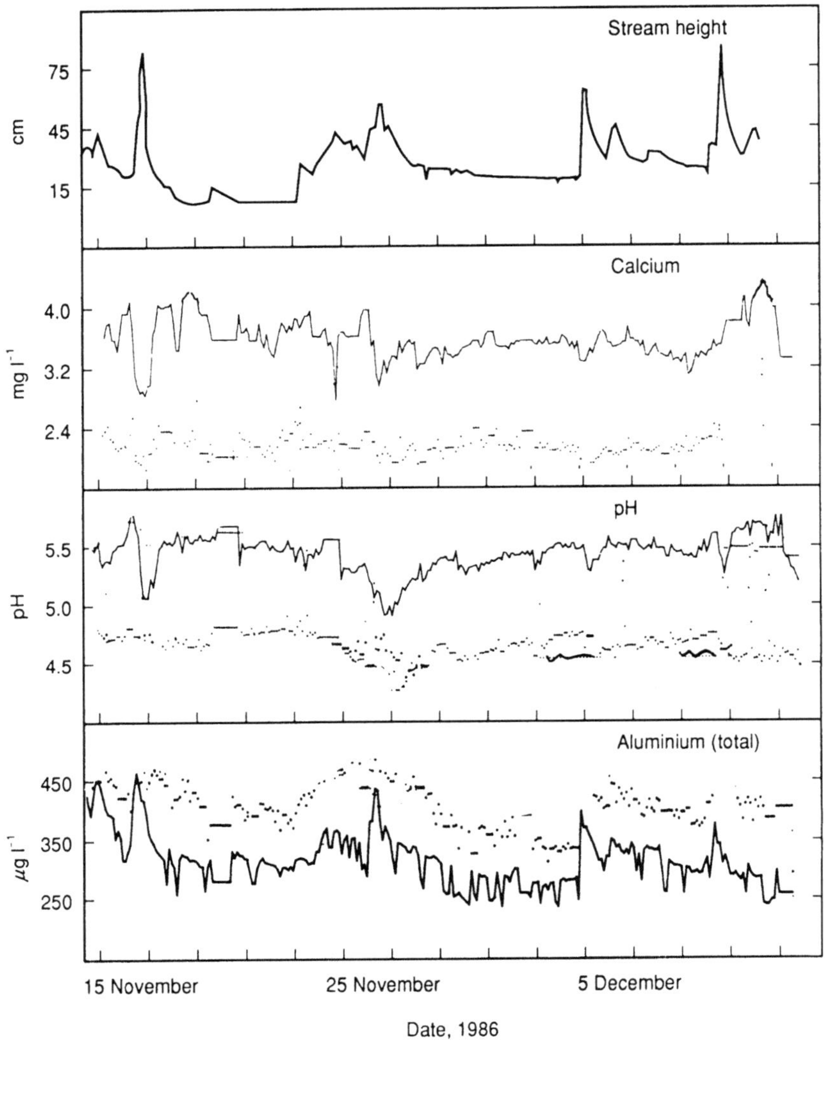

FIG. 10.6 (b) Changes in water quality of surface and intragravel water during flow variations with lime chips in the gravel bed, Fell of Fleet.

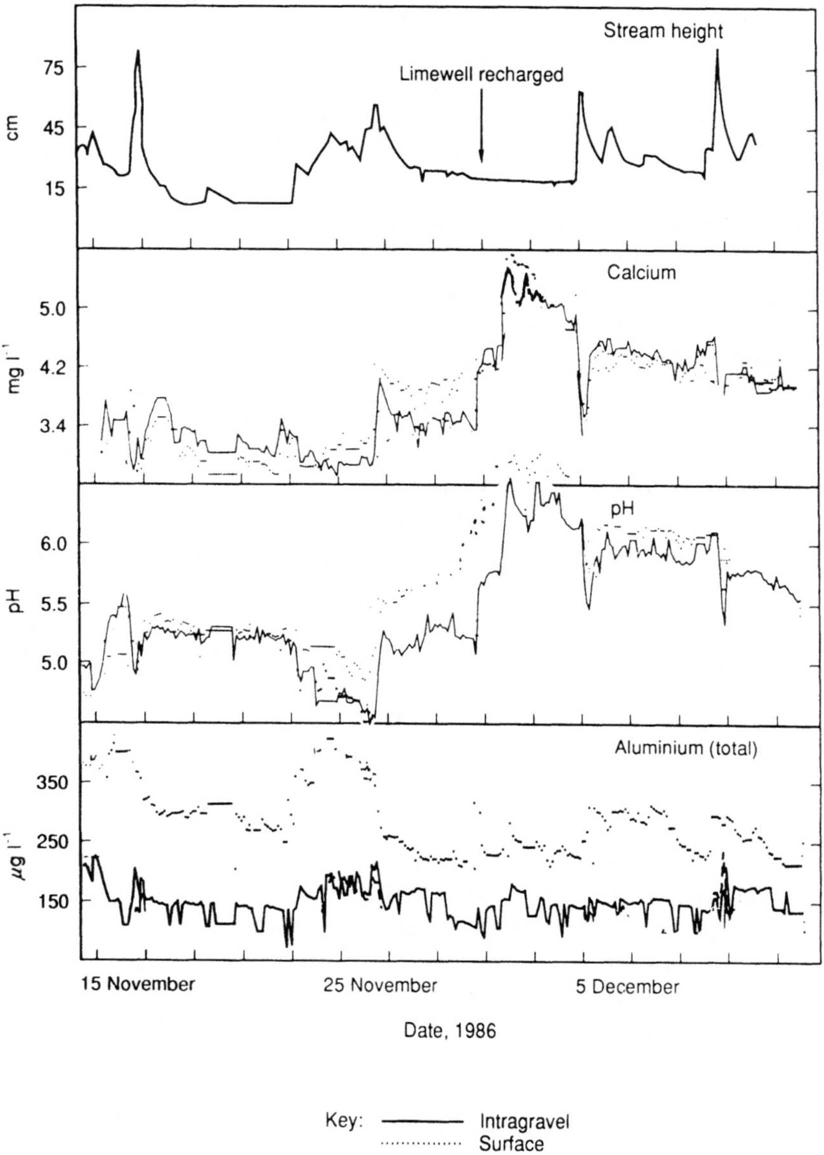

FIG. 10.6 (c) Changes in water quality of surface and intragravel water with lime well dosing, Fell of Fleet.

of the spawning season. Immediately after the application of material in October 1986, the pH differential between surface and interstitial waters was about 1 pH unit; by the end of the incubation period the following April, this differential had declined to 0·25 pH unit. The cause of this decline was not investigated but was probably due to either particle coating or progressive loss of limestone material from the test section.

10.4.4 Conclusions from the Fell of Fleet Study

A more detailed description of the biological findings can be found in Chapter 13 and these aspects will not be addressed further at this point.

The main findings regarding the effectiveness of in-stream treatments indicate that, for spawning ground treatment, surface water dosing and streambed liming are both helpful and, indeed, complementary. Streambed treatments clearly have potential for deacidifying interstitial waters, at least for the first few weeks after application. Since the first few days after fertilisation represent the most critical period of acid sensitivity for salmonid eggs (Brown and Sadler, 1989), application of material just before the spawning season would appear to offer a sensible method of early protection. Improved water quality differentials could be achieved by including finer-graded material (e.g. 1–2 mm) which, if applied during low flow conditions to an open gravel matrix, would largely penetrate into the gravel and be retained.

The best use of lime wells would be to complement streambed treatments to improve water quality in the upper gravel layers and overlying waters. A strategy would be to introduce lime well treatment towards the end of the incubation period when gravel treatments were becoming less effective and when it is required to create a hospitable surface water environment to receive fry as they emerge from the gravel. An important prerequisite for successful lime well function is to determine the likely flow range of the stream to size the capacity of the lime well(s) appropriately. The lime well used in the present study was inadequate to cope with the highest flows, and a parallel set of wells might have performed better.

10.5 COMPARISONS WITH CATCHMENT LIMING

10.5.1 Time-scales of Treatment

Longevity of treatment is one of the chief differences between catchment liming and aquatic liming. Whereas lakes with a retention time of 1 year require reliming every second or third year, and lakes with a retention time of 3 years require retreatment every fourth year, catchment liming, although requiring a higher initial dose, will last many times longer. Loch Fleet, with a retention time of 6 months, would require reliming every one to two years had the lake liming route been chosen. Present predictions are that the catchment liming will remain effective beyond the turn of the century (see Chapter 16). Littoral or lake bed liming would require retreatment at some intermediate interval, although results by these methods are at present somewhat unpredictable.

The benefits of a long-lasting treatment are not confined to reducing the reapplication frequency. All current evidence (see Chapter 14) indicates that aquatic communities take several years to accommodate to changes brought about by liming, and ecological stability would therefore be jeopardised by sharp changes in water quality on a time-scale shorter than this. The slow, steady decay rate of treatment effectiveness produced by catchment liming offers aquatic communities a better chance to stabilise.

10.5.2 Water Chemistry

The differences in water chemistry associated with catchment versus aquatic liming have already been mentioned. The retention of potentially toxic metals within soils after catchment liming is a significant benefit to aquatic biota. Although aquatic liming renders metals largely non-toxic, these metals remain within the aquatic system and are potentially available for re-dissolution once more acidic conditions favour this.

10.5.3 Spawning Grounds

Catchment liming of headwater bogs at Loch Fleet has proved successful in restoring conditions suitable for spawning within the main feeder stream, with a predicted longevity of 7 years (see Chapter 16). Whilst in-stream methods have been shown to be workable at Loch Fleet, no treatment of this type will last more than a single season; indeed, lime wells require almost daily attention.

In-stream treatments may be considered worthwhile at permanently manned locations, especially where bog liming is considered ecologically undesirable.

10.5.4 Ease of Treatment

The liming of lake surfaces is relatively undemanding in terms of resources and the level of pre-treatment planning required. Simple facilities and simple design guidelines put aquatic liming within the realm of the intelligent layman.

Catchment liming, on the other hand, is much more demanding in these respects. To ensure success requires a good understanding of soil and vegetation conditions, and of the hydrological characteristics of the catchment to be treated. A number of catchment liming projects have been initially unsuccessful through lack of understanding of these requirements. Potential side-effects on terrestrial ecosystems must also be fully considered. Catchment liming should therefore be considered the province of the specialist.

10.6 NON-LIMING METHODS

Other novel approaches to restoring fisheries and aquatic ecology in acid waters are currently being investigated.

One such approach involves the injection of alkaline groundwater into lakes to restore pH. The feasibility of using this approach at Loch Fleet has been investigated by Cook *et al.* (1991). Groundwater sampled from a borehole in the catchment has alkaline geochemistry (pH 7·2, HCO_3^- 2000 μeq litre^{-1}) and it is estimated that this contributes 3–4 litre s^{-1} to the loch (equivalent to 5% of the loch outflow). Prior to the recent acidification this was sufficient to buffer the loch water at around pH 6. Titrations of groundwater with loch water (non-limed) indicate that as little as 0·06 litre s^{-1} (1892 m^3 yr^{-1}) of additional groundwater would be required to restore conditions for a self-sustaining fishery. Double this amount is the calculated requirement to restore the loch to the conditions prevailing in the pre-industrial era. Cook *et al.* propose that the groundwater input could be increased at Loch Fleet by sinking additional boreholes, relying on the natural hydraulic gradient to induce artesian flow. The estimated cost would be quite modest, around £6000 plus maintenance. This has not yet been put to the test at Loch Fleet, and is unlikely to in the near future since the loch has already been treated by catchment lime applications predicted to last for many years.

A trial of this procedure has also been undertaken in a spawning stream stocked with brown trout and brook trout in Pennsylvania, USA (Gagen *et al.*, 1989). This has improved water quality (pH increased from 4·9 to 6·0, and aluminium (monomeric) decreased from 360 μg litre^{-1} to 30 μg litre^{-1}), but a large volume runoff episode following restocking led to 25% mortality. At normal flow, no mortality of caged fish was found after groundwater addition. A similar method is being applied for the restoration of a small (9·2 ha) lake in Wisconsin, USA (Garrison and

Rose, 1990). The expected increase in pH, ANC, and calcium (pH 5·1 and ANC 10 μeq litre^{-1} initially) aims to bring pH to 5·6 in the first year and 6·1 in the second. Pumping of groundwater will be about 100 m^3 day^{-1} for a period of 20 days. It is expected that radical or sudden changes in water chemistry will be avoided.

An entirely different approach – to restore fisheries but not water quality – is the subject of a British/Norwegian collaborative study described by Rosseland *et al.* (1990). The REFISH project (Restoring Endangered Fish in Stressed Habitats) seeks to establish the suitability of different brown trout strains to restock marginally acidic lakes. The programme comprises a series of laboratory screening trials and field toxicity experiments to identify tolerant strains, with accompanying lake stockings, followed subsequently by annual test fishings. The experiments have yet to show that self-sustaining populations can be achieved by strain selection.

Two non-liming procedures were trialled at Loch Fleet: the application of nitrogen fertiliser to reduce soil acidification by forest trees, and burning of moorland vegetation. On a pilot scale the former treatment had shown promise at Loch Fleet and elsewhere in Scotland (Miller *et al.*, 1976; Leech and Nisbet, 1987) while the latter strategy is indicated by observation that runoff from areas subject to forest fire in Norway (Rosenqvist, 1981). However, in the conditions found at Loch Fleet improvement in runoff water quality was transitory and failed to meet the specified target water quality.

10.7 CONCLUSIONS

This brief review of alternative techniques for the mitigation of acid water, including direct surface water liming, catchment liming, and some non-liming strategies, suggests that circumstances within each location, as well as the desired target water quality and species to be safeguarded, will determine the method of choice. The predicted duration of effective treatment, and the acceptable costs, will also be necessary components of any treatment strategy.

REFERENCES

Alenäs, I., Andersson, I.B., Hultberg, H. and Romarin, A. (1990). Liming and reacidification reactions of a forest lake ecosystem, Lake Lysevatten, in SW Sweden. *Water Air Soil Poll.* (in press).

Baalsrud, K. (editor) (1987). *Liming of Acid Water.* Norwegian Liming Project. Dept. of Environment and Directorate of Nature Management, Trondheim, Norway, 147 pp.

Bengtsson, B., Bagalius, A. and Thornelof, E. (1984). *Kalkning av sjoar och vattendrag.* Statens naturvardsverk rapport, SNV PM 1786, 140 pp.

Booth, G.M., Hamilton, J.G. and Molot, L.A. (1986). Lake liming in Ontario: short-term biological and chemical changes. *Water Air Soil Poll.* **31**: 709–20.

Brocksen, R.W. and Emler, P.W. (1988). Living lakes: an aquatic liming and fish restoration demonstration programme. *Water Air Soil Poll.* **41**: 85–94.

Brown, D.J.A. and Sadler, K. (1989). Fish survival in acid waters, pp. 31–44 in *Acid Toxicity and Aquatic Animals* (eds. R. Morris, E.W. Taylor, D.J.A. Brown and J.A. Brown), SEB Seminar Series, **34**. Cambridge University Press, Cambridge, 282 pp.

Burns, J.C., Coy, J.S., Tervet, D.J., Harriman, R., Morrison, B.R.S. and Quine, C.P. (1984). The Loch Dee Project: a study of the ecological effects of acid precipitation and forest management on an upland catchment in south-west Scotland. 1. Preliminary investigations. *Fish. Manag.* **15**(4): 145–67.

Cook, J.M., Edmunds, W.M. and Robins, N.S. (1991). Groundwater contribution to an acid upland lake (Loch Fleet, Scotland) and the possibilities for amelioration. *J. Hydrol.*, **125**: 111–128.

Davison, W. (1987). Internal element cycles affecting the long-term alkalinity status of lakes: implications for lake restoration. *Schweiz. Z. Hydrol.* **49**: 186–201.

Dickson, W. (1983). Liming toxicity of aluminium to fish. *Vatten* **39**: 400–4.

Fraser, J.E. and Britt, D.L. (1982). *Liming of Acidified Waters: A Review of Methods and Effects on Acidified Ecosystems.* FWS/OBS-80/40.13. US Fish and Wildlife, 189 pp.

Gagen, G.J., Sharpe, W.E., DeWalle, D.R. and Kimmel, W.G. (1989). Pumping alkaline groundwater to restore a put-and-take fishery in a stream acidified by atmospheric deposition. *N. Amer. J. Fish. Manag.* **9**: 92–100.

Garrison, P.J. and Rose, W.J. (1990). Increasing the pH of acidic lakes by augmenting the ground-water input: an alternative to liming. Internat. Conference *Acidic Deposition: Its Nature and Impacts*, Glasgow, UK, Roy. Soc. Edinburgh, Abstract p. 605.

Gee, A.S. and Stoner, J.H. (1989). A review of the causes and effects of acidification of surface waters in Wales and potential mitigation techniques. *Arch. Environ. Contam. Toxicol.* **18**: 121–30.

Gunn, J.M. and Keller, W. (1980). Enhancement of the survival of rainbow trout (*Salmo gairdneri*) eggs and fry in an acid lake through incubation in limestone. *Can. J. Fish. Aquat. Sci.* **37**: 1522–30.

Gunnerod, T.B. (chairman) and Baalsrud, K. (project manager) (1985). *Liming Acid Water Manual*, Liming Project 1985. Min. Environment and Directorate for Wildlife and Freshwater Fish, Trondheim, Norway, 47 pp.

Hindar, A. (1987). Lime treatment as a countermeasure, pp. 23–39 in: *Liming of Acid Water*, Final Technical Report on the Liming Project (ed. K. Baalsrud) Dept. of the Environment and Directorate of Nature Management, Trondheim, Norway, 147 pp.

Högberg, E. (1983). Kalksten-medicin mot försurning. H. Sverdrup, I. Bjerle og P.

Warfvinge, Kalkning av rinnande vatten (Liming of running waters). Kompendium till symposium. 13 maj, Alvesta. LUTKDH/TKKT-7015/1-112/1983. Lund, Sweden.

Jensen, K.W. and Snekvik, E. (1972). Low pH levels wipe out salmon and trout populations in southernmost Norway. *Ambio* **1**: 223–25.

Johannessen, M. (1985). Description of project executed, pp. 40–56 in *Liming of Acid Water* (ed. K. Baalsrud) Liming Project, Dept. of Environment & Directorate of Nature Management, Trondheim, Norway, 147 pp.

Johannessen, M. and Skogheim, O.K. (1985). Hydrological effects of lime treatment, pp. 57–78 in *Liming of Acid Water*, (ed. K. Baalsrud) Norwegian Liming Project, Dept. of Environment and Directorate of Nature Management, Trondheim, Norway, 147 pp.

Lacroix, G.L. (1992). Mitigation of low pH by substrate manipulation in a stream and its effects on salmonids. *Env. Poll.*, in press.

Leech, A.F. and Nisbet, T.R. (1987). Acid Transfer Through Vegetation and Soils. Final Report to Department of Environment. Macaulay Land Use Research Institute, Craigiebuckler, Aberdeen, 233 pp.

Lessmark, O. and Thornelof, E. (1986). Liming in Sweden. *Water Air Soil Poll.* **30**: 809–16.

Matzow, D., Rosseland, B.O. and Skogheim, O.K. (1985). Effects of lime treatment on fish, pp. 110–30 in *Liming of Acid Water* (ed. K. Baalsrud). Dept. of Environment and Directorate of Nature Management, Trondheim, Norway, 147 pp.

McNeill, W.J. (1962). Variations in the Dissolved Oxygen Content of Intragravel Water in Four Spawning Streams of Southeastern Alaska. US Fish and Wildlife Survey, Spec. Sci. Rep. Fish. No. 402, 15 pp.

Miller, H.G., Miller, J.D. and Pauline, O.J.L. (1976). Effect of nitrogen supply on nutrient uptake of Corsican pine. *J. Appl. Ecol.*, **13**: 955–66.

Olem, H. (1990). *Liming Acidic Surface Waters*, Lewis Publishers, Michigan, US, 331 pp.

Rosenqvist, I.T. (1981). Importance of acid precipitation and acid soil in freshwater lake chemistry, *Vann* **4**: 402–9.

Rosseland, B.O. and Hindar, A. (1988). Liming of lakes, rivers and catchments in Norway. *Water Air Soil Poll.* **41**: 165–88.

Rosseland, B.O. and Skogheim, O.K. (1986). Neutralization of acid brook water by using a shell-sand filter and sea water: effect on eggs, alevins and smolts of salmonids. *Aquaculture* **58**: 99–110.

Rosseland, B.O., Lien, L., Kroglund, F., Sadler, K. and Dalziel, T.R.K. (1990). Strains of brown trout (*Salmo trutta* L.). Stocking and test fishing 1988 and 1989. Field and laboratory toxicity experiments. Norwegian Institute of Water Research (NIVA), Report No. 87178, February 1990, 26 pp.

Schreiber, R.K. (1988). Cooperative federal-state liming research on surface waters impacted by acid deposition. *Water Air Soil Poll.*, **41**: 53–74.

Skogheim, D., Rosseland, B.O., Krogland, F. and Hagenlund, G. (1986). Addition of NaOH, limestone slurry and fine grained limestone to acidified lake water and the effects on smolts of Atlantic salmon (*Salmo salar* L.) *Water Res.* **21**: 435–43.

Sverdrup, H. (1985). Calcite dissolution kinetics and lake neutralization, Doctoral dissertation, Dept. Chem. Engineering, Institute of Technology, Lund, Sweden. LUTKDH/TKKT/1002/1-169, 169 pp.

Sverdrup, H. and Warfvinge, P. (1984). A reacidification model for acidified lakes neutralised with calcite. *Water Res. Res.* **21**(9): 1374–80.

Sverdrup, H., Rasmussen, R. and Bjerle, I. (1984). A simple model for the reacidification of limed lakes, taking the simultaneous deactivation and dissolution of calcite in the lake sediments into account. *Chem. Scripta* **24**: 53–66.

Tidestrom, H. and Moberg, P.O. (1984). *Liming Running Waters. Machine Characteristics Assessed*, SNV PM 1873. Board of Fisheries No. 2.

Turnpenny, A.W.H. and Williams, R. (1982). A conductiometric technique for measuring the water velocity in salmonid spawning beds. *Water Res.* **16**: 1383–90.

Underwood, J., Donald, A.P. and Stoner, J.H. (1987). Investigations into the use of limestone to combat acidification in two lakes in West Wales. *J. Env. Manag.* **24**: 29–40.

Warfvinge, P., Sverdrup, H. and Bjerle, I. (1984). Dissolution of calcite powder sinking in an acidic column of water in relation to a theoretical lake liming model. *Chem. Scripta* **24**(2): 67–72.

Watt, W.D., Farmer, G.J. and White, W.J. (1984). Studies on the use of limestone to restore Atlantic salmon habitat in acidified rivers, in *Lake and Reservoir Management*, Proc. Third Annual Conf. of the North American Lake management Society, Knoxville, Tennessee, US EPA 440/5 84-001, Washington DC, pp. 374–379.

Wickett, W.P. (1958). Review of certain environmental factors affecting the production of chum and pink salmon. *J. Fish. Res. Bd. Can.* **11**: 1103–26.

PART IV

EFFECTS OF LIMING

CHAPTER 11

Soils of Acid Catchments Before and After Liming

E.J. Wilson
National Power plc, Leatherhead, Surrey, UK

G. Hudson and B.F.L. Smith
The Macaulay Land Use Research Institute, Aberdeen, UK

11.1 INTRODUCTION

Soils modify the composition of incoming precipitation passing through them to streams or lakes by interaction between the dissolved constituents and ions adsorbed on the clays and organic matter that make up the soil exchange complex. Hydrogen and aluminium are released when those ions adsorbed on the exchange complex greatly exceed adsorbed nutrient bases. Conversely, high levels of adsorbed bases buffer the soil against acidity in incoming precipitation. Basic ions on the exchange complex can be replenished from easily weatherable soil minerals and afford long-term buffering against acidification.

A recently published map shows Scottish soils grouped into three classes according to their susceptibility to acidification (M.J. Wilson *et al.*, 1988). Soils developed under a cool climate and moderate or high rainfall on nutrient-poor parent materials become acid, and under acid conditions (pH $< 4{\cdot}2$)* aluminium is released to the soil solution. Incoming precipitation and the exchange complex are the principal sources of nutrient bases for peat soils which, because they lack weatherable minerals, are particularly sensitive to acidification.

The soils of Loch Fleet are characterised in the main by peat soil horizons overlying hard unweathered granite rock. Some thicker deposits of peat also occur. These conditions render the soils and the runoff draining from them highly susceptible to acidification.

Chapter 2 discussed the causes from a present and historical viewpoint of soil acidification at Loch Fleet and the requirement for liming in these soil conditions to modify acid runoff. An improvement in water quality is achieved by neutralising soil acidity, precipitating dissolved aluminium as aluminium hydroxides, and providing a source of calcium in runoff. Soil studies at Loch Fleet aimed to assess these changes on soil chemistry and to monitor the partitioning of applied lime between the soil exchanger and the reserve of free lime. Since the effectiveness and duration of liming depends on the presence of free lime in the soil, a measure of its rate of depletion and availability for chemical reaction provides a method for predicting treatment longevity (see also Chapter 16).

The success of catchment liming depends to a large extent on initial soil conditions, hydrology, and climatic factors. The buffer capacity of a limed soil is influenced by chemical and physical properties, namely type of parent material, degree of weathering, cation exchange capacity (CEC) dependent on clay and organic matter content, soil pH, base saturation, exchangeable aluminium, as well as the internal and external inputs of acidity. Following the initial catchment soil survey, soil samples were taken after lime application to sectors IV, VI, and VII, beginning in November 1986, and in 1987 and 1988.

11.2 CHARACTERISTICS OF LOCH FLEET SOILS

11.2.1 Pre-liming Conditions and Methods

At Loch Fleet a preliminary survey (Hudson *et al.*, 1986) was made to assess soil and vegetation resources of the catchment, to establish a

*The pH of water with a soil suspension is the common measure of soil acidity. In some circumstances the pH of a soil suspension is obtained in a weak electrolyte ($0{\cdot}01$ M $CaCl_2$) since in practice the values are more consistent. The pH value in the latter case is typically lower by about $0{\cdot}8$ pH unit than that measured in a water suspension.

baseline for measurements, to provide data on soil properties, and to monitor selected soils after the initial intervention treatments.

Soil properties, vegetation, and landforms were recorded at more than 450 points at 50-m intervals on a squared grid parallel to and conforming with the National Grid (Fig. 11.1). Eleven soil profiles to characterise the principal soil types, and of 40 topsoils to cover a range of site factors, were sampled prior to liming in July 1985. After liming, 12 sites were resampled in 1987.

Soil profiles were described and sampled in pits excavated with a spade to bedrock or 1 m depth, unless prevented by boulders or flooding. The soil horizon thicknesses and depths were measured, soil colours were obtained by comparison with Munsell Color Charts (Munsell Color Company, 1971), field textures accorded with the USDA system (USDA, 1951), and soil structures, fraction of stones, and root numbers were estimated from standard charts. Each profile was classified using the system of the Soil Survey of Scotland (Soil Survey of Scotland, 1984).

Soil samples were air-dried, and loss on ignition, proportion of sand, silt, and clay, exchangeable cations, base saturation, pH, carbon, and nitrogen were determined. The exchangeable cations were measured in a neutral, Normal ammonium acetate leachate; calcium, sodium, and potassium by flame photometry; and magnesium by atomic absorption (Scott *et al.*, 1971). Exchangeable acidity was determined by electrometric titration of a neutral Normal barium acetate leachate (Parker, 1929). The pH was measured in 0·01 M $CaCl_2$ by means of a glass electrode. Total carbon and nitrogen were determined using an Elemental Analyzer (Pella and Colombo, 1973).

11.2.2 Soil Types and Characteristics (Pre-liming Conditions)

Eight soil types were identified in the catchment (Fig. 11.2) each having peaty surface layers overlying thin mineral horizons or rock. Four major soil subgroups, peat, peaty gleys, peaty rankers, and peaty gley rankers, occupy 91·6% of the catchment, with peat being most extensive in the western part of the catchment, and peaty rankers and peaty gley rankers in the eastern part. In sector IV, glacial till is greater than in other sectors, and in sector VIII, peaty gley rankers are greater than elsewhere. Waterlogging predominates in the catchment soils creating anaerobic conditions and gleying of the mineral horizons in which the grey and ochreous mottled appearance of the mineral soil horizons is due to the segregation of oxidised and reduced iron weathering products. Colloidal humus, infiltrated from overlying peaty horizons, often imparts dark coloured

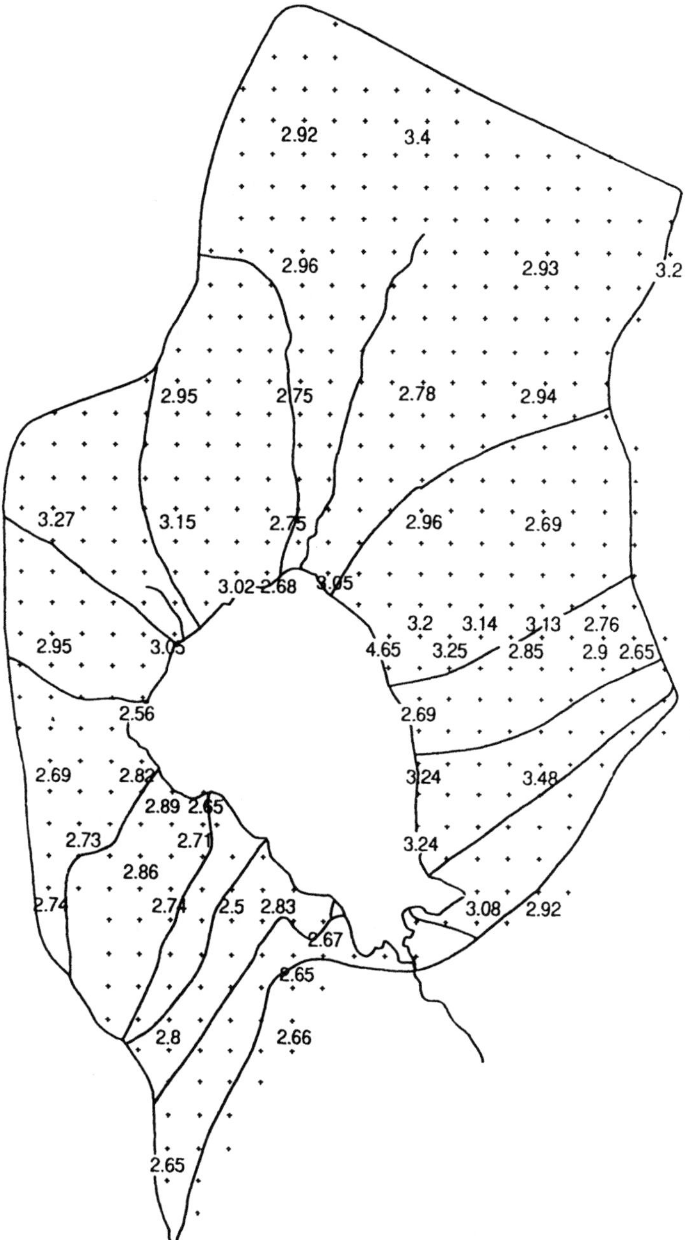

FIG. 11.1 Loch Fleet and its catchment: soil survey sites (+) and pH ($CaCl_2$), prior to liming.

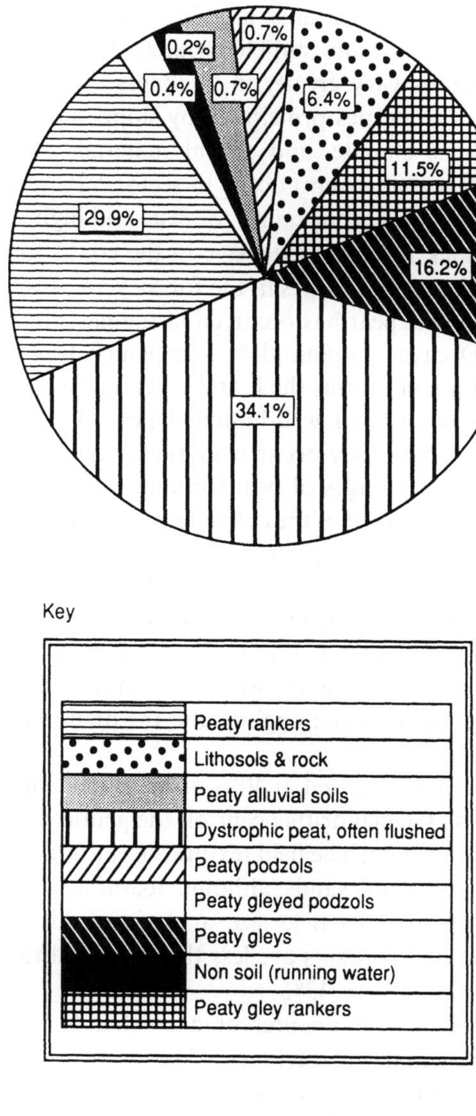

FIG. 11.2 Loch Fleet catchment: proportionate allocation of soil types.

stains which partly mask the gley colours. Waterlogging and the associated gleyed appearance of the mineral horizons is most evident on gentle slopes receiving seepage water and rather less on steep slopes and sites that shed moisture. Three soil features are described separately: organic horizons, mineral horizons, and soil chemical analysis.

The organic horizons vary greatly in thickness and are influenced in their formation and properties principally by slope angle and form, soil hydrologic conditions, climate, and parent materials. Areas with thin organic horizons (less than 26 cm) occur in sector VII, extend along the slopes east of the loch, and occur on ridge crests and high summits. Deep peat over 100 cm thick is locally extensive in the col at the north of the catchment and on the gentler slopes in the forestry areas west of the loch. Erosion (hagging) of the peat has occurred in the past in the col drained by the Altiwhat. Decayed organic debris, in which fragments of plant remains can be recognised, forms the L and F layers overlying the decomposed, black, amorphous and well-humified peat. The basal layers of the peat, especially where they occur over rock, contain mineral grains incorporated in the early stages of peat formation in the catchment.

By far the major component of the catchment soils, the organic horizons are strongly acid and have low contents of nutrient cations (calcium and magnesium) on the exchange complex. Replenishment is probably dependent principally on incoming precipitation rather than by dissolution from the underlying rock or mineral layers. The low permeability of these organic horizons to soil water, allied with the high rainfall, contributes to slow infiltration, persistent waterlogging, and the predominance of water movement by surface flow.

Mineral subsoils were recorded in 40% of the catchment and are often less than 20 cm thick. Glacial drifts occur on the gentle slopes west of the loch, where peat is prevalent. The field textures of the mineral horizons are mainly sandy loams or loamy sands, although loam, silty loam, and sandy clay loam were recorded. Particle size analysis of 11 mineral subsoil samples shows that clay contents are less than 6% and the silt fractions (2–50 μm) range from 12% to 43%.

These thin mineral horizons are present in less than 50% of the soils and are only a minor element in the catchment. They have low contents of clay and the silt and sand fractions comprise mainly nutrient-poor slow weathering minerals derived from granite. There is little provision of nutrient bases to the exchange complex or to buffer the effects of incoming acid precipitation.

The soils were sampled prior to liming over a short period in 1985 to minimise seasonal or weather-dependent variation in chemistry and to allow comparison between results. Throughout the catchment, pH values (in 0·01 M $CaCl_2$), were low, ranging from 2·5 to 4·6 (Fig. 11.1), with variations explained partially by land use, flushing, type of organic horizon, and soil type. Samples collected from afforested soils generally had lower pH than comparable soils under semi-natural moorland vegetation. Soil acidity also appears to be linked to position and altitude. On the eastern slopes of the catchment, pH values increase downslope, probably reflecting an increase in flushing. LF and H horizons, occurring at or near the surface, are generally less acid than the underlying O horizons. Exchangeable bases and percentage base saturation* were low throughout the catchment prior to the intervention treatments.

The chemistry of the soils, soil waters, tributary streams, and loch waters were monitored on a regular basis before and for a year after liming. The results indicate that soil seepage waters do not play a significant role in the determination of the chemistry of the loch waters, and that under conditions of persistent waterlogging surface runoff must be all important.

11.2.3 Post-liming Changes

As anticipated, there were immediate changes in the chemical properties of the topsoil horizons (5–10 cm) sampled after lime treatments, notably increases in pH, exchangeable calcium, and base saturation* and a decrease in exchangeable acidity. However, there was little evidence that liming had any effect on sub-surface soils. Particular attention was paid to the composition of soil seepage waters which were sampled at a 30 cm

*A standard method for determining cation exchange capacity (CEC) and base saturation of a soil (MAFF, 1986) employs 1·0 M ammonium acetate or barium acetate as an extractant; this produces conditions in the sample of pH 7·0. Base cations (Na, K, Ca, Mg) in the leachate are analysed, as well as H^+ (hydrolytic acidity). The sum of these values provides a "derived" CEC at pH 7. The percent base saturation (%BS) is calculated as the ratio, exchangeable cations/CEC × 100. A problem arises in that CEC is strongly pH dependent, and is also influenced by the soil organic content. An alternative extractant, such as ammonium chloride or potassium chloride, provides a more realistic measurement at the pH of the soil sample, extracting similar amounts of exchangeable base cations, plus exchangeable hydrogen and aluminium ions. The sum of these constituents is the "effective cation exchange capacity" and represents exchangeable base cations + exchangeable acidity; %BS is then exchangeable cations/ECEC × 100, and thus will be much larger than by the previous method (Bache, 1979). For Loch Fleet soils, the former method provides values of %BS of about 10% in unlimed soils, while the latter provides values of greater than 45%.

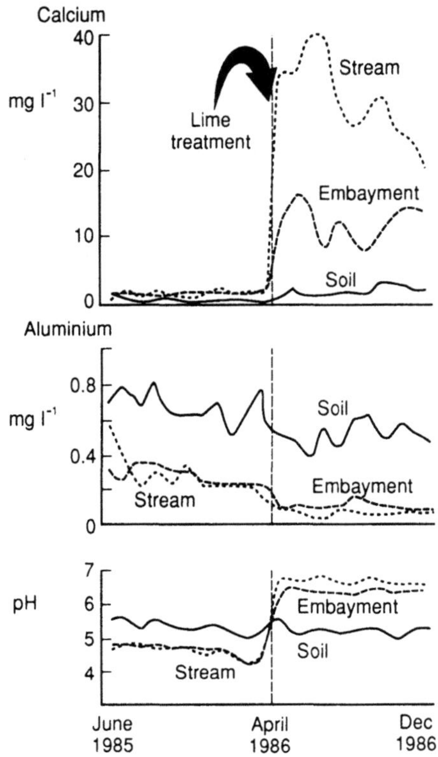

FIG. 11.3 Changes in soil pore water chemistry in the period prior to and after liming.

depth at the interface between the organic and mineral soils around the perimeter of the loch. The acidity of the soil waters was not affected within the first year by the lime treatment although the effect on embayment and stream waters was rapid and dramatic (Fig. 11.3). However, a notable difference is that aluminium, generally high prior to liming, is decreased within soil waters with a consequent drop in aluminium in stream waters and loch embayments.

The climate in the Loch Fleet catchment is euoceanic, perhumid, and mainly oroboreal with some orohemiarctic on the highest ridges of the Fell of Fleet (Birse, 1971). Land having this type of climate occupies around 6% of Scotland and, where it occurs, the major soil groups and subgroups are similar to those recorded at Loch Fleet. They are, however, developed on a broader spectrum of parent materials than the Cairnsmore of Fleet granites.

11.3 ROLE OF SOIL IN CATCHMENT LIMING

11.3.1 General Aspects

In the UK, where upland lakes are characterised by rapid turnover times, application of lime to the soil facilitates a slow release of calcium and alkalinity, increasing treatment duration compared with lake liming. In contrast to lime applications to surface waters, relatively few studies of catchment (soil) liming treatments are reported. However, current projects are likely to provide relevant data in the next few years (Section 11.7).

Land use and hydrological conditions in the limed areas are crucially important. The lime requirement will be influenced by the types of vegetation in the catchment which modify the acidity of the soil (and hence runoff) to differing extents (e.g. Skeffington, 1983; Miles, 1985). In upland Wales, the replacement of moorland by forest plantations has resulted in an increase in the concentration and flux of mobile anions into the soils. This has led to the mobilisation of organic acids in peats and in soils with organic horizons, providing runoff of low pH (about 4·0), and of aluminium and silicate from soil mineral material (Hornung *et al.*, 1990a). This is particularly evident in the drainage from coniferous forest (eg. L. Brianne catchment) which is substantially more acid than that from moorland within the same hardness category; significantly higher concentrations of aluminium and sulphate are also found (Bird *et al.*, 1990).

The optimum depth to which liming needs to penetrate to increase base saturation will depend upon the region of hydrochemical reaction. At Loch Fleet, hydrological flow of those waters influencing loch chemistry appears to be more than 90% by surface rather than sub-surface pathways (see Chapter 8, but also Chapter 7): if correct, only lime in surface soils will improve the quality of runoff.

11.3.2 Target Water Quality and Soil Lime Requirement

In agricultural use, "lime requirement" is defined as the amount of limestone required to neutralise an acid soil to a desired pH. At Loch Fleet, it can be defined as the amount of limestone required to provide drainage or runoff water quality satisfactory for brown trout (*Salmo trutta*) survival (see Chapter 3). The definition should also include a time dimension such that the lime requirement provides an economic duration of improvement while minimising any adverse effects on catchment vegetation.

There are two sources of soil acidity:

- ionised hydrogen in the soil solution, referred to as the "intensity factor" and measured as soil pH, and

- the "capacity factor" or exchangeable acidity, comprising exchangeable hydrogen and exchangeable aluminium adsorbed on the soil surface. Aluminium is a source of acidity since hydrogen ions are generated on hydrolysis:

$$Al^{3+} + 2H_2O \longrightarrow Al(OH)^{2+} + H_3O^+ \qquad (1)$$

In a very acid soil (e.g. pH $< 4{\cdot}0$) such as at Loch Fleet, most of the exchangeable acidity will be Al^{3+} rather than H^+. In samples from sectors IV and VI at Loch Fleet, exchangeable acidity was 62% and exchangeable aluminium 75% (Roberts, pers. comm.). Hydrogen and aluminium are in equilibrium in acid soil, the latter being released when hydrogen ions react with soil minerals.

The addition of limestone to soil results in hydrogen ions in the soil water being neutralised:

$$\begin{aligned} CaCO_3 + 2H^+ &\rightarrow Ca^{2+} + H_2CO_3 \\ &\rightarrow Ca^{2+} + CO_2 + H_2O \qquad (2) \end{aligned}$$

and calcium exchanges for H^+ on the soil exchanger:

$$\text{soil}\Big]^{-H}_{-H} + Ca^{2+} + 2HCO_3^- \rightarrow \text{soil}\Big]^{-}_{-} Ca + H_2CO_3 + CO_2 + 2H_2O \qquad (3)$$

The result will be an increase in soil pH (equation (2)) and base saturation (equation (3)). Less aluminium will be released from soil minerals, thereby reducing the concentration of Al^{3+} in the soil water. The potential for aluminium–base cation exchange and loss of bases by leaching will diminish. If the limestone addition is sufficient to bring soil pH into the range 4·0–5·2, aluminium will be precipitated:

$$Al^{3+} + 3OH^- \longrightarrow Al(OH)_3 \qquad (4)$$

Hydrogen ions are removed from solution by exchange/mineral dissolution, and base cations are brought into solution. Buffering will be efficient in soils such as Loch Fleet which have a high organic matter content and hence a high CEC. Saturation of the soil with calcium (equation (3)) will buffer internal and external acid inputs, and provide a reserve of calcium for release in runoff.

Equations (2), (3), and (4) show how catchment liming can counter low pH, base cation deficiency, and aluminium mobilisation in acid soils. The rapid translation of these reactions into improved water quality has been demonstrated at Loch Fleet (see Chapter 12).

Having identified soil parameters central to drainage/runoff water quality improvement, it should be possible to determine a "target soil composition" which will produce surface waters to meet the "target water quality" for fish survival. In practice, this target is a more complex and dynamic concept for soils than that applied to water chemistry. A starting point would be the determination of a threshold base saturation and soil pH which buffers acid inputs by $H^{+}-Ca^{2+}$ exchange, causing aluminium to be retained producing surface waters with pH >6 and calcium and aluminium concentrations of $100\,\mu eq\,litre^{-1}$ and $<100\,\mu g\,litre^{-1}$, respectively. From ion exchange buffer curves and $Ca^{2+}-Al^{3+}$ exchange isotherms, Reuss and Walthall (1989) suggest a base saturation threshold greater than 15%–20% at which pH changes and aluminium mobilisation are small. At Loch Fleet, the "target base saturation" (45%–53% when measured at soil pH)* would have to exceed pre-liming values in the top 10 cm (Hudson *et al.*, 1986; E.J. Wilson *et al.*, 1988). Adams *et al.* (1990), working on acidic upland soils of mid-Wales, found that the release of inorganic aluminium to soil water and runoff was negligible when soil pH was greater than 5·0, which may be a rough guide to a target pH. The criteria for the "target soil composition" should be applied to a specified depth down the soil profile, determined by the region of hydrochemical response.

It is too early in the Loch Fleet soil sampling programme to determine a "target soil composition" other than the fact that pH values and base saturation must lie between those pre-treatment values when the loch could not support fish, and current values (Wilson *et al.*, 1988) which meet the "target water quality". The target soil composition can be determined in practice as those soil conditions at which stream water chemistry falls below target quality.

11.4 POST-LIMING SOIL CHARACTERISTICS

11.4.1 Programme of Sampling and Analysis

Soil sampling in the three sectors treated in 1986 (see Chapter 9) was initiated to investigate the effect of liming on soil chemistry and to monitor the changes in free lime and exchangeable calcium pools. A knowledge of the dynamics of these soil pools provides a method to estimate the efficacy of catchment liming. Soils from sectors IV, VI, and VII were sampled in

*See footnote on p. 200.

October 1986, November 1987, and November 1988. The 1986 results showed high inter-sample variability, masking significant differences in various parameters. However, it was apparent that most of the applied lime remained in the litter and upper soil layers in the short term and subsequent sampling regimes were modified to include more samples from shallower depths. This greatly reduced the variability observed between samples from a given sector. In 1987 and 1988, samples from sectors IV and VI were taken systematically along transects running from the loch edge up-slope, using a corer of known surface area. On sector VII core samples were taken from the limed *Sphagnum* bog. The soil cores were subsectioned down the profile, dried to constant weight at 35°C, and ground (2 mm sieve) prior to analysis.

Samples were analysed for pH, exchangeable acidity, exchangeable calcium, base saturation, and free lime (Wilson, 1989). In samples that did not contain free lime, exchangeable bases were determined by the single extraction method of Trüby and Aldinger (1984), using a solution of ammonium chloride (0·5 M). Base saturation was calculated as the ratio of exchangeable bases to the sum of exchangeable acidity and exchangeable bases (or CEC), multiplied by 100. Exchangeable acidity was determined by titration of the ammonium chloride extract to pH 7·0 by sodium hydroxide. In the presence of free lime, exchangeable calcium (and hence base saturation) cannot be determined by conventional single exchange methods owing to wash-out of calcium into the leachate. The CEC was determined by a double exchange method, i.e. saturating cation exchange sites with sodium using a sodium acetate–sodium chloride solution, followed by extraction with magnesium nitrate (Gupta *et al.*, 1985). Exchangeable cations other than calcium were determined in a single extraction using ammonium chloride. The fraction of the CEC not occupied by the exchangeable cations (including exchangeable acidity) determined in the single extract was assumed to be occupied by calcium. Hence exchangeable calcium was calculated as the CEC (from the double exchange method) minus the sum of exchangeable cations (from the single extraction). This is only valid if the applied lime can be considered as pure $CaCO_3$. Chemical analysis of the lime used at Loch Fleet (EAU, 1986; see Chapter 9) shows this assumption to be justified. Free lime was determined using a pressure calcimeter (Page *et al.*, 1982). The 1986 and 1987 values for pH, exchangeable acidity, base saturation exchangeable calcium, and free lime are shown graphically in Figs. 11.4–11.8. The differences between data for the two years have been analysed statistically (*t*-test).

The total lime remaining on the catchment in December 1987 is cal-

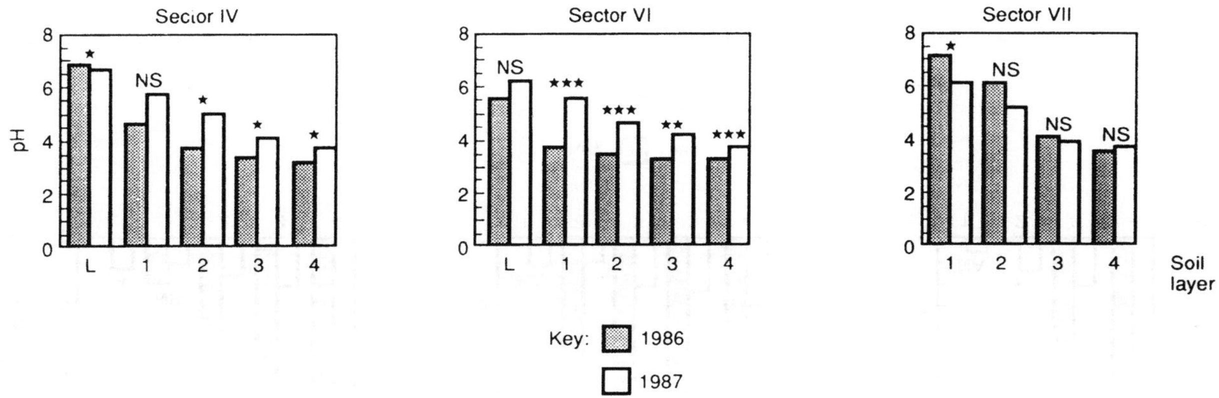

FIG. 11.4 Mean soil pH in treated sectors in 1986 and 1987). (*** $p < 0{\cdot}001$, ** $p < 0{\cdot}01$, * $p < 0{\cdot}05$; NS = not significant, NA = not applicable).

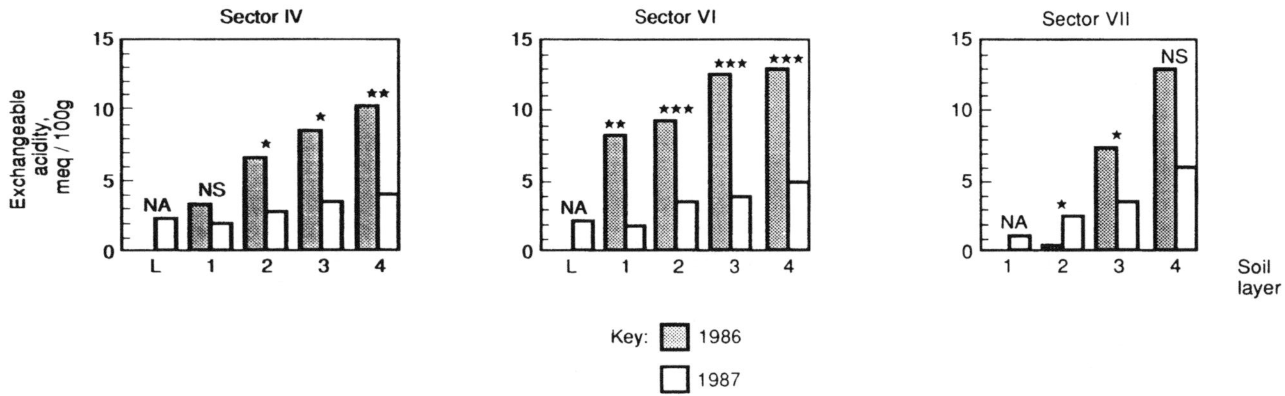

FIG. 11.5 Mean exchangeable acidity of soils in treated sectors, 1986 and 1987. (Significance as Fig. 11.4.)

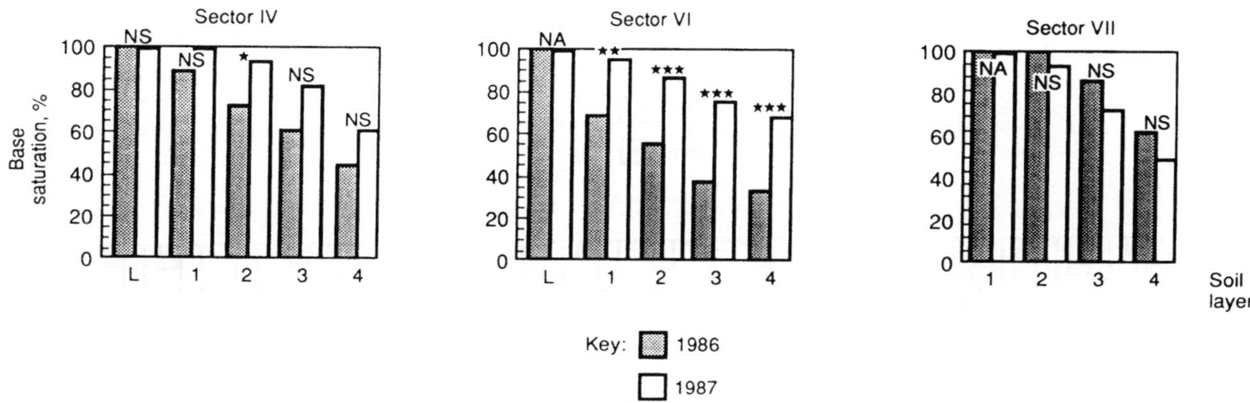

FIG. 11.6 Mean base saturation in soils in treated sectors, 1986 and 1987. (Significance as in Fig. 11.4.)

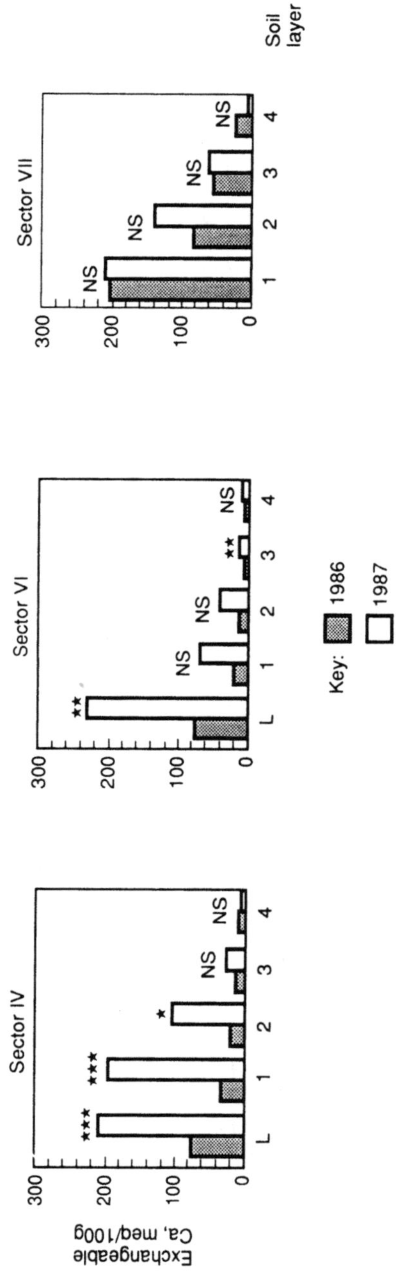

FIG. 11.7 Mean exchangeable calcium in soils in treated sectors, 1986 and 1987. (Significance as in Fig. 11.4.)

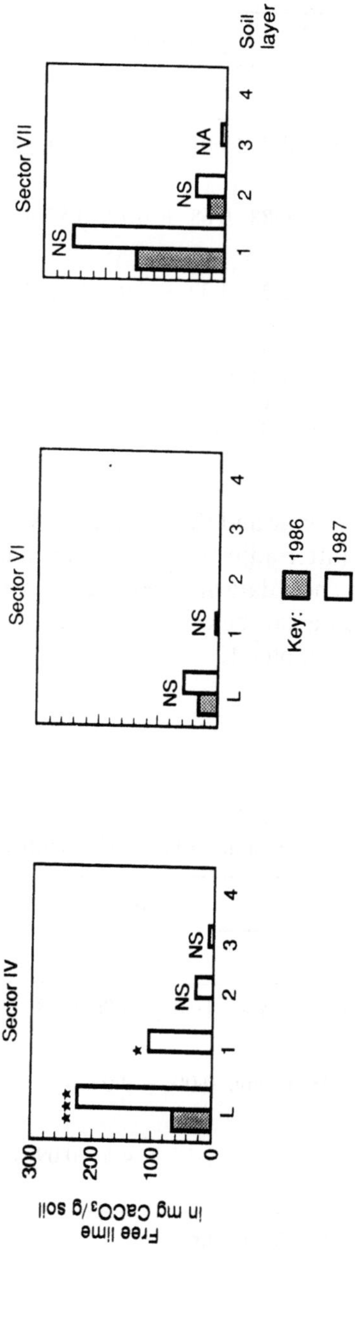

FIG. 11.8 Mean free lime soils in treated sectors, 1986 and 1987. (Significance as in Fig. 11.4.)

TABLE 11.1
Remaining exchangeable calcium expressed as $CaCO_3$ in t ha^{-1} (December 1987)[a]

Sector	*Soil depth*				
	L	*1*	*2*	*3*	*4*
IV (forest)	5·81 ± 0·73	4·48 ± 1·7	3·01 ± 2·3	1·74 ± 1·9	0·42 ± 0·30
VI (moorland)	6·86 ± 2·5	1·02 ± 0·88	0·68 ± 0·72	0·57 ± 0·27	0·26 ± 0·14
VII (wetland)		4·43 ± 2·2	3·00 ± 3·9	1·13 ± 2·3	0·16 ± 0·48

[a]Figures are means with 95% confidence limits.
Notes: L = litter; depth 1 = 0–2·5 cm, 2 = 2·5–5 cm, 3 = 5–10 cm, 4 = 10–15 cm (forest and moorland soil); 1 = 0–5 cm, 2 = 5–10 cm, 3 = 10–15 cm, 4 = 15–20 cm (wetland).

culated as the sum of the free lime and that lime from which the exchangeable calcium was derived (after adjustment for exchangeable calcium in untreated Loch Fleet soil). These data are presented for each treated sector in Tables 11.1–11.3. The data can be compared with those derived from the stream budget data in Chapter 12.

TABLE 11.2
$CaCO_3$ remaining as free lime in t ha^{-1} (December 1987)[a]

Sector	*Soil depth*				
	L	*1*	*2*	*3*	*4*
IV (forest)	12·47 ± 3·7	4·92 ± 4·9	1·58 ± 0·86	0·38 ± 1·7	0·00 ± 0·0
VI (moorland)	2·10 ± 2·4	0·05 ± 0·06	0·00 ± 0·0	0·00 ± 0·0	0·00 ± 0·0
VII (wetland)		10·36 ± 7·5	2·15 ± 4·5	0·09 ± 0·22	0·00 ± 0·0

[a]Figures are means with 95% confidence limits.
See notes to Table 11.1.

TABLE 11.3
Calcium remaining on each sector expressed as $CaCO_3$ in t ha^{-1} (December 1987)[a]

Sector	*Ex. Ca (as $CaCO_3$)*	*Free lime*	*Total lime remaining*
IV (forest)	15·46 ± 3·13	19·35 ± 6·06	34·81 ± 6·98
VI (moorland)	9·39 ± 2·45	2·15 ± 2·17	11·54 ± 3·42
VII (wetland)	8·72 ± 10·36	12·55 ± 6·24	21·27 ± 9·21

[a]Figures are means with 95% confidence limits.

11.4.2 Post-liming Changes

11.4.2.1 pH and Exchangeable Acidity

Liming has increased the pH to greater than 5 in surface soils in all three sectors compared with pre-treatment values of approximately 3·0 (E.J. Wilson *et al.*, 1988). In sectors IV and VI soil pH has increased since liming in all layers below the litter layer to a depth of 20 cm (Fig. 11.4). This is paralleled by a decrease in exchangeable acidity down the profile (Fig. 11.5) reflecting the movement of lime out of the litter layer where most was present in 1986, and with a more even distribution to lower depths.

11.4.2.2 Exchangeable Cations, Base Saturation, and Free Lime

By 1987 liming had doubled the base saturation in the litter or surface layers in sectors IV, VI, and VII (Fig. 11.6) compared with pre-liming values (E.J. Wilson *et al.*, 1988). Liming has increased base saturation considerably even at a depth of 10–15 cm (Fig. 11.6). Analysis of free lime shows an increase from 1986 to 1987 down the profile in sectors IV and VI. Exchangeable calcium increased in the upper three layers (to 15 cm) on sector VII (Fig. 11.7). Evidence of the migration of free lime down the profile supports the increase in exchangeable calcium and base saturation (1986 to 1987) at lower depths. Redistribution of free lime from the litter layer has been more uniform in sector IV compared with sector VI, reaching depths of 10 cm and 5 cm, respectively (Fig. 11.8). The fraction of incoming precipitation penetrating the soil profile compared with that flowing over the surface was greater on sector IV than VI, thereby increas-

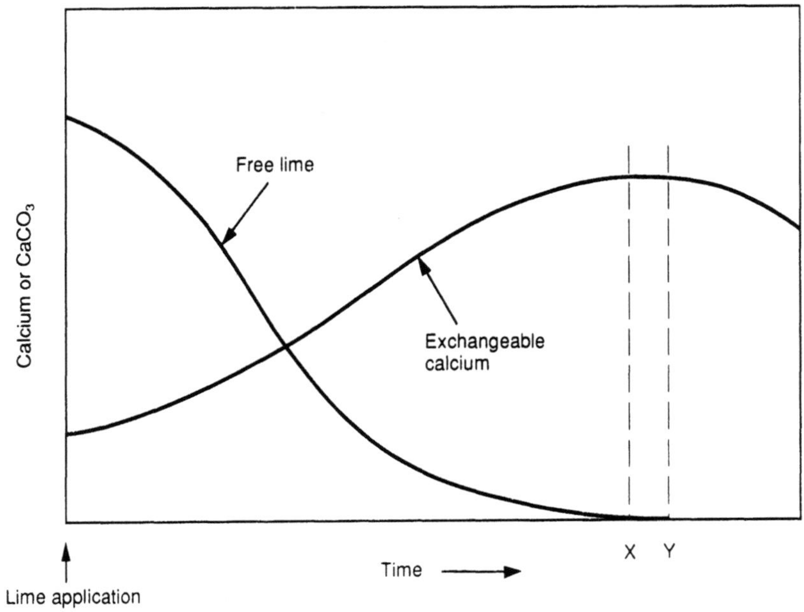

FIG. 11.9 Schematic time-scale of treatment effectiveness, expressed as availability of free lime and exchangeable calcium.

ing the percolation of lime to lower depths. This results from a combination of a shallower gradient, interception by the forest canopy, and soil channels created by roots. On sector VII (*Sphagnum* bogs) high variability as a result of small sample size makes it difficult to reach confident conclusions. However, this sector showed the most uniform distribution of lime with depth due to the porous structure of the *Sphagnum*.

11.5 DISCUSSION

Immediately following liming, free lime in the soil will be at a maximum and will gradually decrease with time due to dissolution and removal in drainage and stream water or on to the exchanger (see Fig. 11.9). There will be a concomitant increase of calcium on the exchanger, which will reach a maximum (point X in Fig. 11.9). Calcium removed from the exchanger by incoming cations in precipitation or internal acid generation will be replenished continually by the dissolution of free lime until this calcium source is exhausted or deactivated. When this point is reached (point Y in Fig. 11.9), the treatment effectiveness time is limited by the amount of calcium remaining on the exchanger, the calcium concentration desired in runoff, and the rate of supply from weathering.

On all three sectors liming has brought soil pH into the range at which aluminium is likely to precipitate as aluminium hydroxides. This was achieved as early as 6 months after liming and was maintained during the following year. The change in soil pH is reflected in a decreased flux of aluminium to the loch (see Chapter 12). The increase of soil pH to 5 in the upper layers produced runoff with pH greater than 7 on all three sectors (see Chapter 12), thus exceeding the target pH of greater than 6.

Measurements of base saturation and exchangeable acidity show that the movement of calcium on to the exchanger and the removal of hydrogen ion has been rapid. Critically, base saturation in the region of hydrochemical reaction (litter layer and upper 5 cm) approached 100% 6 months after liming, considerably exceeding the threshold value at which the soil will buffer acid inputs. The increase in exchangeable calcium and base saturation from 1986 to 1987 indicates a rapid rate of limestone dissolution and movement of calcium on to the exchanger. It appears that the lime has not been deactivated by some form of coating. This is corroborated by investigations using energy dispersive X-ray analysis (Allison, 1989, pers. comm.). A source of free lime is required to maintain this high base saturation and the efficient buffering of internal and external sources of acidity. Free lime measurements show that most of the limestone still remains in the litter layer and upper 2·5 cm of soil after liming, within the region of hydrochemical reactions in the soils at Loch Fleet. A significant increase in free lime at lower depths indicates movement down the soil profile from 1986 to 1987. If vertical migration of lime continues, it may become unavailable for improving water quality and limit treatment effectiveness.

Calcium flux calculations show that calcium export from the treated sectors as a proportion of remaining lime is decreasing (see Chapter 16). This could be a result of depletion of the free lime or an insufficient rate of dissolution of the remaining lime. However, the movement of lime down the soil profile below the main hydrological access is most likely to be responsible for the observed decline in calcium export. This factor would be most significant on sector VII where penetration of lime to greater depths was most rapid. This is consistent with predictions, from the extrapolation of calcium flux data, that calcium will first fall below the target value on sector VII.

More soil data are obviously required to determine the longevity of the increased soil pH and base saturation attained so rapidly after lime application. The distribution of free lime down the soil profile and its subsequent rate of depletion appears to play a central role in treatment effectiveness. This will continue to be monitored at Loch Fleet.

11.6 SOIL CHEMISTRY DATA TO ESTIMATE TREATMENT EFFECTIVENESS TIMES

Treatment effectiveness times can be determined by the balance between "available" limestone remaining and the rate of dissolution (Fig. 11.10) and replenishment of exchangeable calcium. From a knowledge of the amount of free lime and exchangeable calcium (Fig. 11.9), preliminary predictions can be made. Such calculations are inevitably based on a number of assumptions which may not be reproduced in the field. At present, predictions made in this way are limited by the number of years, post-liming, for which soil chemistry data are known.

A primary assumption in predicting treatment effectiveness times is that all the limestone remaining on the catchment in December 1987 is available for water quality improvement. If all the free lime remaining at this time passed into the loch on dissolution, and was succeeded by calcium removal from the exchanger, the treatment longevity would equal the sum of the time taken to deplete the free lime pool to zero and exchangeable calcium to below the "target base saturation". For the purpose of these calculations, the latter has been estimated at 60% on the basis of pre-treatment values of 45%–53% (measured at approximately soil pH). The exchangeable calcium available to maintain water quality after December 1987 is that in excess of 60% base saturation (Wilson, 1989).

Certain assumptions may lead to an over-estimation of treatment effectiveness. First, all the remaining free lime is assumed to pass into the loch, although in reality it may become unavailable due to migration in the soil below the region of hydrochemical reaction. Secondly, the export rate of exchangeable calcium is likely to be less than from free lime, and may be insufficient when free lime has been depleted to zero.

A prediction of treatment effectiveness time for sector IV has been calculated from a measure of calcium remaining (as free lime and exchangeable calcium in excess of 60% base saturation) divided by annual calcium export derived from calcium flux data for October 1986 to October 1987 (see Chapter 12). This gives the following results:

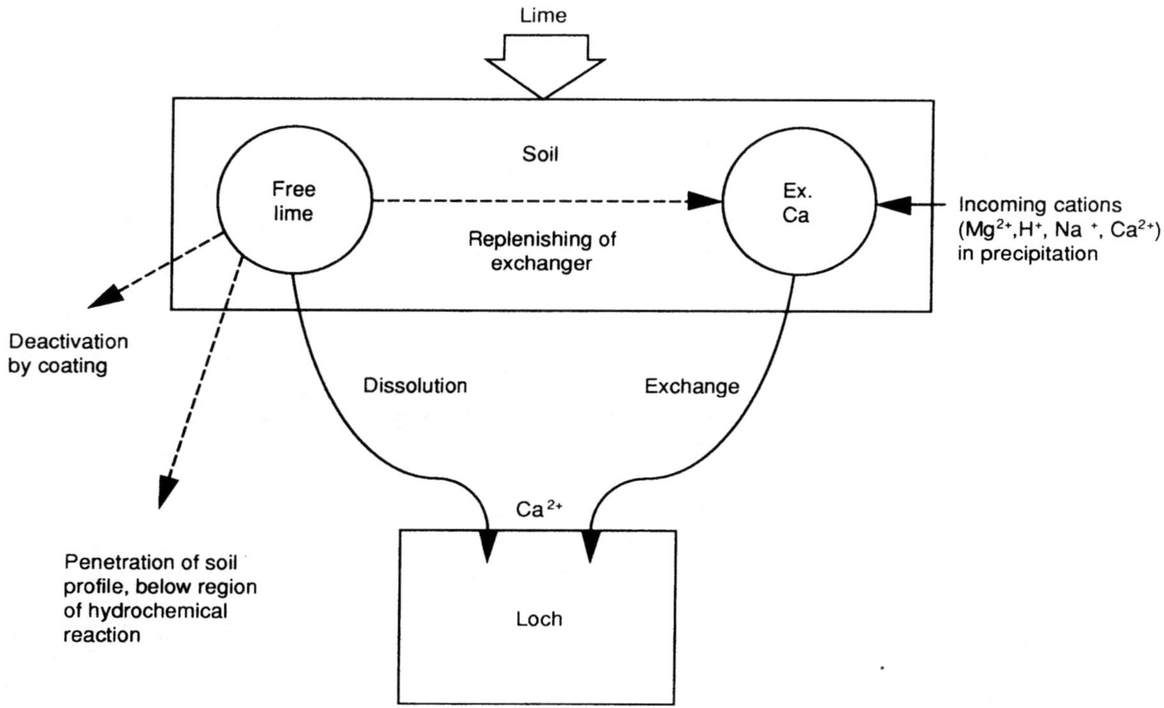

FIG. 11.10 Possible fates of limestone applied to catchment soils.

- calcium remaining: 495 keq ha^{-1};
- calcium export: 15·23 keq ha^{-1} yr^{-1};
- land, treatment effectiveness time: 32·5 years.

Similar calculations have not been made for sectors VI and VII at this stage owing to the high variability of free lime and exchangeable calcium measurements on these sectors.

A predicted treatment effectiveness time of 32·5 years for sector IV does not appear unrealistic in view of data from Welsh catchments (Adams and Evans, 1989) which show that liming at the lower rate of 15 t ha^{-1} still has effects on water quality 40 years after application. This prediction is also in line with those based on modelled stream chemistry data, which indicate that stream water quality will be maintained on this sector well into the next century (see Chapter 16).

11.7 SOIL STUDIES AT OTHER CATCHMENT LIMING SITES

At present, data from different sites on the effects of liming on soil chemistry are sparse. Although liming has been employed to counter surface water acidification in Scandinavia for many years, this has generally been undertaken on a pragmatic basis. Furthermore, since liming is usually directly to surface waters, even detailed scientific studies such as those at Lake Gårdsjön in Sweden do not provide comparable soil data.

11.7.1 Llyn Brianne and Other Studies in Wales

The Llyn Brianne project in south-west Wales was initiated by the Welsh Water Authority in response to declining fisheries at about the same time as the Loch Fleet study. The Llyn Brianne area has peaty acid soils similar to those at Fleet and comparable rainfall (1807 mm, 30-yr mean; Welsh Water Authority, 1987). A detailed soil studies programme in limed areas is being undertaken at Llyn Brianne which will provide data comparable with Loch Fleet where liming is on similar vegetation types and at similar rates. Soil chemistry data from one catchment a year after liming shows a very similar soil response to Loch Fleet. This site, dominated by *Molinia* and *Sphagnum* and limed at a rate of 25 t ha^{-1}, is comparable with the moorland sector VI at Loch Fleet which was limed at 21 t ha^{-1}. At Llyn Brianne liming increased pH in the surface horizons (peat) by 0·5 unit to pH 4·3 (Reynolds and Emmett, 1990), while at Loch Fleet pH in the top 10 cm on sector VI increased from pH 3·1–3·3 to 3·54 and 5·16, 6 months and 1·5 years after liming, respectively. Assuming a linear increase in pH, this suggests a value of 4·35 at Fleet one year after liming, comparable with Llyn Brianne. Since Loch Fleet soils are more acid, the rate of lime dissolution and neutralisation must be greater than at Brianne to achieve the same target soil pH over the same time period. As at Loch Fleet, there was very little vertical penetration of lime down the profile 1 year after liming. Since vertical migration will be influenced by rainfall volume and soil structure, similar patterns of movement might be expected at the two locations. Exchangeable aluminium decreased in the surface layers with a concomitant increase in exchangeable calcium and base saturation as at Loch Fleet. Different analytical methods used by the two projects prevent more detailed comparison. Furthermore, some areas at Brianne with stagnopodsols and brown podsolic soils failed to show improved runoff quality (Hornung *et al.*, 1990b).

Studies on upland pasture catchments in mid-Wales with unimproved soils and soils that had been limed during the last 20 years, provide useful data on the longevity of treatment effects on stream chemistry (Adams and Evans, 1989). The soils of the catchments included acidic peaty podzols with low base saturation, similar to those at Fleet. In streams draining from land limed as long as 20 years ago, water quality was found to meet the criteria for fish survival. pH was in the range 5·9–6·7, calcium 100–200 μeq litre^{-1}, and total monomeric aluminium 7–49 μg litre^{-1}, compared with a pH range of 4·3–5·9, calcium greater than 50–90 μeq litre^{-1}, and total monomeric aluminium 60–121 μg litre^{-1} in streams draining unlimed catchments. All areas (except one for which the application rate is unknown) were limed at 5 t ha^{-1}. This study indicates not only that target water quality may be maintained many years after liming, but that it may be achieved even at such low application rates. In the soils with higher pH values (5·9 and 6·1), base saturation was still about 90%, which may be an indication of the "target base saturation threshold" of soils required to provide water able to sustain a trout population. From the considerations of Reuss and Walthall (1989) this seems a very conservative estimate. It will, however, depend on sulphate inputs from deposition and internal anion production. Evidently more data are required from the simultaneous monitoring of soil and water chemistry to determine the base saturation at which water quality fails to meet the criteria for fish survival.

Where agricultural liming has been carried out at higher application rates, there is evidence that effects on water quality may last for as long as 40 years. Studies of solution chemistry of stagnopodzol soils were made at four sites in mid-Wales which had received different pasture improvement treatments in the past. A lime application of 15 t ha^{-1} about 40 years ago resulted in a soil pH greater than 6 in upper horizons, and calcium greater than 100 μeq litre^{-1} to a depth of 50–70 cm, differing significantly from soils of untreated areas (Hornung *et al.*, 1986). Grazing may have contributed to the long duration of treatment effects in these examples by enhancing nutrient cycling, thereby releasing bases normally locked up in organic matter.

11.7.2 Woods Lake, USA

Preliminary data on catchment soils has established that the soil pH is 3·29 (surface) rising to 4·13 in the B horizon. Exchangeable calcium is 8320 μeq per 100 g in the O horizon, 950 μeq per 100 g in the A horizon, and 220 μeq per 100 g in the B horizon. CEC is highest in the A_1 horizon (27 530 μeq per 100 g) and somewhat lower in the O and B horizons. Percent base saturation is 43% in the humus (O), but only 5% and 2% in the A and B horizons (Cronan, 1984).

Liming (1000 t as pellets at 1 t ha^{-1}) on two subcatchments was undertaken in September 1989. After 1 year some surface calcium is still evident but no calcium has penetrated to the B horizon (Porcella, pers. comm.).

11.7.3 Scandinavia

The long-term (20–24 years) effects of liming and fertiliser application on soil properties has been demonstrated at six forest sites in Finland. Limestone was added on a single occasion at 2 t ha^{-1} and nitrogen as ammonium sulphate and ammonium nitrate at varying doses (82–180 kg N ha^{-1}) on a number of occasions.

Calcium is retained for a long period; 65·75% of limestone applied was still present in the humus and mineral soil almost 25 years after liming. Liming was associated with a considerable increase in organic content in the humus layer, and a corresponding decrease in the mineral soil. It also increased the CEC of the humus layer and decreased it in the mineral soil, following the change in organic matter. Base saturation in the humus layer increased by 15%, and in the mineral soil by 5%–15%. Exchangeable acidity of soils below pine was reduced by about 10%, but did not change below spruce. Monomeric aluminium was dramatically reduced in the humus (by 50%) and mineral soils (by 20%) (Derome *et al.*, 1986).

Liming with calcium carbonate of 16 forest stands in Sweden was undertaken as long as 70 years ago. Doses were 3–20 t ha^{-1} to the forest floor. At all sites soil changes were an increase in pH and base saturation, especially in the humus layer, but some long-term effects have also been seen at 10–20 cm depth. In addition, exchangeable aluminium decreased in humus and mineral soils. The exchange capacity (CEC) increased in the mineral soil after liming, although changes were not dramatic. Transfer of calcium from the upper soils to deeper soils varied from 50 to 400 kg $CaCO_3$ ha^{-1} yr^{-1}. The loss has decreased with time, but the applied lime has remained within the soil profile – 16% is retained at one site even after 68 years (Popovic and Andersson, 1984; Hallbacken and Popovic, 1985).

Further liming experiments on five meadow sites in soils in southern and eastern Sweden tested a variety of alkaline materials at 1–30 t ha^{-1}. Observations over 4 years demonstrated that soil liming has resulted in higher pH, increased base saturation, decreased aluminium, and increased available calcium (with calcitic limestone or dolomite) and of magnesium (with olivine or dolomite), mostly at the 0–5 cm depth. Insignificant effects were found at lower soil depths. Effects were related to dose and the material employed: calcitic limestone > wood ash > dolomite > olivine. Some observations on meadow plant species suggest that acidophilic species were affected, but the number of species was not significantly changed (Blom and Wincent, 1990).

11.8 SUMMARY AND CONCLUSIONS

The soil studies programme at Loch Fleet was initiated to

- assess the chemical and physical characteristics of soils of the catchment prior to liming;
- monitor the changes in soil chemistry as a result of liming; and
- relate these to changes in drainage and runoff water quality.

Sufficient data should provide, in due course, an independent method of estimating treatment effectiveness times to that developed in Chapter 16.

The soil types of the Loch Fleet catchment fall into four major soil sub-groups: peats, peaty-gleys, peaty rankers, and peaty-gley rankers. They are characterised by a thick organic horizon (> 26 cm) and sparse, thin mineral soils. High rainfall at Loch Fleet leads to waterlogging and slow decomposition of plant matter. The high flux of anions leaching bases from the upper layers, combined with low base reserves for release by mineral weathering, result in very acid soils (pH 2·5–4·6) of low base saturation (see Chapter 2). At Loch Fleet, where the hydrological pathway appears to be near the surface, water flow will be confined to these acid, peaty horizons.

A limed soil improves drainage water quality by neutralising soil acidity, precipitating aluminium as aluminium hydroxides, and increasing base saturation to provide a reserve of calcium for release into surface waters. Soils sampled from sectors IV, VI, and VII at 6 months, 1·5 and 2·5 years after liming are being used to assess changes in soil chemistry. Lime application neutralised acidity, increasing pH in surface layers from 3·0 to 5·0, as early as 6 months after liming. An increase in pH occurred in lower layers in the soil profile from 1986 to 1987. This was matched by a decrease in exchangeable acidity and an increase in exchangeable calcium and base saturation over the same period, indicating a migration of lime down the profile. This was corroborated by free lime measurements, although most of the applied lime still remained in the surface layers 1·5 years after treatment.

It appears that a pH of 5·0 and base saturation greater than 90% produces water of "target quality" and may represent "target soil composition". From the 1986 and 1987 soil sampling data, treatment effectiveness seems unlikely to be limited in the short term by a depletion of free lime or an insufficient rate of dissolution. Where lime migrates down the soil profile below the region of hydrochemical reaction (>5 cm), it may become unavailable for buffering acidity and this may prove to be the factor limiting treatment effectiveness.

A preliminary prediction of treatment longevity using the 1986 and 1987 soil sampling data shows that water quality will be maintained until approximately the year 2020 on sector IV. This is longer than, but not inconsistent with, predictions derived from streamwater chemistry data and calcium flux calculations (see Chapter 16). Stream chemistry for sites in mid-Wales which were limed for agricultural purposes up to 40 years ago also show that treatment still has significant effects on pH and calcium concentration of runoff water. Since treatment effectiveness will depend on initial soil conditions, soil and water chemistry data from a range of sites are required before optimum liming rates and longevity of water improvement can confidently be predicted for other sites.

REFERENCES

Adams, W.A., Ali, A.W. and Lewis, P.J. (1990). Release of cationic aluminium from acidic soils into drainage water and relationships with land use. *J. Soil Sci.* **41**: 255–68.

Adams, W.A. and Evans, G.M. (1989). Effects of lime applications to parts of an upland catchment on soil properties and the chemistry of drainage waters. *J. Soil Sci.* **40**: 585–97.

Bache, B.W. (1979). Base saturation, pp. 38–42 in *The Encyclopedia of Soil Science* (eds. R.W. Fairbridge and C.W. Finkel). Dourden, Hutchinson & Ross, Stroudsburg, Pa, USA, 646 pp.

Bird, S.C., Brown, S.J. and Vaughan, E. (1990). The influence of land management on stream water chemistry, pp. 241–53 in *Acid Waters in Wales* (eds. R.W. Edwards, A.S. Gee and J.H. Stoner). Kluwer, Dordrecht, Netherlands, 328 pp.

Birse, E.L. and Dry, F.T. (1971). *Assessment of Climatic Conditions in Scotland. 3. The Bioclimate Sub Regions*. Macaulay Institute for Soil Research, Aberdeen.

Blom, G. and Wincent, H. (1990). *Effects of Liming on Meadow Vegetation*. SNV PM 3605, 123 pp (and abstract in *Acidification Research*, INFORMS pp. 2–3, June 1990).

Cronan, C.S. (1984). Vegetation and soil chemistry of the ILWAS watersheds, pp. 3-1 to 3-31 in *The Integrated Lake – Watershed Acidification Study* 4 (ed. R.A. Goldstein). Electric Power Research Institute, Palo Alto, Ca, USA, 222 pp.

Derome, J., Kukkola, M. and Malkonen, E. (1986). Forest Liming on Mineral Soils: Results of Finnish Experiments, Report 3084. National Swedish Environmental Protection Board (SVN), 107 pp.

Environmental Advisory Unit (EAU) (1986). The Loch Fleet Study Land Intervention Treatment. Report No. 6, University of Liverpool.

Gupta, R.K., Singh, C.P. and Abrol, I.P. (1985). Dissolution of gypsum in alkali soils. *Soil Sci.* **140**: 382–86.

Hallbacken, L. and Popovic, B. (1985). Effects of Forest Liming on Soil Chemistry. Review of Earlier Swedish Liming Experiments. SNV PM 1880 (and abstract in Acidification Research INFORMS, pp. 5–7, May 1985).

Hornung, M., Stevens, P.A. and Reynolds, B. (1986). The impact of pasture improvement on the soil solution chemistry of some stagnopodzols in mid-Wales. *Soil Use Manag.* **2**(1): 18–26.

Hornung, M., Reynolds, B., Stevens, P.A. and Hughes, S. (1990a). Water quality changes from input to stream, pp. 223–40 in *Acid Waters in Wales* (eds. R.W. Edwards, A.S. Gee and J.H. Stoner). Kluwer, Dordrecht, Netherlands, 335 pp.

Hornung, M., Brown, S.J. and Ranson, A. (1990b). Amelioration of surface water acidity by catchment management, pp. 311–28 in *Acid Waters in Wales* (eds. R.W. Edwards, A.S. Gee and J.H. Stoner). Kluwer, Dordrecht, Netherlands, 335 pp.

Hudson, G., Robertson, J.S., Campbell, C.G.B. and Henderson, D.J. (1986). *Soils and Vegetation of Loch Fleet Catchment – Baseline Studies*. Macaulay Institute for Soil Research, Aberdeen. 137 pp.

MAFF (1986). *The Analysis of Agricultural Materials* (A manual of the analytical methods used by the Agricultural Development and Advisory Service), 3rd edn. Ref. Book 427. HMSO, London, 248 pp.

Miles, J. (1985). The pedogenic effects of different species and vegetation types and the implications of succession. *J. Soil Sci.* **36**: 571–87.

Munsell Color Company, Inc. (1971). *Munsell Soil Color Charts*. Munsell Color Company, Inc., Baltimore.

Page, A.L., Miller, R.H. and Keeney, D.R. (editors) (1982). *Methods of Soil Analysis*. Part 2: *Chemical and Biological Properties*. American Society of Agronomy Inc., Soil Science Society of America Inc. Publisher, Madison, Wisconsin, USA.

Parker, F.W. (1929). The determination of exchangeable hydrogen in soils. *J. Amer. Soc. Agronomy* **21**: 1030–39.
Pella, E. and Colombo, B. (1973). Study of carbon, hydrogen and nitrogen determination by combustion-gas chromatography. *Mikrochimica Acta* **5**: 679–719.
Popovic, B. and Andersson, F. (1984). Markkalkning och skogsproduction (English translation: Liming and Forest Production). SNV PM 1792, 107 pp.
Reuss, J.D. and Walthall, P.M. (1989). Soil reaction and acidic deposition, pp. 1–33 in *Acidic Precipitation* (eds. S.A. Norton, S.E. Lindberg and A.L. Page). Springer-Verlag, New York, Berlin, Heidelberg.
Reynolds, B. and Emmett, B.A. (1990). Annual report on the soil and soil water chemistry of selected sites at Llyn Brianne. DOE Contract PECD 7/12/049.
Scott, R.O., Mitchell, R.L., Purves, D. and Voss, R.C. (1971). Spectrochemical methods for the analysis of soils, plants and other agricultural materials. *Consultative Committee for Development of Spectrochemical Work*. Bulletin 2.
Skeffington, R.A. (1983). Soil properties under three species of tree in southern England in relation to acid deposition in throughfall, in *Effects of Accumulation of Air Pollution in Forest Ecosystems* (eds. B. Ulrich and J. Pankrath). Reidel, Dordrecht, Boston, London, pp. 219–231.
Soil Survey of Scotland (1984).*Organization and Methods: Soil and Land Capability for Agriculture 1 : 250 000 Survey*. The Macaulay Institute for Soil Research, Aberdeen.
Trüby, P. and Aldinger, E. (1984). Eine Methode zur schnellen Bestimmung der effectiv austuschbaren Kationen, in *Düngung von Waldstandorten – eine Möglichkeit zur Milderung der Waldschäden*? (ed. P. Trüby). *Allg. Forstz*, **39**: 1302–1304.
USDA (United States Dept. of Agriculture) (1951). *Soil Survey Manual*. Handbook of US Dept Agriculture No. 18. Govt. Printing Office, Washington DC, 503 pp.
Welsh Water Authority (1987). Llyn Brianne Acid Waters Project. An Investigation into the Effects of Afforestation and Land Management on Stream Acidity. First Technical Summary Report, 91 pp.
Wilson, E.J. (1989). The Effect of Lime Application on Soils at Loch Fleet – 1987 sampling. CERL Report No. ESTD/L/0010/R89, 18 pp.
Wilson, E.J., Roberts, D.J. and Skeffington, R.A. (1988). The effect of lime application to the Loch Fleet catchment on tree and soil chemical properties – an initial report. CERL Report No. TPRD/L/3283/R88, 25 pp.
Wilson, M.J., Lilly, A., Nolan, A.J. and Cresser, M.S. (1988). Vulnerable soils and their distribution, pp. 59–70 in *Acidification in Scotland 1988*, Symposium Proceedings. Scottish Development Dept, Edinburgh, 128 pp.

CHAPTER 12

Water Quality of Surface Waters Before and After Liming

T.R.K. Dalziel
PowerGen plc, Nottingham, UK

M.V. Proctor
National Power plc, Leatherhead, Surrey, UK

K. Paterson
ScottishPower plc, Technology Centre, Glasgow, UK

12.1 INTRODUCTION

A target water quality for fish and other aquatic fauna of a liming project is discussed in Chapter 3. In brief, the pH and calcium concentration need to be raised and the toxic inorganic fraction of total aluminium reduced. As mentioned in Chapter 3, the specific target water quality of a liming project will be dependent on a number of factors. At Loch Fleet, since the objective was to improve the water quality so that brown trout survival and reproduction was possible, the target water quality was set at pH greater than 6, calcium concentration greater than $100\,\mu eq\,litre^{-1}$, and monomeric inorganic labile aluminium less than $30\,\mu g\,litre^{-1}$. A regular programme of sampling of surface waters started early in the project to assess the extent to which the water quality would need to be improved and, since liming, to establish the water quality changes that followed. Water quality data for pH, calcium, and aluminium are obviously crucial to judging the success of an acid mitigation project. Sampling and analysis of surface waters, of course, can extend further, and in many instances a more comprehensive programme is implemented to accompany a programme of analysis for deposition inputs (see Chapter 5). This allows hydrochemical input/output budgets or flux calculations to be made. These can often help to understand acidification processes in a catchment and provide a way of predicting the duration of treatment effectiveness (see Chapter 16).

At Loch Fleet, samples of stream and loch water before and after catchment liming were subjected to extensive chemical analyses. This chapter outlines the sampling and analysis programme and describes the development and calculation of simple hydrochemical budgets. The results of studies on groundwater chemistry and sediment chemistry are also presented. Comparisons are made with other regional, national, and international data.

12.2 SAMPLING AND ANALYSIS

Throughout the project daily manual sampling (Monday to Friday) of surface waters has been carried out at a number of scheduled sampling sites within the catchment (Plate II). The locations of these are shown in Fig. 12.1. Additional sampling points were established within the loch (inside and outside embayments, for example) and downstream of the loch outlet. Samples were transferred to the SSEB (now ScottishPower) chemistry laboratory at East Kilbride, near Glasgow, on a weekly basis for analysis. Major ions were analysed throughout, while minor element analyses, including trace metals, continued until August 1987, when it was judged that no further significant changes following liming would be seen. The methods of analysis for surface water samples are summarised in Table 12.1.

In addition to measuring total aluminium, aluminium fractionation was also carried out based on the ion exchange fractionation procedure developed by Driscoll (1984) (modified by Lynam and Sadler, 1987) to determine three aluminium fractions;

- total reactive aluminium following acid digestion;
- total monomeric aluminium; and
- non-labile monomeric aluminium (not retained by ion exchange).

Total reactive aluminium was measured by the pyrocatechol violet colorimetric method (Dougan and Wilson, 1974) following digestion with 0.1 M

PLATE II Water sampling at the loch outlet showing the trapezoidal flume used to measure stream water flow.

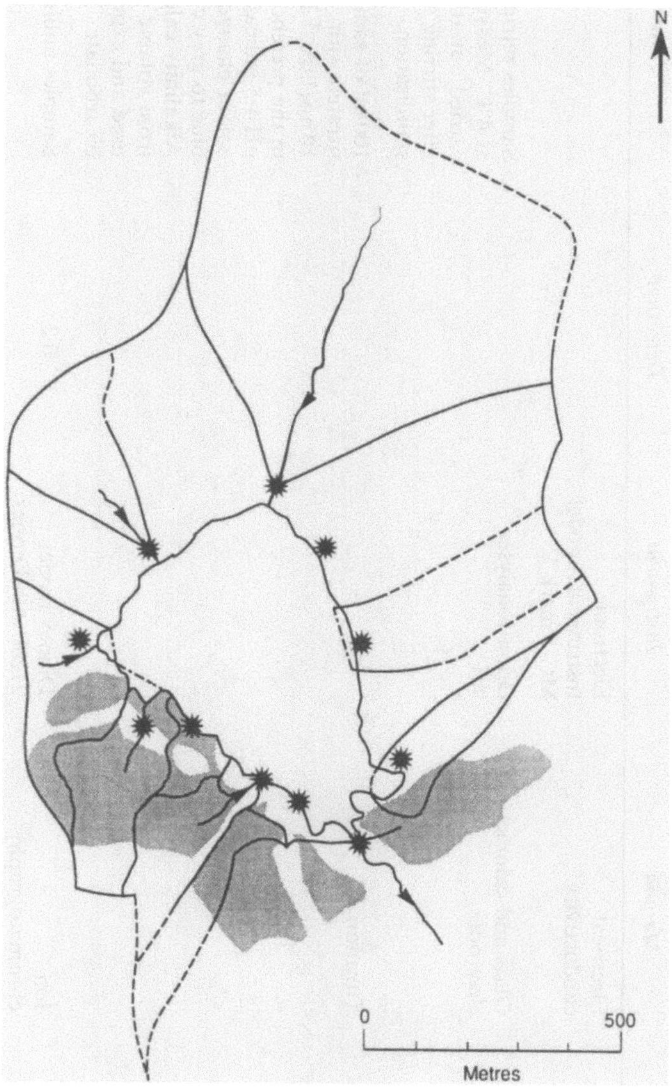

FIG. 12.1 Loch Fleet and its catchment surface water sampling sites.

TABLE 12.1
Details of analytical methods used to measure major ions and trace elements in surface water samples

Determinand	*Method*	*Instrument*	*Detection*[a]	*Notes*
Conductivity	Electrical conductivity	Electronic Instruments model MC-1 mark V		
pH	Glass and calomel electrodes	Orion Ionalyser 901		Samples refrigerated at 4°C. Measurement carried out at 20°C after stirring and stabilisation.
Alkalinity	Titration	–		100 ml of sample titrated with 10 mM litre^{-1} H_2SO_4 in the presence of pH 4·5 indicator until colour change from blue to grey complete. Alkalinity calculated from volume of acid used and expressed as bicarbonate.
Chloride, sulphate, nitrate, sodium, ammonium, potassium	Ion chromatography	Dionex model 2120i with Dionex columns	0·2	Samples unfiltered

Sulphur	Inductively coupled plasma emission spectrophotometry	Instrumentation Laboratory Plasma Emission Spectrophotometer	0·032	Samples unfiltered
Phosphorus	Inductively coupled plasma emission spectrophotometry	Instrumentation Laboratory Plasma Emission Spectrophotometer	0·0078	Samples unfiltered
Zinc	Inductively coupled plasma emission spectrophotometry	Instrumentation Laboratory Plasma Emission Spectrophotometer	0·006	Samples unfiltered
Lead	Inductively coupled plasma emission spectrophotometry	Instrumentation Laboratory Plasma Emission Spectrophotometer	0·072	Samples unfiltered
Nickel	Inductively coupled plasma emission spectrophotometry	Instrumentation Laboratory Plasma Emission Spectrophotometer	0·072	Samples unfiltered
Silica	Inductively coupled plasma emission spectrophotometry	Instrumentation Laboratory Plasma Emission Spectrophotometer	0·086	Samples unfiltered

(Continued)

TABLE 12.1 – *Contd.*

Determinand	*Method*	*Instrument*	*Detection*[a]	*Notes*
Manganese	Inductively coupled plasma emission spectrophotometry	Instrumentation Laboratory Plasma Emission Spectrophotometer	0·007	Samples unfiltered
Magnesium	Inductively coupled plasma emission spectrophotometry	Instrumentation Laboratory Plasma Emission Spectrophotometer	0·009	Samples unfiltered
Calcium	Inductively coupled plasma emission spectrophotometry	Instrumentation Laboratory Plasma Emission Spectrophotometer	0·086	Samples unfiltered
Copper	Inductively coupled plasma emission spectrophotometry	Instrumentation Laboratory Plasma Emission Spectrophotometer	0·031	Samples unfiltered
Aluminium	Inductively coupled plasma emission spectrophotometry	Instrumentation Laboratory Plasma Emission Spectrophotometer	0·037	Samples unfiltered
Sodium	Inductively coupled plasma emission spectrophotometry	Instrumentation Laboratory Plasma Emission Spectrophotometer	0·155	Samples unfiltered

Potassium	Inductively coupled plasma emission spectrophotometry	Instrumentation Laboratory Plasma Emission Spectrophotometer	0·126	Samples unfiltered
Iron	Inductively coupled plasma emission spectrophotometry	Instrumentation Laboratory Plasma Emission Spectrophotometer	0·014	Sample acidified to 1% solution with nitric acid and heated to 85°C for a minimum of 1 h.
Fluoride	Ion selective electrode	Orion 901 I analyser and Orion 94 09 electrode	0·002	Sample unfiltered.
Total organic carbon	Flame ionisation detection	Phase Separation Tocsin II Aqueous Carbon Analyser	1·5	Sample acidified to remove inorganic carbon. Allowed to mix with air and heated to 920°C for complete oxidation to CO_2. CO_2 mixed with H_2 and mixture reacted over Ni catalyst. CO_2 reduced quantitatively to CH_4 and then dried and passed through silica gel tube and measured.

[a]Detection limit is defined as the concentration for which there is a small probability that the analysis result would be excluded as insignificant if the sample contained slightly above this value. The detection limit is therefore taken as 3 times standard deviation of n blank samples. For the above table $n = 10$. (Units = $mg\,l^{-1}$.)

hydrochloric acid for at least 24 h. Absorbance was measured at 580 nm using a battery powered Corning 252 Colorimeter on site. For total monomeric aluminium determination the samples were not acidified. Phenanthroline and catechol violet solutions were added to the sample followed by the simultaneous addition of hexamine buffer and hydrochloric acid. Absorbance was measured after 4 min. The labile and non-labile aluminium fractions were separated by passing the sample through a cation exchange resin (Amberlite IR 120 regenerated in the sodium form). The resin-treated sample was analysed by the method described for the analysis of total monomeric aluminium (Table 12.1).

12.3 WATER CHEMISTRY BEFORE AND AFTER LIMING

Sectors IV, VI, and VII were limed in April 1986 (see Chapter 9). Drainage from each sector showed almost immediate and marked changes in chemistry following the lime applications. Five further sectors received liming or other treatments in 1987. Some of these did not prove as successful in improving water quality as the 1986 treatments. The water quality changes resulting from the different treatments are described below.

12.3.1 Calcium, pH, and Aluminium Concentrations on Sectors IV, VI, and VII Treated in 1986

Drainage from sector IV before liming had a calcium concentration of about 100 μeq litre^{-1} and a pH that fluctuated considerably, depending on flow conditions. When flows were greater than 1 litre s^{-1} (e.g. February 1986) the pH was around 4, but during prolonged dry periods it rose to around 6. The calcium concentration also rose during dry periods and total aluminium fell from values between 150 and 200 μg litre^{-1} to around 50 μg litre^{-1}. Following liming, the pH rose quickly to around 7·5, the calcium concentration to greater than 30 mg l^{-1} (1500 μeq litre^{-1}), and labile monomeric aluminium fell to around 10 μg litre^{-1}. Since liming in April 1986, the calcium concentration has declined somewhat, but there is still sufficient alkalinity present to maintain the pH at around 7 and the labile monomeric aluminium at non-toxic levels (see Fig. 12.2).

Before liming, sector VI had drainage water with a pH of about 4·2 and a calcium concentration around 1 mg l^{-1} (50 μeq litre^{-1}). The total aluminium concentration exceeded 200 μg litre^{-1}, about half in the labile form. Following liming, the pH rose to almost 8 along with a calcium

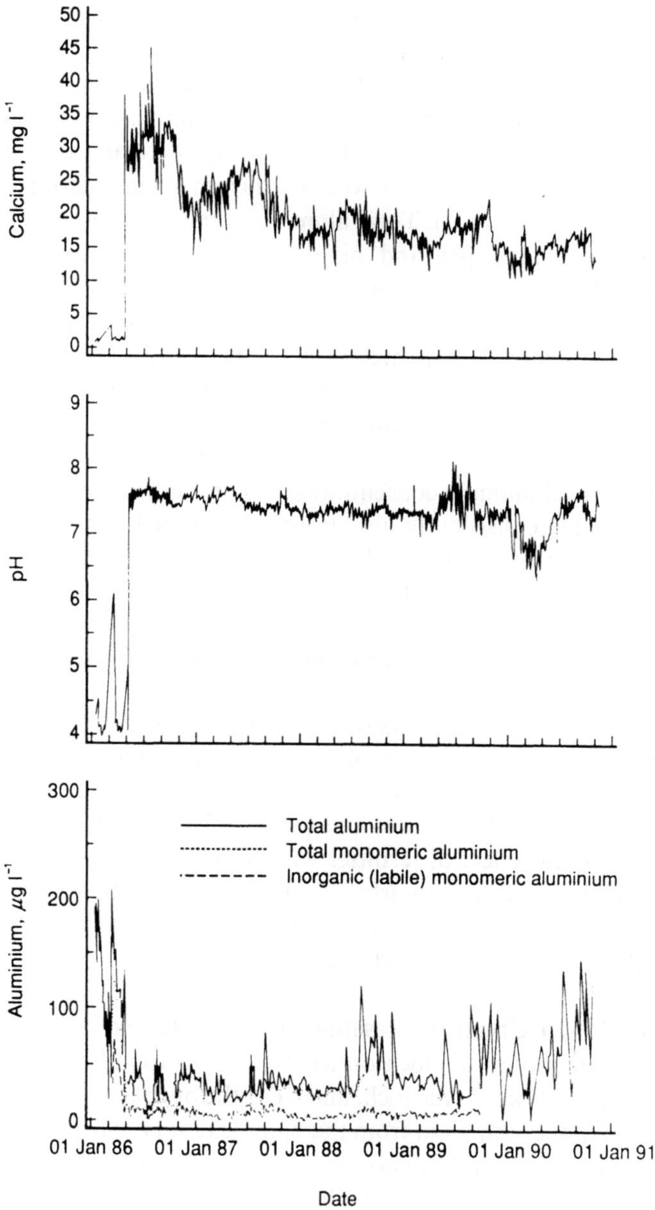

FIG. 12.2 Time series changes in stream water quality, sector IV, January 1986 to December 1990.

concentration increase to almost 30 mgl^{-1} (1500 μeq $litre^{-1}$). Labile monomeric aluminium fell from 100 $\mu g\,litre^{-1}$ to around 10 $\mu g\,litre^{-1}$. The calcium concentration has dropped since then to a value (in 1990) of around 10 mgl^{-1} (500 μeq $litre^{-1}$) (see Fig. 12.3) but labile aluminium remains low.

A somewhat different pattern of response to liming was recorded for sector VII. Before liming, the pH of the main inlet stream, the Altiwhat, fluctuated between 4·1 and 5·0 with a calcium concentration of 1·0–1·4 mgl^{-1} (50–70 μeq $litre^{-1}$). Total aluminium ranged from 120 $\mu g\,litre^{-1}$ to 290 $\mu g\,litre^{-1}$. Low pH and high aluminium concentrations were associated with high flows. Labile monomeric aluminium was generally around 100 $\mu g\,litre^{-1}$. Liming caused a rise in pH to above 7 and a calcium concentration greater than 14 mgl^{-1} (700 μeq $litre^{-1}$). Labile monomeric aluminium fell to levels less than 20 $\mu g\,litre^{-1}$. Post-liming calcium concentration, in particular, was influenced by flow conditions with low flows reducing in concentration. There was a tendency under these conditions also for the total aluminium concentration to rise, but the labile monomeric fraction remained consistently low. Four years after liming, the calcium concentration has dropped considerably but the pH and aluminium concentrations are still adequate to support fish (see Fig. 12.4).

12.3.2 Calcium, pH and Aluminium Concentrations on Sectors Treated in 1987

Table 12.2 summarises the changes in water quality recorded on the sectors treated in 1987. A variety of treatments were applied to sectors of the catchment in 1987. In some cases (sectors Z_1 and IX) these explored responses to lower applications of lime (5 and 7·5 t ha^{-1}), in others (sectors I and II) to test the effect of fertiliser with or without lime, and in sector VIII, to test the effect of "muirburn". With the exception of the pellet application of lime on sector IX, effects were insufficient to meet the target water quality for the runoff, and were transient.

12.3.3 Loch Water Chemistry Resulting from Catchment Treatments

The water chemistry of the loch outlet stream was found to be representative of the main body of the loch, and since the outlet was sampled more frequently than the loch, data from this sample point are used to show the changes in chemistry that have occurred in the loch since liming. As would be expected for the loch outlet which represents an integrated sample of catchment drainage, pre- and post-liming chemistry has been less variable than sectoral drainage, and the response to catchment treatments less immediate. The pH rose from 4·5 to 6·3 over a period of about 2 months

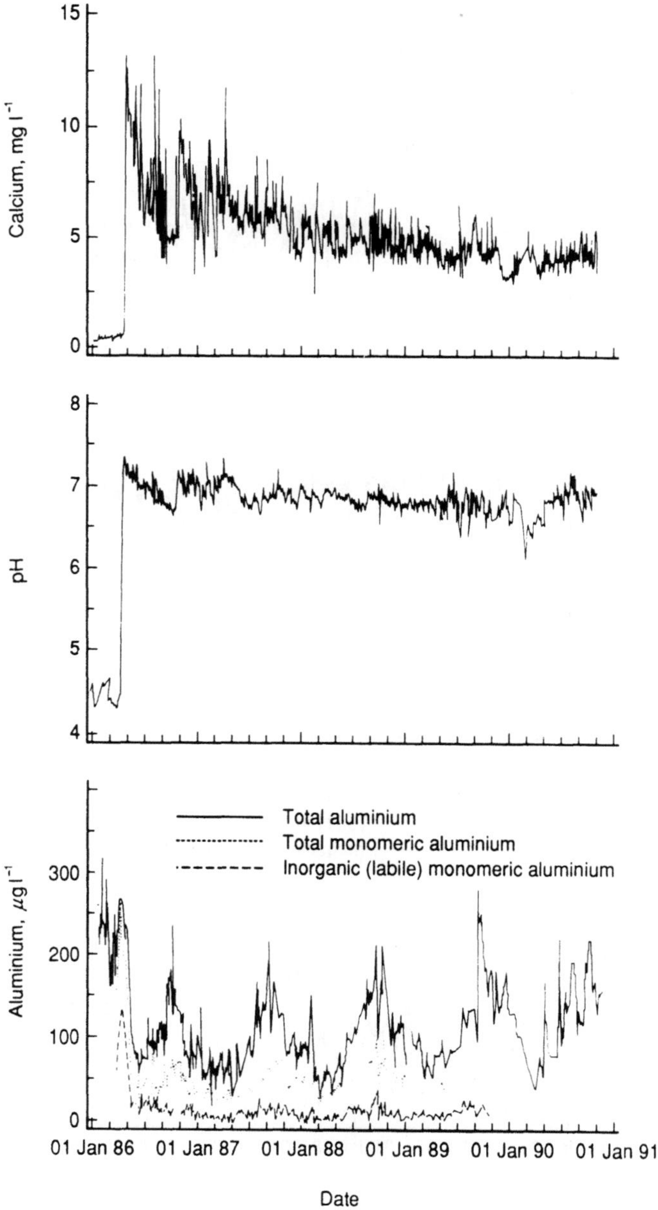

FIG. 12.3 Time series changes in stream water quality, sector VI, January 1986 to December 1990.

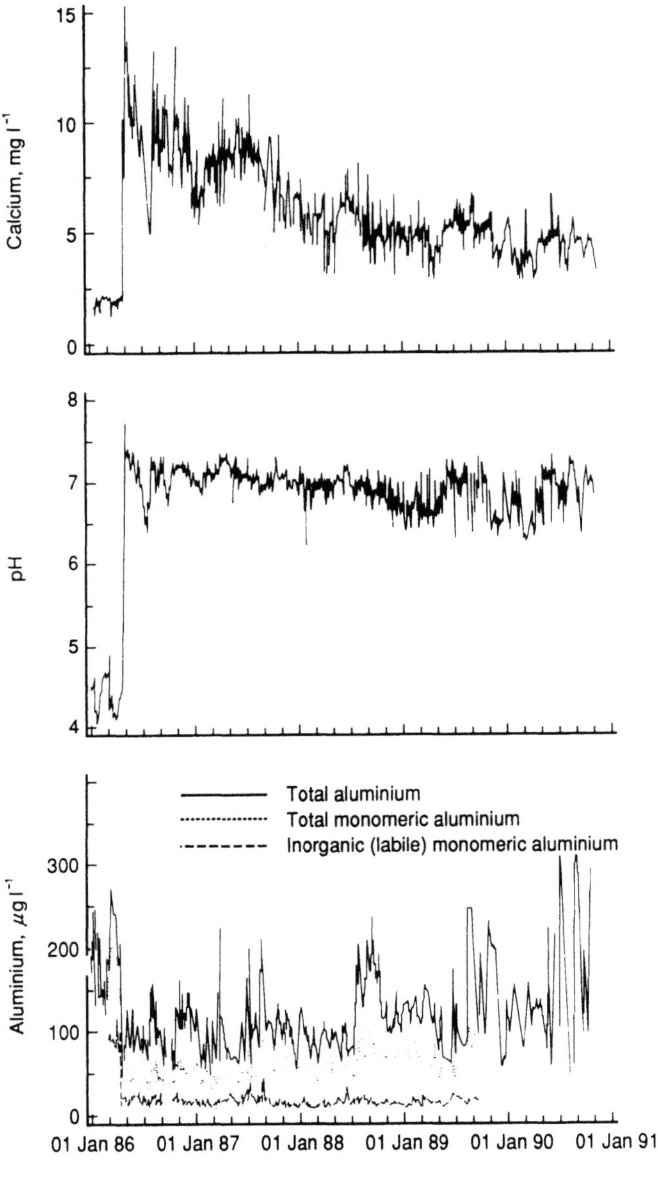

FIG. 12.4 Time series changes in Altiwhat water quality, sector VII, January 1986 to December 1990.

TABLE 12.2
Summary of water quality changes on sectors that received liming or other treatments in 1987

Sector	*Treatment*	*Pre-treatment water quality*	*Post-treatment water quality*
I	Fertiliser (1 t ha^{-1}) under the forest canopy	pH 4–4·2. Ca less than 1 mg litre^{-1}. Lab. mono. Al as high as 100 μg litre^{-1}.	Rise in pH to above 5·4 but only sustained for a month. No change in Ca or Al.
II	Limestone (5 t ha^{-1}) and fertiliser (1 t ha^{-1}) under the forest canopy	Around pH 4. Ca less than 50 μeq litre^{-1}. Lab. mono. Al as high as 100 μg litre^{-1}.	Rise to pH 4·5 following liming with some additional increase to as high as pH 5 after applying fertiliser. Ca rose to between 100 and 200 μeq litre^{-1}. Al increased after fertiliser applied.
Z_1	Limestone (5 t ha^{-1}) uniformly on moorland	pH 4. Ca around 50 μeq litre^{-1}. Lab. mono. Al as high as 200 μg litre^{-1}.	pH rose to 7 but then declined to pH 6 within 6 months. Ca peaked at 600 μeq litre^{-1} but within 4 months had declined to between 50 and 200 μeq litre^{-1}. Lab. mono. Al dropped to 10–20 μg litre^{-1}. Overall, only a brief improvement in water quality.
VIII	Muirburn (burning of heather moorland)	pH around 4·1. Ca around 50 μeq litre^{-1}. Total Al as high as 300 μg litre^{-1}.	No change in pH or Ca. Al increased in the summer following burning, but this was not attributed to the treatment.
IX	Limestone pellets (7·5 t ha^{-1})	pH around 4. Ca less than 50 μeq litre^{-1}. Total Al greater than 100 μg litre^{-1}.	Rapid rise to pH 6·5. Rise in Ca to fluctuate between 100 and 400 μeq litre^{-1}. Rise in total Al, in the summer after treatment, consistent with rest of catchment. No data on Al fractionation.

following the 1986 treatments and the calcium concentration increased from 1–3 mgl^{-1} (50 μeq $litre^{-1}$ to 150 μeq $litre^{-1}$) over the same time. The decrease in total aluminium was less dramatic, with a decline in total concentrations from just below 200 μg $litre^{-1}$ to values typically between 100 and 150 μg $litre^{-1}$. However, the labile monomeric fraction dropped from about 50–60 μg $litre^{-1}$ to around 10 μg $litre^{-1}$. Improved water quality has been maintained since limestone was first applied to parts of the catchment in April 1986 (see Fig. 12.5). The additional treatments in 1987, comprising only a further 93 t of lime, appeared to have little effect on the loch chemistry.

12.3.4 Changes in Other Major Ions and Trace Elements Following Liming

The emphasis so far has been on changes in pH, calcium, and aluminium concentrations elicited by the various treatments. Analysis results for other major ions show that there was little change in the concentrations of these since liming. In general, a good ion balance for major elements was observed, although no account was taken of the contribution of organic ions or trace elements to the ion balance. Sea-water-derived salts make a significant contribution to the total input (see Chapters 5 and 8) and this is reflected in stream water samples. Potassium, ammonium, and nitrate concentrations are low, reflecting the nutrient-poor status of the site. Minor element concentrations, including trace metals, at principal sampling sites are shown in Table 12.3. There was a significant decline of trace metals in drainage from limed sectors – all concentrations declined significantly following liming ($p < 0{\cdot}001$, Mann–Whitney *U*-test; Sokal and Rohlf, 1981); silica and total organic carbon (TOC) concentrations were unaffected by liming. In sufficient concentration, silicon, present as $Si(OH)_4$, is known to be an effective complexan for toxic monomeric aluminium (molar ratio Al : Si = 1 : 13; Birchall *et al.*, 1989). This ratio was never reached before or after liming at Loch Fleet and thus aluminium toxicity there could not be ameliorated by silicon. Similarly, citrate and fluoride complexes detoxify aluminium (Baker and Schofield, 1980), but the effective concentrations to do this in the field were not observed. Fluoride was undetectable prior to liming and present at only very low concentrations after. Dissolved organic materials, however, are also well demonstrated as detoxifying in natural waters in Nova Scotia (Lacroix and Townsend, 1987), where concentrations of about 10 mgl^{-1} are associated with very low concentrations of monomeric aluminium species, even after liming (Lacroix, 1991).

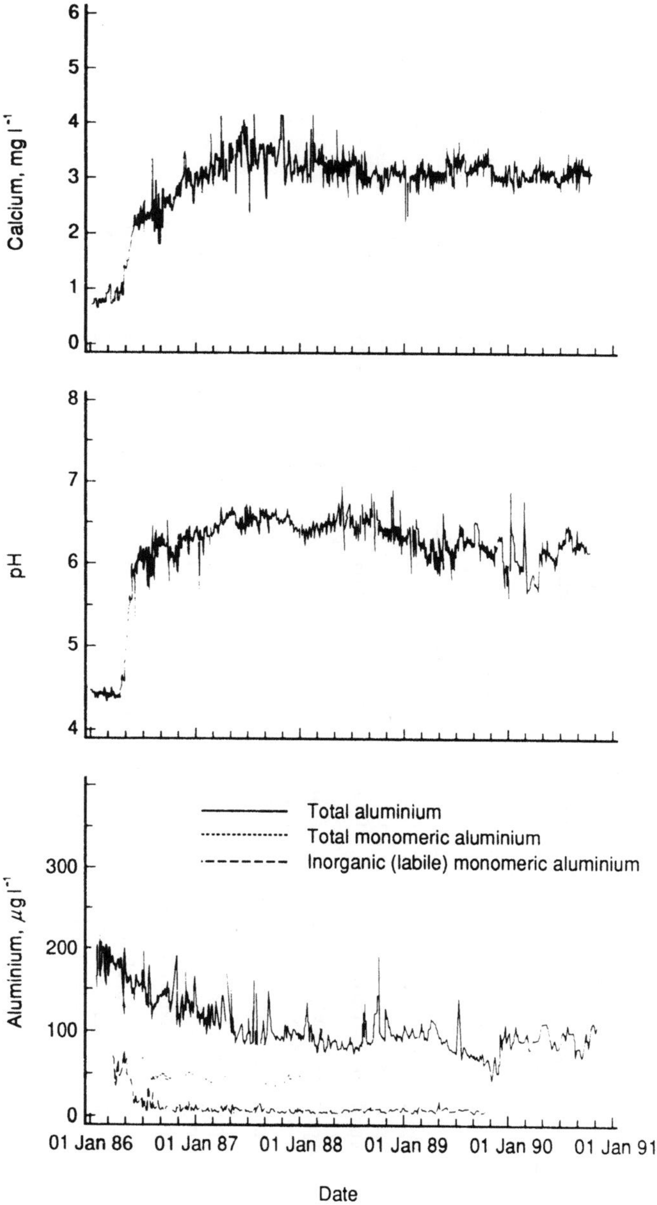

FIG. 12.5 Time series changes in water quality at the loch outlet, January 1986 to December 1990.

TABLE 12.3

Minor element concentrations at principal sampling sites, before and after liming (concentrations mg litre^{-1}, annual means)

Site	Sample period	Minor element concentrations							
		Fe	*Cu*	*Zn*	*Mn*	Al_{tot}	SiO_2	*TOC*	*F*
Inflow stream	April 1985–March 1986	0·09	0·01	0·05	0·04	0·26	2·33	6·3	0
	April 1986–March 1987	0·05	0	0·01	0	0·07	2·56	5·3	0·03
	April 1987–March 1988	0·06	0	0·01	0	0·08	2·09	5·5	0·03
Stream 4	April 1985–March 1986	0·48	0·01	0·08	0·11	0·23	4·01	10·3	0
	April 1986–March 1987	0·14	0	0·01	0·04	0·04	6·10	7·7	0·04
	April 1987–March 1988	0·19	0	0·01	0·02	0·02	5·13	7·9	0·03
Stream 6	April 1985–March 1986	0·20	0·01	0·04	0·05	0·25	2·44	7·1	0
	April 1986–March 1987	0·11	0	0·01	0·01	0·07	5·52	8·1	0·02
	April 1987–March 1988	0·14	0	0·02	0	0·06	3·77	7·8	0·02
Outlet	April 1985–March 1986	0·15	0·01	0·01	0·17	0·20	0·76	6·1	0
	April 1986–March 1987	0·10	0	0	0·04	0·12	1·27	3·7	0·02
	April 1987–March 1988	0·08	0	0·03	0	0·08	1·13	4·8	0·08

12.3.5 Acid Pulses in Stream Water Coincident with High Sea Salt Inputs

The dominance of sea salt components in deposition inputs is documented in Chapters 5 and 8. During 5 years of observation sodium and chloride input in deposition has averaged 2·62 and 3·07 keq ha^{-1}, respectively, with significantly higher values in 1989–90. Soil ion exchange reactions (see Chapter 10) can occur and result in a proportion of sea salt cation input (principally sodium) exchanging for H^+ ions on soil exchange sites, with the result that some of the sodium is retained in the soil. The associated anion (chloride) is not retained and passes into drainage or stream water with the electrical balance being provided partly by H^+ ions. The acid pulses that result can be potentially very harmful to aquatic biota. The phenomenon of sea-salt-induced acid episodes in surface waters is documented for remote sites in Norway (Wright *et al.*, 1988), Wales (Edwards *et al.*, 1990), and Scotland (Langan, 1987, 1989). The contribution of marine ions to deposition is substantial in western Britain, especially in coastal areas, accounting for as much as 90% of the total ionic strength (Irwin *et al.*, 1990). This has been associated with the frequency of westerly weather patterns, bringing in materials of Atlantic origin. Prior to liming at Loch Fleet, acid stream water pulses were also recorded coincident with high sea salt inputs.

One such episode occurred in January 1986 when 23 mm rain fell over two days and had very high sea salt concentrations. Rainfall on 14 January had 2138 μeq Cl $litre^{-1}$ and 1383 μeq Na $litre^{-1}$, and on 15 January 3220 μeq Cl $litre^{-1}$ and 2340 μeq Na $litre^{-1}$. (The rainfall pH was 5·51 on both days.) This coincided with the pH of the Altiwhat decreasing from pH 4·7 to 4·1. Other tributary streams to the loch also experienced similar pH depressions. Empirical evidence that cation exchange reactions occur comes from stream water concentrations of sodium relative to chloride. When the equivalent ratio of Na : Cl (or sometimes Mg : Cl) becomes less than that of sea water, then it is assumed that ion exchange has occurred. Amounts of non-marine (or excess) sodium and magnesium were calculated using the factors 0·857 and 0·199 for the ratios of sodium and magnesium, respectively, relative to chloride in sea water. Large reductions (to negative values) of excess sodium (Fig. 12.6) and magnesium are apparent over the time of the acid episode. Similar acid episodes attributable to ion exchange following high sea salt inputs were recorded at other times at Loch Fleet prior to liming. Since liming, despite several instances of high sea salt input, no stream water acid pulses have been recorded.

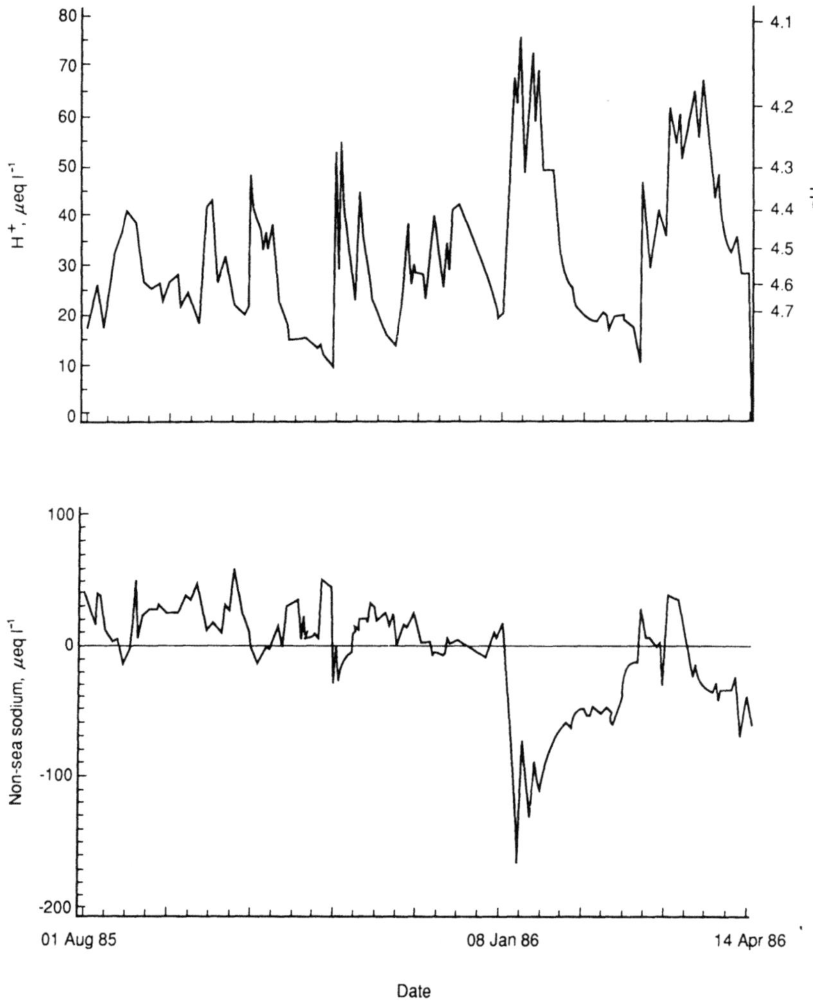

FIG. 12.6 Hydrogen ion and non-sea salt sodium concentrations in the Altiwhat during a season of high marine inputs (August 1985 to April 1986).

12.4 HYDROCHEMICAL BUDGET CALCULATIONS

Hydrochemical budget (or flux) calculations show whether a catchment is a net source or a net sink for various chemical constituents (Likens *et al.*, 1977). These calculations also help to clarify acidification processes.

Assessment can be made of the relative contributions to acidity by nitrogen and sulphur species in deposition and the ways in which acidification processes vary between areas with different land use (e.g. forest and moorland), soil types, rainfall, and other catchment characteristics. So far as the Loch Fleet project is concerned, the calcium budgets are the most important to assess the rate at which the applied limestone in liming treatments is being mobilised. These data are then used to make predictions of the duration of treatment effectiveness. This procedure is described in detail in Chapter 16. Here, a description is given of how the budgets of all major ions, including calcium, were calculated.

A water budget should satisfy the equation:

$$\text{input} = \text{output} \pm \Delta\text{storage}$$

or

$$\text{deposition} = \text{runoff (streamflow)} + \text{evapotranspiration} \pm \text{soil moisture (or lake) storage}$$

Sectoral streamflow contains components of both surface flow, i.e. that over and through the soil surface, and deeper sub-surface pathways, although it is possible that some sub-surface flow may not be included in streamflow measurements. On the basis of hydrological studies undertaken at Loch Fleet (see Chapter 7), however, it has been assumed that any sub-surface flow not accounted for in streamflow measurements was small enough to be ignored, although interpretations of isotope separations (see Chapter 8) do not support this.

A further assumption in the budget estimates was that sectoral soil moisture storage at the start of the budget period is essentially equal to that at the beginning of the period and thus that storage in the terrestrial component is negligible. Differences in lake water height, however, were used to account for Δstorage in the aquatic component over each year (for budgets involving the whole catchment), since this can fluctuate more rapidly than soil moisture.

Records from automatic streamflow monitoring (see Chapter 6) were found occasionally to be erroneous or interrupted, mainly because the recording apparatus malfunctioned, e.g. during frost or drought periods. To produce a complete continuous flow record the missing flow periods were infilled employing a cross-correlation relationship derived between the flows from adjacent sectors (200–300 m apart) near to the period of data infilling required for one of the sectors. Missing data were then synthesised or infilled from the flow records of other sites. In general, a very good correlation between flows from different sectors was found (correlation coefficient, $r^2 > 0{\cdot}9$ in most cases) and so a reliable estimate was provided for any missing flow records.

Following rainfall chemistry analyses, as described in Chapter 5, ion inputs were computed by multiplying rain ion concentrations in the samples from the bulk collectors by the corresponding rain volumes over the time of sample collection and totalling these over the year. (It is acknowledged that this procedure must underestimate total inputs, which would include those from dry deposition and mist/fog interception.) Ion outputs were computed in the same way, this time multiplying ion concentrations from stream water samples by the volume of water having flowed between the sample times (from streamflow measurements).

Table 12.4 shows the results of budget calculations performed for the year before any liming (mid-April 1985 to mid-April 1986) and for the subsequent four years. The dominance of sea salt components is obvious, in particular sodium and chloride. The outputs of these ions were always greater than the inputs. Without a significant catchment source, this indicates a contribution by other forms of deposition not fully accounted for in bulk rainfall chemical analysis (see Chapter 5 and 8). The budgets of these ions were similar for each sector despite the 86% forest cover on sector IV from where it might have been expected that ion outputs would be greater than from the other sectors due to atmospheric scavenging by the trees. The mean sulphate and nitrate influxes for the five years were 1·06 and 0·36 keq $ha^{-1} yr^{-1}$, respectively. Mean ammonium influx (adding to the nitrogen input, but not shown in Table 12.4) was 0·67 keq $ha^{-1} yr^{-1}$. Mean effluxes from the catchment for these ions were calculated as: sulphate, 1·66 keq $ha^{-1} yr^{-1}$, and nitrate, 0·37 keq $ha^{-1} yr^{-1}$; ammonium efflux was negligible. Thus, sulphate efflux considerably exceeds influx while nitrate and ammonium are undoubtedly utilised in biological activity, accounting for less nitrate and negligible ammonium efflux. In oligotrophic acidified waters sulphate is usually the major anionic species (excluding sea salts) and thus that linked with acidity, with other anions absent or negligible. In limed waters, either by direct limestone addition to the water bodies or by application to the catchment, sulphate concentrations are unaffected, but the limestone addition provides carbonate as a source of alkalinity, thus reducing the H^+ concentrations and providing a bicarbonate reserve of alkalinity. The effect of liming on calcium (and H^+) outputs from the sectors and the whole catchment is clear (Table 12.4). This will be dealt with in more detail in Chapter 16.

12.5 GROUNDWATER CHEMISTRY

Contract work was undertaken by the British Geological Survey in spring 1986 to investigate the groundwater contribution to the loch, in terms of both volume and chemistry. This involved the drilling of three shallow boreholes to provide samples of groundwater for analysis. The concentrations of a wide range of major and trace elements were determined using inductively coupled plasma-optical emission spectroscopy (ICP-OES) after 20-fold preconcentration by evaporation under infra-red lamps, carried out in laminar flow hoods to minimise contamination. Automated colorimetry was used to determine chloride, bromide, iodide, fluoride, and nitrate-nitrogen, and automated titrimetry to measure total alkalinity.

The principal findings of this study were that a linear geological feature along the Altiwhat in sector VII, passing through the centre of the loch, is believed to represent a zone of shattered, probably faulted, rock. Most groundwater flow in the catchment is associated with this feature and the borehole drilled to a depth of 10 m adjacent to it overflowed at 1×10^{-2} litre s^{-1}. The results of chemical analyses show that there was a net gain by borehole water compared with rain water in sodium, potassium, calcium, bicarbonate, sulphate, chloride, silicon, strontium, iron, manganese, lithium, and fluoride and net losses in nitrate, aluminium, zinc, and boron. These changes are in accordance with the known mineralogy and bulk rock chemistry of the granite and the secondary fracture mineralogy infills. The average groundwater alkalinity was found to be 2329 μeq litre^{-1}. The main conclusion of this work was that, although groundwater does not contribute significantly to the quantity, it does contribute significantly to the ionic composition of the loch water and has an important role in buffering the loch chemistry.

The favourable chemistry of the groundwater has suggested that its contribution ($< 5\%$ of inflow) was sufficient to buffer the loch at around pH 6·0 until the late 1960s (Cook *et al.*, 1991) and that a modest increase could restore conditions suitable for a self-sustaining fish population.

12.6 LOCH SEDIMENTS – THEIR ROLE AS A SOURCE OR SINK OF CALCIUM

Studies on this topic were undertaken by Northern Environmental Consultants. A series of grab samples were taken from fixed positions before

TABLE 12.4
Hydrochemical budgets for major ions, 1985–90 (units are: water, litres × 10^7 ha^{-1}; ions, keq ha^{-1})

1985–86	*In*	*Out*			
		IV	*VI*	*VII*	*Whole*
Water	2·34	1·84	2·09	1·98	1·89
Cl^-	2·78	3·92	3·56	3·45	3·42
SO_4^{2-}	1·15	1·81	1·80	1·86	1·84
NO_3^-	0·28	0·20	0·32	0·30	0·30
Na^+	2·23	3·42	3·23	3·17	3·18
Mg^{2+}	0·50	1·05	0·90	0·91	0·88
Ca^{2+}	0·18	1·42	0·75	0·99	0·86
H^+	0·41	0·91	0·98	0·67	0·55
1986–87	*In*	*Out*			
		IV	*VI*	*VII*	*Whole*
Water	2·41	1·67	1·88	2·08	1·88
Cl^-	2·83	4·31	3·82	4·16	3·93
SO_4^{2-}	1·08	2·40	2·22	2·00	1·75
NO_3^-	0·42	0·54	0·68	0·56	0·44
Na^+	2·47	3·86	3·49	3·83	3·43
Mg^{2+}	0·48	1·37	0·96	0·97	0·83
Ca^{2+}	0·27	20·20	15·77	8·84	2·67
H^+	0·50	0·003	0·002	0·007	0·04
1987–88	*In*	*Out*			
		IV	*VI*	*VII*	*Whole*
Water	2·35	1·64	1·82	1·77	1·93
Cl^-	2·54	3·92	2·92	2·73	2·77
SO_4^{2-}	0·90	1·64	1·42	1·31	1·59
NO_3^-	0·37	0·35	0·39	0·22	0·46
Na^+	2·10	3·15	2·59	2·56	2·68
Mg^{2+}	0·40	1·12	0·72	0·67	0·69
Ca^{2+}	0·24	16·47	11·07	5·74	3·34
H^+	0·46	0·001	0·001	0·002	0·005

(*Continued*)

TABLE 12.4 – *Contd.*

1988–89	*In*	*Out*			
		IV	*VI*	*VII*	*Whole*
Water	2·56	1·66	2·01	1·87	1·94
Cl^-	3·24	4·59	3·75	3·50	3·30
SO_4^{2-}	1·21	1·78	1·74	1·51	1·63
NO_3^-	0·43	0·23	0·48	0·30	0·40
Na^+	2·79	3·51	3·19	3·10	2·90
Mg^{2+}	0·52	1·09	0·81	0·74	0·68
Ca^{2+}	0·34	14·04	10·43	4·19	3·01
H^+	0·49	0·001	0·001	0·003	0·007

1989–90	*In*	*Out*			
		IV	*VI*	*VII*	*Whole*
Water	2·06	1·47	1·83	1·40	1·34
Cl^-	3·96	5·64	5·24	4·09	3·37
SO_4^{2-}	1·34	1·87	2·18	1·43	1·46
NO_3^-	0·32	0·19	0·36	0·23	0·27
Na^+	3·53	4·10	4·06	3·37	2·85
Mg^+	0·60	1·14	0·90	0·73	0·60
Ca^{2+}	0·30	11·36	7·89	2·84	2·07
H^+	0·29	0·001	0·002	0·003	0·010

parts of the catchment were limed (July 1985) to provide a baseline data set. Repeat sampling surveys were carried out following the first set of liming treatments in 1986 and after the second set in 1987, taking samples at the same sites. Immediately after collection, samples were transported back to the laboratory where they were air dried, ground, and sieved through a 200 μm mesh. A known amount of sieved sediment was shaken for 1 h with either 1, 5, and 10 mg l^{-1} calcium solutions ($CaCl_2$). These were then centrifuged and filtered through a 0·45 μm membrane filter and analysed by atomic absorption spectrophotometry for calcium.

The principal findings were that at sites in the north-east of the loch there was little significant change in the relatively high concentrations of total calcium in the sediments, following liming. At sites in the southern end of the loch there were either slight or significant increases in total calcium following liming, in particular after the second set of treatments. Concentrations of water-soluble calcium started to decrease 1 year after the first set of liming treatments and 2 months following the second set of treatments. In some areas, total and exchangeable calcium had significantly increased, but in the central part and southern end of the loch the changes, although quite significant initially, tended to plateau even after the second set of treatments. After the two sets of liming treatments, the majority of sediments still showed a tendency to remove (or take up) calcium from the surrounding water, especially at the higher concentrations of calcium in the experimental solutions (NEC, 1986, 1987).

12.7 COMPARISON WITH REGIONAL, NATIONAL, AND INTERNATIONAL DATA AND OTHER LIMED WATERS

12.7.1 Sweden

At the outlet of limed Lake Gardsjon (Hultberg and Nystrom, 1988), acidity was reduced to zero at liming, and maintained for about 3 years, after which some reacidification developed as the subsequent runoff progressively diluted the limed body of water. Sulphate concentrations over pre- and post-liming years were virtually constant (annual weighted means 64·5–70·7 μeq litre^{-1}) as were magnesium and potassium. Some significant increase in sodium and chloride concentrations were observed, suggesting an increased maritime component during the 6 years of monitoring (1979–1985). Calcium was less than 2 mgl^{-1} (100 μeq litre^{-1}) prior to liming, which then raised calcium concentrations to 10 mgl^{-1} (500 μeq litre^{-1}), reducing to about 5 mgl^{-1} (250 μeq litre^{-1}) by the end of 1985 (with 2·5 changes of lake volume). Alkalinity, however, changed little, i.e. from about 180 μeq litre^{-1} prior to liming to 200 μeq litre^{-1}.

12.7.2 Norway

Store Hovvatn in Norway was limed (directly to the lake shoreline) in 1981 and relimed (directly to the lake) in 1987 (Rosseland and Hindar, 1988); 200 t of powdered limestone ($CaCO_3$) was spread on the lake ice cover, and a smaller application was made to a headwater lake. Prior to liming, pH was 4·4, having declined from a pH of 5·2 over the previous 40 years; calcium was 0·4 mgl^{-1} (20 μeq litre^{-1}) and reactive aluminium 150–300 μg litre^{-1}. Liming in winter 1981 led to a sharp increase in pH (> 7·0) and calcium (30 mgl^{-1}, ~600 μeq litre^{-1}) at ice-out, although only a calculated 14% of the applied lime was dissolved in the first year. No increases in sediment calcium were reported. Further dissolution of the applied limestone was evident in the following years, with 50% utilised by

1984. The later liming (1987) directly to the lake water was better utilised, and pH was restored to 6·8. Aluminium concentrations in Hovvatn were reduced to 50 μg litre^{-1} in 1983, increasing with reacidification to 150 μg litre^{-1} in 1985–86, and to 175 μg litre^{-1} prior to reliming in 1987.

12.7.3 North America

A catchment and lake liming experiment is underway at Woods Lake in the Adirondack Mountains of New York State. Limestone was applied to the lake as a slurry, with high dissolution efficiency (86% in 4 weeks), producing an immediate increase in pH (from less than 5·0 to more than 9·0), and in calcium, alkalinity, and dissolved inorganic carbon (DIC). Furthermore, aluminium speciation was shifted to less toxic forms, and total aluminium and trace metals were reduced in concentration, depositing to lake sediments (Olem, 1990). In this small lake, retention time is less than 6 months, and reacidification was indicated after about 16 months.

In October 1989, limestone pellets were applied by helicopter to the forested slopes above the two main tributary streams flowing into the lake (D. Porcella, pers. comm.). Following this limestone application, rain showers helped to dissolve the pellets and similar stream water quality changes to those recorded at Loch Fleet were observed. pH rose from around 4·5 to greater than pH 7, along with a calcium concentration increase from about 2 mgl^{-1} (100 μeq litre^{-1}) to a peak of 20 mgl^{-1} (1 000 μeq litre^{-1}), before declining to around 10 mgl^{-1} (500 μeq litre^{-1}). Organic and inorganic aluminium concentrations also declined following catchment liming (C. Driscoll, pers. comm.).

12.7.4 United Kingdom

Moorland and forest streams in the Llyn Brianne have been the subject of a variety of treatments designed to improve their water quality (Welsh Water Authority, 1987, 1988, 1989; Edwards *et al.*, 1990). Moorland streams in the area have an average pH of 5·3, compared with forest streams of pH about 4·75. Sulphate levels in moorland streams are slightly higher than concentration in bulk deposition (93 μeq litre^{-1}), while forest streams reflect canopy scavenging to have about twice the level of bulk deposition ($\sim$ 160 μeq litre^{-1}). Aluminium concentrations parallel the characteristic pH levels, with moorland streams about 130 μg litre^{-1} and forest streams about 400 μg litre^{-1}.

Agricultural improvement techniques (ploughing, liming, reseeding), applying approximately 10 t ha^{-1} of magnesian limestone, fertiliser application, and reseeding, produced only slight upward trends in pH and water hardness in the drainage stream water. Acidic episodes with elevated aluminium still occur (Welsh Water Authority, 1988). On another catchment at Llyn Brianne, source area liming, similar to the treatment at Loch Fleet to the wetland at the head of the Altiwhat, was employed. About 25 t ha^{-1} of fine powdered limestone was applied to the unplanted catchment headwaters. Stream water pH and calcium concentration rose shortly after liming from around pH 5·0 to around pH 6·5 and from 2 mgl^{-1} (100 μeq litre^{-1}) to around 3 mgl^{-1} (300 μeq litre^{-1}), respectively. Filterable aluminium dropped from levels as high as 400 μg litre^{-1} to less than 100 μg litre^{-1}. Dry periods following liming produced declines in pH to around 5·5, although increases to pH 7 followed heavy rain. A further source area liming on another catchment involved the application of 15–20 t ha^{-1} powdered limestone (50% of source area treated) and caused the calcium concentration to increase from less than 1 mgl^{-1} (50 μeq litre^{-1}) to 6–10 mgl^{-1} (300–500 μeq litre^{-1}) and the pH to rise from 5·0 to 6·9.

Moorland liming at Llyn Brianne using about 10 t ha^{-1} fine powdered limestone caused a more sustained water quality improvement. The stream water pH rose from 5·2 to 6·4 after liming and this has been sustained, along with parallel improvements in calcium and aluminium.

At Loch Dee, where stream water quality is similar to that at Loch Fleet, applications of sea shells and limestone chips were made directly to one of the catchment streams and limestone powder was applied along the stream banks. These treatments are recorded to have had little effect on loch water chemistry. The treatments remained effective in increasing calcium concentrations and pH above pretreatment levels for only a maximum of 1·5 years (Tervet and Harriman, 1988).

Liming of stream-side land alongside the River Esk, Cumbria (3200 t to 640 ha) in September 1987 and June 1988 led to expected improvements in stream pH (from pH 6·0 to pH 6·6) and a reduction in total aluminium, but the measured calcium fell from 1·7 mgl^{-1} (85 μeq litre^{-1}) to 1·2 mgl^{-1} (60 μeq litre^{-1}), possibly because of the selection of high flow conditions for sampling. A small but consistent increase in soil pH (in 40 riparian fields) was reported (Diamond *et al.*, 1992).

12.8 CONCLUSIONS

Comparison of the record of water quality improvement at Loch Fleet can be made with direct lake liming and catchment liming elsewhere. It is evident that direct lake liming, where lake retention time is short, provides only short-term changes in pH, calcium, and aluminium levels before reacidification. Catchment liming, in contrast, has usually led to more sustained water quality improvement.

Catchment liming at Loch Fleet brought about almost immediate and sustained changes in water quality in streams draining treated sectors. The most obvious and important changes were in the greatly increased calcium concentrations and decreased acidity and aluminium concentrations. Importantly, too, trace metal concentrations decreased following liming from potentially toxic to tolerable levels. Liming less than 20% of the total catchment area in 1986 was sufficient to improve the loch water quality within about 2 months to a point suitable for fish survival. Satisfactory water quality has been maintained on these sectors and in the loch four and a half years following liming. The liming treatments made to other sectors in 1987 served to demonstrate that a limestone application rate greater than $5\,t\,ha^{-1}$ was necessary to elicit a satisfactory runoff water quality change and also that fertiliser application and muirburn were unable to affect runoff water quality significantly.

It is of interest that, as an alternative acid amelioration method, it is estimated that only a modest increase in the groundwater input to the loch would be sufficient to improve the loch water quality for fish survival.

REFERENCES

Baker, J.P. and Schofield, C.L. (1980). Aluminium toxicity to fish as related to acid precipitation and Adirondack surface water quality, pp. 292–93 in *Ecological Impact of Acid Precipitation* (eds. D. Drablos and A. Tollan). Proc. International Conference, Sandefjord, Norway, 383 pp.

Birchall, J.D., Exley, C., Chappell, J.S. and Phillips, M.J. (1989). Acute toxicity of aluminium to fish eliminated in silicon-rich acid waters. *Nature* **338**: 146–48.

Cook, J.M., Edmunds, W.M. and Robins, N.S. (1991). Groundwater contribution to an acid upland lake (Loch Fleet, Scotland) and the possibilities for amelioration. *J. Hydrol.*, **125**: 111–128.

Dalziel, T.R.K., Paterson, K. and Procter, M.V. (1991). Loch Fleet Project—trace metal concentrations before and after catchment liming; significance for fish survival. PowerGen report no. TR/91/23113/R.

Diamond, M., Hirst, D., Winder, L., Crawshaw, D.H. and Prigg, R.F. (1992). The effect of liming agricultural land on the water quality of the River Esk, Cumbria. Internat. Conference on Acidic Deposition: Its Nature and Impacts, Glasgow, UK. *Env. Poll.*, in press.

Dougan, W.K. and Wilson, A.L. (1974). The absorptiometric determination of aluminium in water. A comparison of some chromogenic reagents and the development of an improved method. *Analyst* **99**: 413–30.

Driscoll, C.T. (1984). Procedure for the fractionation of aqueous aluminium in dilute acidic waters. *Int. J. Env. Anal. Chem.* **16**: 267–83.

Edwards, R.W., Stoner, J.H. and Gee, A.S. (editors) (1990). *Acid Waters in Wales.* Kluwer, Dordrecht, Netherlands, 337 pp.

Hultberg, H. and Nystrom, U. (1988). The role of hydrology in treatment duration and reacidification of the limed Lake Gadsjon, pp. 95–134 in *Liming of Lake Gårdsjön* (ed. W. Dickson). National Swedish Environmental Protection Board, Solna, Sweden, 327 pp.

Irwin, J.G., Campbell, G.W., Cape, J.N., Clark, P.A., Davies, T.G., Derwent, R.G., Fisher, B.E.A., Kallend, A.S., Longhurst, J.W.S., Martin, A., Smith, F.B. and Warrilow, D.A. (1990). *Acid Deposition in the United Kingdom 1986–1988.* Dept. Environment, Warren Spring Laboratory, 124 pp.

Lacroix, G. (1992). Mitigation of low pH by substrate manipulation in a stream and its effects on salmonids. Internat. Conference Acidic Deposition: Its Nature and Impacts, Glasgow, UK. *Env. Poll.*, in press.

Lacroix, G. and Townsend, D.R. (1987). Responses of juvenile Atlantic salmon (*Salmo salar*) to episodic increases in acidity of Nova Scotia rivers. *Can. J. Fish Aquat. Sci.* **45**: 1475–84.

Langan, S.J. (1987). Episodic acidification of streams at Loch Dee, SW Scotland. *Trans. Roy. Soc. Edin., Earth Sci.* **78**: 393–97.

Langan, S.J. (1989). Sea-salt induced streamwater acidification. *Hydrol. Proc.* **3**: 25–41.

Likens, G.E., Bormann, F.H., Pierce, R.S., Easton, J.S. and Johnson, N.M. (1977). *Biogeochemistry of a Forested Ecosystem.* Springer-Verlag, New York, Heidelberg, Berlin, 146 pp.

Lynam, S. and Sadler, K. (1987). Evaluation of a Method for the Measurement of Aluminium in Freshwaters Permitting the Fractionation into Monomeric and Polymeric Forms and Organic Complexes. CERL Research Report No. TPRD/L/3082/R86, 11 pp. + annex.

NEC (Northern Environmental Consultants) (1986). Sediment and Water Interactions in Loch Fleet catchment. Report to Loch Fleet Project.

NEC (Northern Environmental Consultants) (1987). Monitoring pH and Calcium in the Sediment of Loch Fleet Pre- and Post-intervention. Report to Loch Fleet Project, 41 pp.

Olem, H. (1990). *Liming Acidic Surface Waters*, Lewis Publishers, Michigan, USA, 331 pp.

Rosseland, B.O. and Hindar, A. (1988). Liming of lakes, rivers and catchments in Norway. *Water Air Soil Poll.* **41**: 165–88.

Sokal, R.R. and Rohlf, F.G. (1981). *Biometry*, 2nd edn. W.H. Freeman and Co., 859 pp.

Tervet, D.J. & Harriman, R. (1988). Changes in pH and calcium after selective liming in the catchment of Loch Dee, a sensitive and rapid-turnover loch in south-west Scotland. *Aquaculture Fish. Manag.* **19**: 191–203.

Welsh Water Authority (WWA) (1987). Llyn Brianne Acid Waters Project. First Technical Summary Report 1987, Dept. of Environment and Welsh Office, 91 pp.

Welsh Water Authority (1988). Llyn Brianne Acid Waters Project. Second Technical Summary Report. Welsh Water Authority, Llanelli, Wales, 32 pp.

Welsh Water Authority (1989). Llyn Brianne Acid Waters Project. Progress Report, Welsh Water Regional Rivers Division, Cardiff, 21 pp.

Wright, R.F., Norton, S.A., Brakke, D.F. and Frogner, T. (1988). Experimental verification of episodic acidification of freshwaters by sea salts. *Nature* **334**: 423–24.

CHAPTER 13

Fishery Restoration After Liming

A.W.H. Turnpenny
National Power plc, Southampton, UK

13.1 INTRODUCTION

An earlier section of this book (see Chapter 1 and Chapter 3) provides a review of the loss of fisheries attributable to acidification of their freshwater habitat, both in general and at Loch Fleet. This chapter will review the progress of fishery restoration at Loch Fleet subsequent to the liming applications in 1986 and 1987. This programme has involved confirmation of the water quality targets, the reintroduction of stock to the loch, and surveillance of the successful spawning and fry development in the improved water conditions.

13.2 POST-LIMING SURVIVAL TESTS

The liming treatments applied to various parts of the Loch Fleet catchment in April 1986 and supplementary treatments applied in 1987 produced an almost immediate water quality response in the Altiwhat feeder stream and a slightly slower response in the body of the loch. The circumneutral pH values, raised calcium levels, and greatly reduced inorganic aluminium and heavy metal concentrations (see Chapter 12) appeared suitable for fish within a few months, and survival trials similar to those carried out before liming were repeated to confirm this. The tests were also extended to measure long-term survival of adult fish in the loch itself.

13.2.1 Egg Survival and Hatching

In the Altiwhat the liming of bog areas in the upper stream catchment reversed the previous pattern of increased acidity in spate flow conditions as well as improving steady-state conditions (see Chapter 12). Freshly-fertilised eggs were planted out in the streambed gravels in November 1986 exactly as in the pre-liming trials described in Chapter 3 (Section 3.3.4), and batches were again lifted at the eyed and hatched stages (Fig. 13.1). Survival through to hatching was 64% (compared with 10–20% pre-liming), comparable with hatchery-held control levels. Sea trout (*Salmo trutta*) and salmon (*S. salar*) eggs planted alongside the brown trout eggs gave similarly good hatching rates in response to the improved water chemistry.

To check that streambed waters were not more acidic than surface waters, measurements of intra-gravel water chemistry were made by means of ceramic sampling points buried alongside the egg boxes at depths of 7·5, 15·0 and 22·5 cm (egg depth about 15·0 cm). The results (Fig. 13.2) indicate the reverse effect: pH and calcium levels were higher in intra-gravel than in surface water. This is thought to have been the result of calcium dissolution from deposited lime particles within the slower moving streambed waters.

13.2.2 Brown Trout Fry and Fingerling Survival

The spring timing of the lime applications in 1986 gave the opportunity to test fry survival at the loch outlet in June 1986 and to make a direct comparison with the test carried out 1 year earlier, prior to liming (see Chapter 3). The 1986 test differed in that it used only raw water from the loch outlet, with no additional treatments for comparison. An 18-day exposure period resulted in no deaths of fry (compared with 20–50% mortality over 9 days in raw water in the 1985 test: see Chapter 3, Fig. 3.13).

Trials with fingerling fish held in stream tanks at the loch inlet and outlet were repeated using the same protocol as in the 1985–86 tests. Fresh stocks of yearling fish (8–10 cm) were placed in the tanks in April 1986 and their subsequent survival and condition monitored over a 180-day period. No deaths occurred over this period, and fish remained in a healthy condition

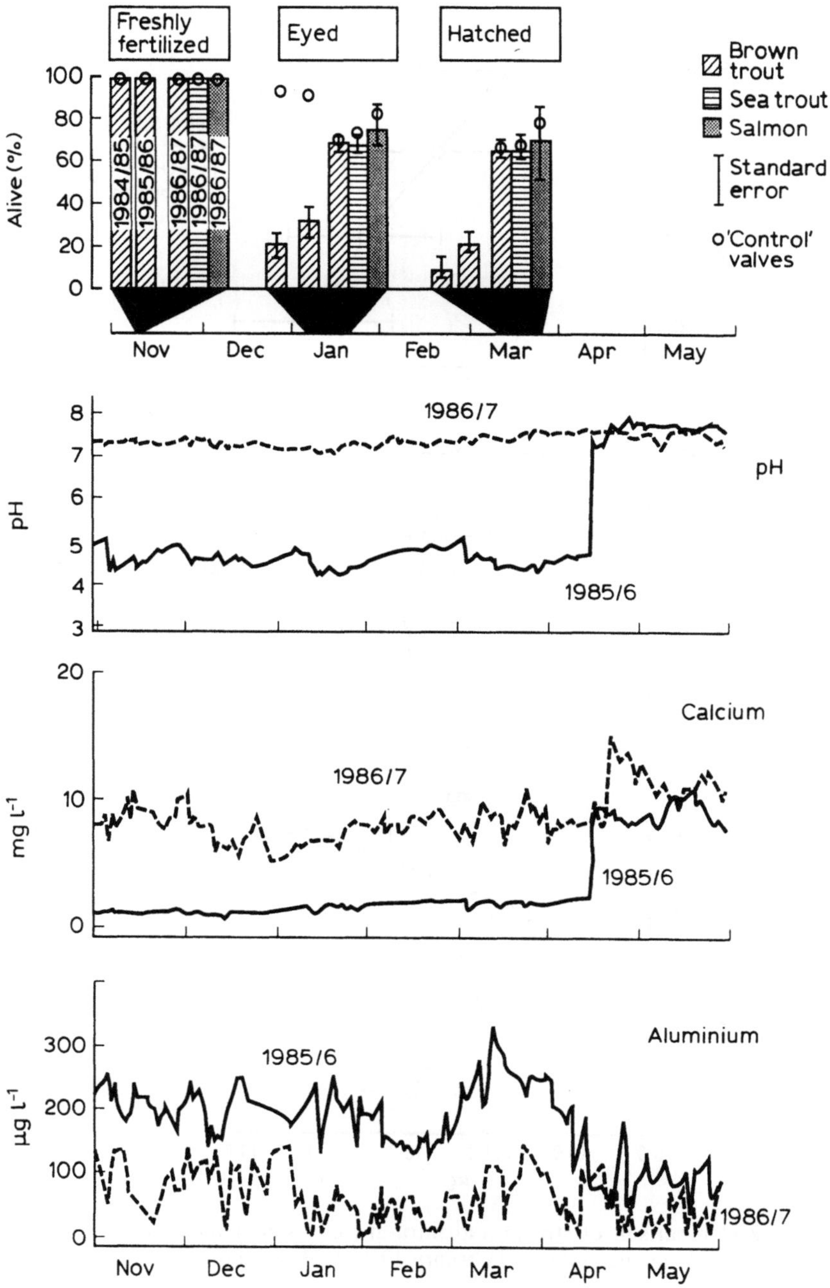

FIG. 13.1 Survival of salmonid eggs in relation to stream water chemistry (Altiwhat) before and after liming.

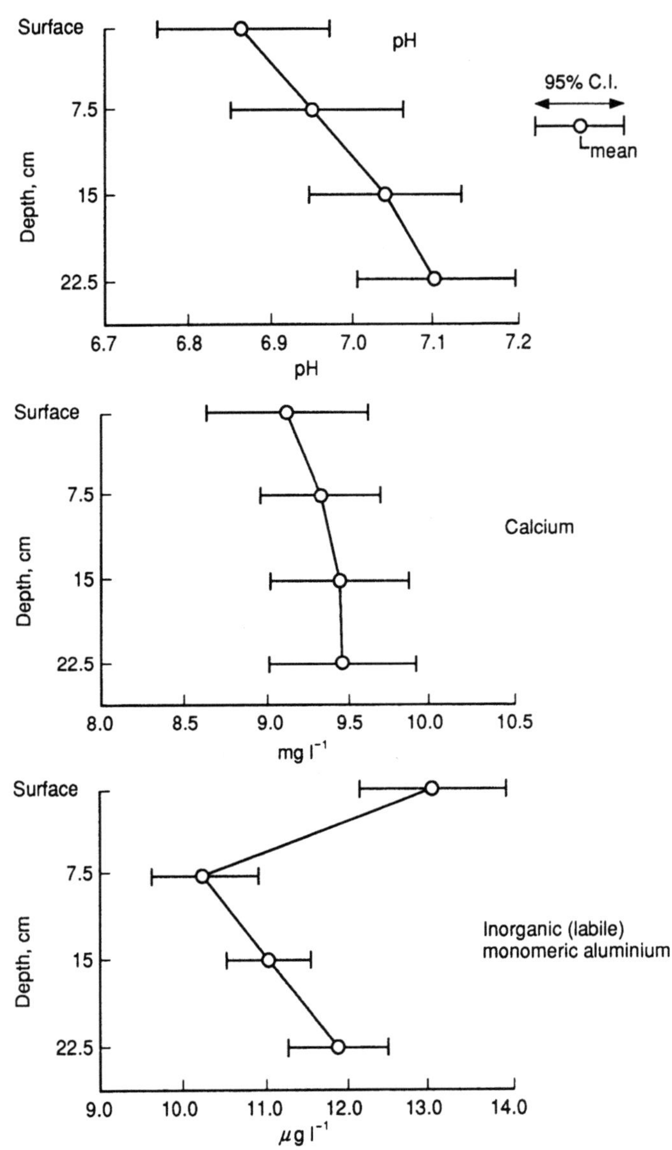

FIG. 13.2 Comparison of stream and intra-gravel water quality (means) in the Altiwhat, November 1988 to February 1989. (C.I. = confidence interval).

PLATE III Successful test fishing, September 1988.

TABLE 13.1
Mean lengths and weights of Loch Dee fish from in floating fish cage (S.D. in parentheses)

Date	*n*	*Length* (cm)	*Weight* (g)	*Condition factor*[a]
28 December 1986	14	19·9 (1·6)	130·1 (35·7)	1·63 (0·14)
12 March 1987	14	20·9 (17)	111·6 (29·9)	1·20 (0·16)
6 October 1987	12[b]	20·9 (2·3)	103·1 (49·4)	1·06 (0·14)

[a]Condition factor $= \frac{\text{weight}}{\text{length}^3} \times 100$.

[b]Two fish escaped.

with a mean condition factor ($100\ W/L^3$, W = weight in g, L = length in cm) of 1·54 at the end of this period (compared with 1·50 at the start).

13.2.3 Survival of Adult Fish

In November 1986, a floating cage was installed in the loch to study the survival and conditions of adult brown trout.

The cage used was a simple cubic nylon net of dimensions 2 m × 2 m × 2 m, suspended from a floating collar to provide about 40 cm freeboard. The top of the cage was then covered by a mesh panel to prevent fish escape or predation, the whole structure being anchored in position toward the centre of the loch.

The cage was stocked with a dozen 2-year-old brown trout taken from floating stock cages at Loch Dee (wild stock). Fish were marked using numbered fingerling tags allowing individual recognition. The fish were not fed, but had to rely upon natural food sources drifting into the cage or colonising the cage surface. The fish were inspected at intervals, and their weights, lengths, and numbers surviving recorded (Table 13.1).

No fish died over the 9-month period, although the fish lost weight due to restricted foraging.

13.3 REINTRODUCTION OF FISH TO LOCH FLEET

With the successful outcome of fish survival trials over a full year from the time of the first liming treatments (applied April 1986), fish were once again introduced directly into Loch Fleet in May 1987. This was a restricted stocking of only 300 brown trout in the first instance because the recovery of the autochthonous fauna was at this stage still rather limited (see Chapter 14). It was considered that a heavier stocking would jeopardise both the recovery of invertebrates in the loch, and the success of the fish themselves. The main purpose of the exercise was to obtain information on the performance of free-living fish in the loch, and especially on their spawning success.

The choice of stock was governed by a number of factors. The ideal choice would have been to re-introduce genetic material from the original Loch Fleet stock, since that stock had presumably evolved to perform well in that particular environment, at least until the onset of recent acidification. That not being possible, the next best source of fish was considered to be stock from the lower reaches of the Little Water of Fleet, which in the past would have had some genetic contiguity with the Loch Fleet stock by downstream loss. A difficulty with this source was the possibility that many, if not all, of the fish in this region of the Fleet were the anadromous (sea trout) form of *S. trutta*, which on smoltification would have been lost from the loch. In the event, approximately 300 fish from this and two other local sources were used for the stocking. The other sources were Loch Dee (wild stock, as used in all the survival trials: see Chapter 3) and Solway Fishery, a local hatchery stock from near Dumfries. This strategy met the recommendations of Thorpe (1986) who proposed that restocking programmes for lost populations of Atlantic salmon (*S. salar*) should, wherever possible, use: (1) genetic material from sources local to the lost population to retain some match to the local geology and climate; (2) a mixture of stock from three or more such populations to increase the size of the gene pool for natural selection; and finally, (3) at least 30 parents from each stock to avoid inbreeding.

The size distributions of the three source stocks are shown in Fig. 13.3. The intention was to use a mixture of one-, two-, and thee-year-old fish in a ratio of 4 : 2 : 1 to approximate a natural age structure. In practice, this was achieved for Dee and Solway stock, but the Fleet fish caught were all yearlings. The fish were batch marked by fin clipping to identify stock origins prior to their introduction.

A second batch of 220 fish, comprising approximately equal numbers of the same three stock types, was introduced into the loch just over one year later, in July 1988. These fish were individually weighed, measured, and marked with serially numbered fingerling tags to provide further infor-

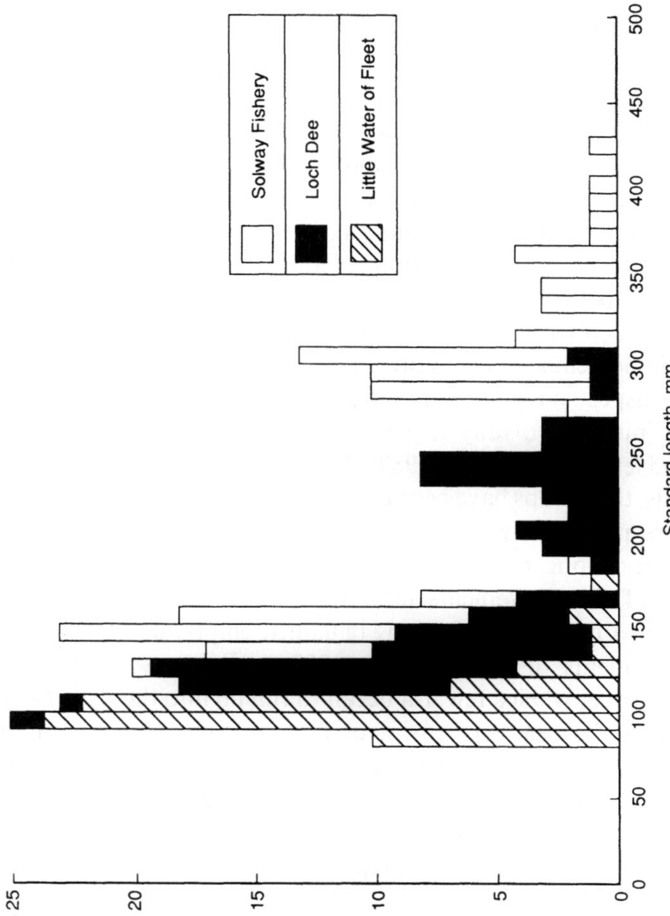

FIG. 13.3 Source and size distribution of brown trout stocked in May 1987.

mation on growth upon recapture. The size/age distribution was essentially the same as for the 1987 stocking.

13.4 SPAWNING AT LOCH FLEET

According to Mr Ian Murray of the Forestry Commission (pers. comm.), spawning runs of brown trout in the Galloway region take place in late October or early November when the water temperature drops below 6°C. The fish will then run from lakes into spawning streams with the first spate flows. This has provided a regular and reliable schedule for trapping fish for stripping in the neighbouring Loch Dee catchment for some years.

To monitor spawning movements out of Loch Fleet, two fish traps were installed, one just above the mouth of the Altiwhat (main inflow), the other at 150 m below the loch outlet. The outlet trap was fitted at the time of fish introductions to check losses from the loch, and a second trap was fitted in September 1987 in preparation for the spawning season. The traps were inspected regularly by site staff, who noted dates of fish movements, tag numbers of any tagged fish, and (when time allowed) fish weights and lengths. Since mid-1989 when the site ceased to be manned on a daily basis, these traps have been replaced by electronic fish counters and data loggers, powered by aerogenerators. These now provide daily counts of fish 20 cm or longer in length.

Records of fish movements into the Altiwhat for the 1987 season are shown in Fig. 13.4. Both traps remained empty, apart from downstream movements of a few fish in 1987 immediately after stocking, until mid-October when a few fish began to run up into the Altiwhat trap. In both years the peak movement of fish took place in early to mid-November, just as water temperature fell below 6°C and high flows occurred. Very few fish entered the outfall trap even at this time of year.

The sizes of fish comprising the spawning run into the Altiwhat are shown in Fig. 13.5. In 1987, the majority of fish were 3 year olds of the Solway stock, then representing the majority of mature fish present in the loch. By 1988, the larger Dee stock fish had reached 3 years of age and had become the predominant spawners.

Inspection of the Altiwhat gravels after the spawning runs in both years revealed no classic redd structures, but by disturbing gravel patches, eggs could be dislodged and collected in a handnet held in the downstream flow. This proved that eggs had been deposited by the trout, but to check hatch rate, additional batches of captive eggs were planted alongside the

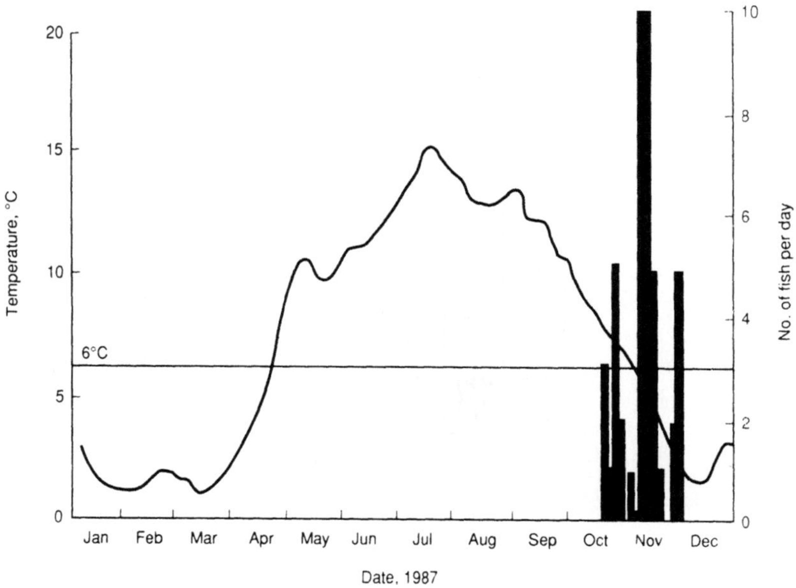

FIG. 13.4 Autumn influx of spawning brown trout into the Altiwhat in relation to stream temperature, 1987.

naturally deposited eggs. When lifted in the spring, these showed a 60–70% hatching rate, not significantly different from hatchery-held controls. By May 1988, and again in May 1989, naturally-spawned fry were evident in large numbers, both along the Altiwhat and in the outlet stream within a 100 m or so of the loch.

Population density estimates of 0-group fry were made in May 1989 by electrofishing the Altiwhat and outlet streams. Fry densities reached a maximum of 1600 fry per 100 m^2 in the Altiwhat and 800 fry per 100 m^2 in the outflow, within 100 m of the loch exit. Minnows (*Phoxinus phoxinus*) were also captured from the loch for the first time since the project began.

With the introduction of automatic fish counters in 1989, no permanent trap was installed, but a trap was fitted temporarily in the lower Altiwhat when the site was manned in early November. This was after the main spawning run in late October (as shown by counter records) and only three fish were intercepted. All were around 250 mm in length and were neither tagged nor batch marked. Examination of scales indicated that they were spawned in 1987 and thus must have been the progeny of fish planted in the loch in May 1987. All three fish were in spawning condition, yielding

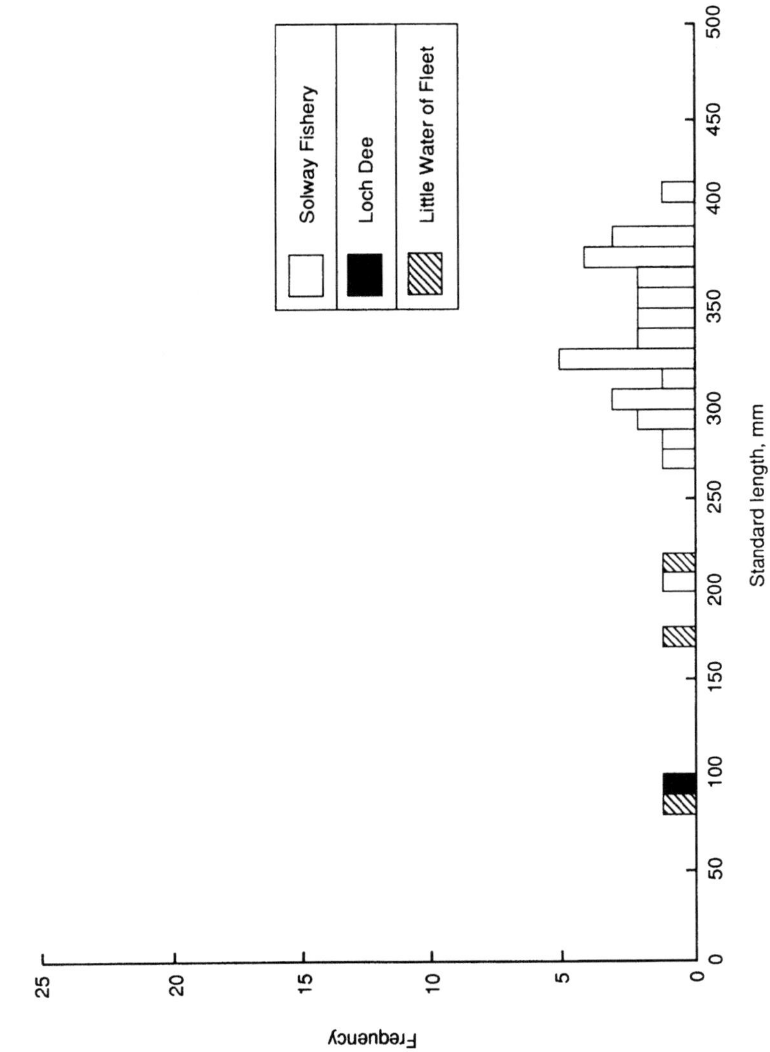

FIG. 13.5 Source and size distribution of brown trout spawning in the Altiwhat, 1987.

TABLE 13.2
Condition factors of brown trout stocked into and recaptured from Loch Fleet

Date	*Source*	*n*	*Length range* (mm)	*Weight range* (g)	*Condition factor* $(K)^a$
July 1988	As stocked	300	95–480	10–1050	1·59 (0·25)
Sept. 1988	Loch, rod and line	11	200–450	118–1400	1·58 (0·22)
Nov. 1988	Altiwhat trap	13	230–430	170–1020	1·32 (0·10)
June 1989	Loch, gill nets	7	175–316	87–386	1·46 (0·15)
Sept. 1989	Loch, rod and line	8	167–245	71–255	1·60 (0·18)
June 1990	Loch, gill nets	7	182–315	95–444	1·32 (0·20)

$^aK = \dfrac{\text{weight}}{\text{length}^3} \times 100.$

eggs or milt when handled. Thus, the life cycle of trout in Loch Fleet had once again been completed.

13.5 GROWTH AND CONDITION OF TROUT IN LOCH FLEET

In view of the rather slow recovery of benthic invertebrates reported in Loch Fleet from samples taken following liming (Chapter 14), and the depression in numbers following the 1987 fish restocking, it was clearly of interest to assess the condition of fish and the growth performance subsequent to stocking. From the first stocking, although fish lengths were measured at the time of introduction, no weights were recorded and fish were only batch marked so that it was not possible to monitor individual growth.

More information was gained from the second stocking of individually tagged fish in July 1988. The first attempts to recapture these were made in September 1988 when a fly-fishing exercise was carried out (Plate III). Four rods were fished for 5 h yielding a total bag of 31 fish. Eleven of these were tagged fish from the previous stocking. On average, these showed a 20% mean (range 0·6%–106%) weight increase over the 79-day interval since introduction. Average values of the condition factor were virtually the same on the date of recapture as at the time of introduction (Table 13.2). The fish were also all in excellent physical condition and were judged by the anglers to have offered good sport. In view of the fact that, prior to introduction, these fish had been fed on an artificial, pelleted diet and had to switch to natural foraging, this rate of growth was considered very encouraging.

Further data on fish growth and condition were obtained from subsequent recaptures by trapping during the 1988 spawning run and from gill netting carried in 1989 and 1990. Condition factors remained within a range of 1·3–1·6 (Table 13.2).

A better indication of fish growth in Loch Fleet is given by growth analysis of naturally-spawned fish which have spent their entire life in the loch. A growth curve for recaptured fish is shown in Fig. 13.6, based on length-at-age back-calculated from scale readings. This is compared with similarly derived values for trout from other Scottish lochs given by Frost and Brown (1967). Growth rates in Loch Fleet are seen to be about average in this context, 3-year-old fish attaining a mean weight of 184 g at a standard length of 23 cm.

It is of interest to know whether the fish in Loch Fleet have adequate food. Elliott (1975) has shown that growth performance of trout (*S. trutta*) where food is not limiting is controlled predominantly by temperature, according to the following model:

$$W_t = [b_1(a + b_2T)t + W_0^{b_1}]^{1/b_1}$$

where W_0 is the initial wet weight and W_t is the final wet weight (in grams) after t days at T°C; values of the constants a, b_1, and b_2 for different temperature ranges are given in Elliott (1975, Table 2). The model has also been used successfully to predict optimum growth rates of wild trout populations exposed to natural temperature regimes (Edwards *et al.*, 1979; Elliott, 1984, 1985, 1988; Allen, 1985; Jensen, 1990). A corollary of the model is that the attainment of optimum growth rates for a particular temperature regime implies that food is not limiting fish growth.

Weekly water temperatures measured at the outlet of Loch Fleet in 1987 are shown in Fig. 13.7. These values have been used to compute weekly growth increments for each year class using the initial weights shown in Fig. 13.6. The model prediction lines plotted alongside observations in Fig. 13.6 mark the start and end of year values only. Predicted growth rates are very close to, and slightly below, observed values for each year class, and hence it is concluded that the fish have adequate food and are growing optimally for that temperature regime.

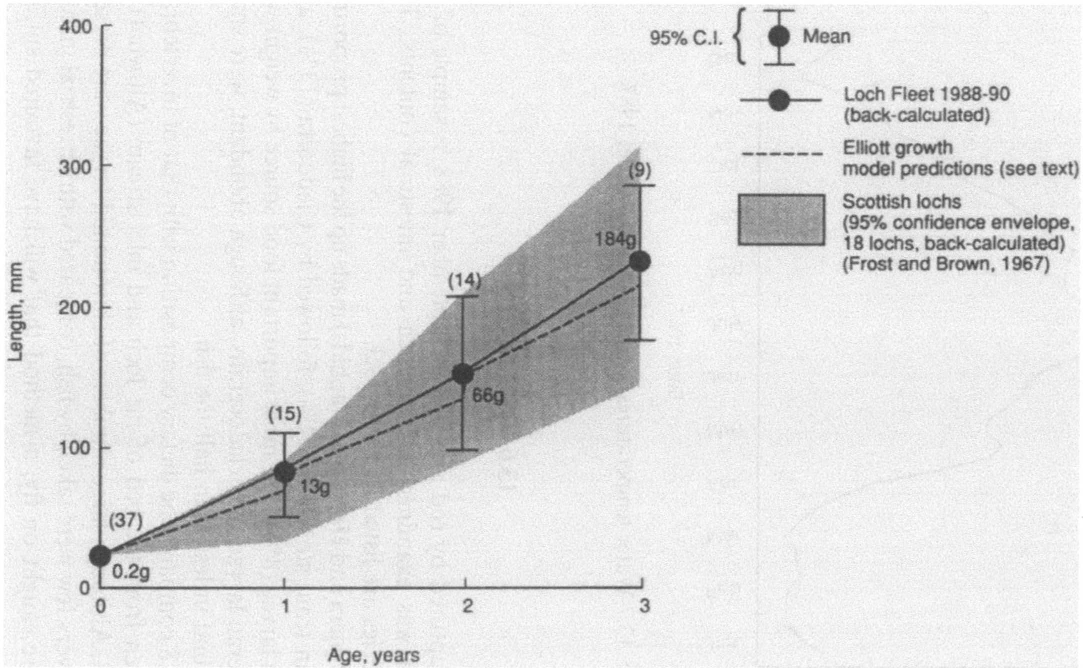

FIG. 13.6 Growth of brown trout in Loch Fleet in comparison with the Elliot growth model and in other Scottish lakes. (C.I. = confidence interval). The bracketed numbers and weights refer to Loch Fleet fish samples.

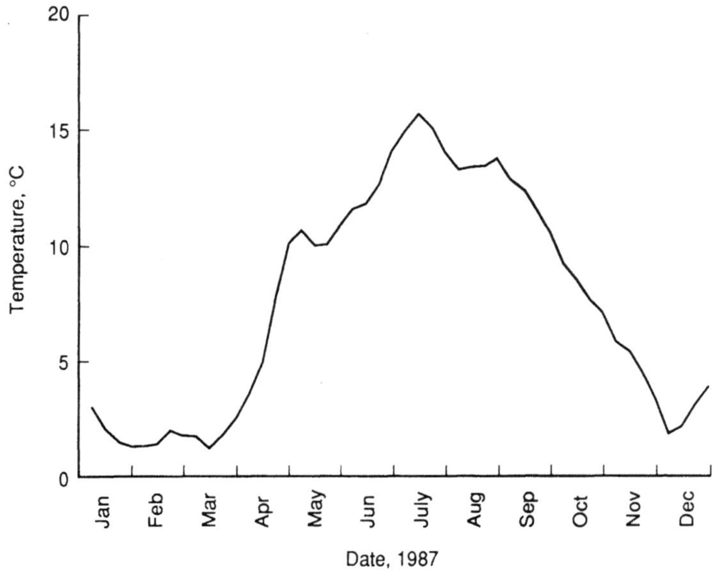

FIG. 13.7 Water temperatures at the Loch Fleet outlet 1987.

13.6 FISH DIETS

From fish captured by rod and line in September 1988, a sample of 22 brown trout was examined for stomach and intestinal contents. The recorded species are listed in Table 13.3.

Winged Diptera and Homoptera (aphids) made up the highest proportion by number in trout stomachs (74%), followed by Cladocera (13%), and trichopteran larvae (8%). The most important food source by weight was the trichopteran larvae. The cladocerans, although abundant, were small and were found in less than half the fish.

Figure 13.8 compares the dietary composition with that of invertebrate samples taken from the bed of the loch and inlet stream (Altiwhat) in August 1988. Although large numbers of chironomid larvae were present in the loch, very few were taken by fish, compared with terrestrial insects. The fish were caught on fly, a method that will have favoured surface feeders.

A second sampling of fish diets, using alternative methods of capture, was made in June 1989. Twenty-five fish were sampled: 8 from the main body of the loch by gill net, 1 from the Altiwhat, and 16 from the loch

TABLE 13.3
Species composition of trout diets (A. Milner, pers. comm., 1989)

Crustacea:	Cladocera:	*Bosmina* sp. *Bythotrephes* sp.
Ephemeroptera larvae:		*Leptophlebia* sp.
Hemiptera:	Heteroptera:	Corixidae
	Homoptera:	Aphidae
		Adelgidae (*A. abietis* = Spruce aphid)
		Pemphigidae
		Cercopidae (Froghoppers)
Trichoptera larvae:		*Plectrocnemia conspersa*
		Phryganeidae
Neuroptera:	Hemerobiidae:	*Hemerobius* sp. (Lacewing)
	Sialidae:	*Sialis* sp. (Larvae)
Coleoptera adults:		*Helophorus* sp. (Aquatic)
		Hydroporus palustris (Aquatic)
		Curculonidae (Weevil)
Diptera adults		Chironomidae: *Chironomus* sp.
		Culicidae
		Phoridae: *Phora* sp.
		Dixidae
		Bibionidae: *Dilophus* sp.
		Mycetophilidae: *Sciara* sp.
		Psychodidae: *Psychoda* sp. (Mothflies)
Hymenoptera:	Apocrita	Proctotrupidae: *Platygaster* sp.
	Formicoidae:	*Formica rufa*
	Apoidae:	*Bombus* sp.
	Sphecidae, Ichneumidae, Cypnipoidex, Braconidae	
	Symphyta: Diprionidae: *Diprion pini* (Pine Sawfly)	

outlet and Little Water of Fleet by electro-fishing. A stomach flushing method (Meehan and Miller, 1978) was used to avoid killing the fish. On this occasion only 7% (by number) of food items found in loch fish were of terrestrial origin, compared with 76% in the previous survey. In stream fish, 24% of food items were of terrestrial origin.

Overall, the most abundant aquatic components of the diet (by number) were chironomid larvae (53%) and trichopteran larvae (20%). This composition much more closely resembled that found in loch and stream samples (Fig. 13.8).

It is not yet clear to what extent the differences in diet shown by the two sets of data were due to seasonal variations or, alternatively, to sampling selectivity.

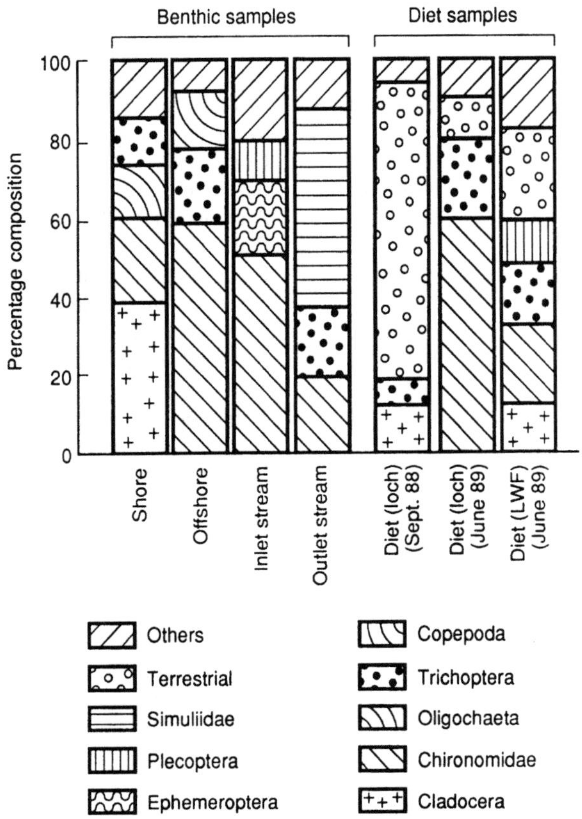

FIG. 13.8 Composition of benthic invertebrate samples from Loch Fleet and fish diet analyses for fish caught in the Loch and Little Water of Fleet (LWF). (A. Milner, pers. comm., 1989).

13.7 EFFECTS ON FISHERIES DOWNSTREAM OF LOCH FLEET

In May 1989 and May 1990 the Fleet system was re-surveyed for fish, repeating exactly the pattern of 1985. It was expected that some effect of both the liming and the introduction of fish at Loch Fleet would be detectable downstream, although the chemical changes would be minimised by dilution from the larger catchment within a few kilometres of passage. Trout and eel (*Anguilla anguilla* L.) population biomass and density values are compared in Table 13.4.

The patterns of trout densities recorded along the course of the Little Water in 1989 are shown in Fig. 13.9 in comparison with 1985 values.

TABLE 13.4

Population biomass of brown trout and eel, Fleet system (excluding 0-group) ($g\,m^{-2}$)

Site[a]	Trout			Eel		
	1985	*1989*	*1990*	*1985*	*1989*	*1990*
LF outflow 0·05 km	0	4·3	0·38	0	0·41	0·96
Lime tanks 1·0 km	0	6·8	2·9	0	2·2	0·77
1st bridge 2·1 km	0	0·39	2·9	0·80	0·50	2·3
Viaduct 4·6 km	0	0·82	0·43	2·0	0·74	1·06
C. Linn above 7·1 km	0	0·15	N.F.	1·5	2·0	N.F.
C. Linn below 7·2 km	0·09	0·47	N.F.	1·0	0·72	N.F.
Drumruck 9·1 km	0·47	2·9	1·15	1·2	8·6	6·6
Culreoch 11·5 km	0·08	3·0	0·66	0·98	2·6	0·45

Population densities of brown trout and eel, Fleet system (excluding 0-group) ($number\,m^{-2}$)

Site	Trout			Eel		
	1985	*1989*	*1990*	*1985*	*1989*	*1990*
LF outflow 0·05 km	0	0·040	0·044	0	0·02	0·18
Lime tanks 1·0 km	0	0·16	0·17	0	0·03	0·03
1st bridge 2·1 km	0	0·021	0·14	0·04	0·01	0·09
Viaduct 4·6 km	0	0·025	0·011	0·05	0·11	0·03
C. Linn above 7·1 km	0	0·0034	N.F.	0·03	0·01	N.F.
C. Linn below 7·2 km	0·01	0·03	N.F.	0·05	0·04	N.F.
Drumruck 9·1 km	0·05	0·19	0·08	0·07	0·40	0·34
Culreoch 11·5 km	0·02	0·30	0·08	0·06	0·18	0·028

N.F. = not fished.

[a]Distances from loch outlet.

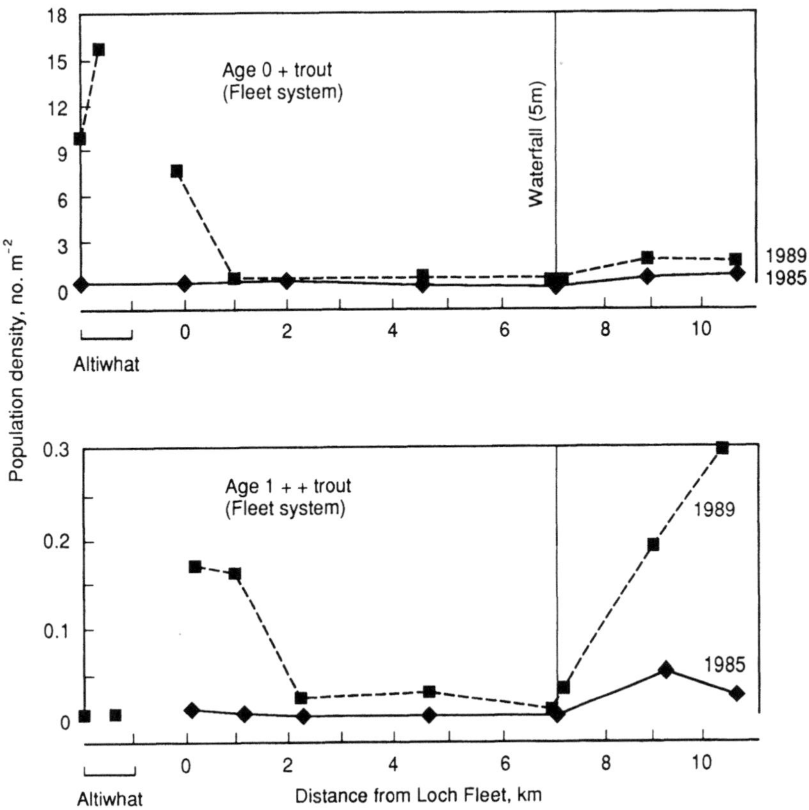

FIG. 13.9 Population density of 0+ and 1++ group trout in the Altiwhat and Little Water of Fleet, before and after liming.

0-group trout fry were prominent in the streams around the loch, but quickly faded downstream, indicating that no spawning had taken place in the Little Water of Fleet below the loch outlet. Fry were again detected below the Craigie Linn, the products of spawning grounds in the lower reaches of the river.

Population densities of yearling and older trout were up to 16 fish per 100 m^2 in the top reaches of the Little Water of Fleet (cf. other British upland waters in Table 3.2). The numbers then diminished rapidly downstream towards the Craigie Linn waterfall, but at least one trout was caught at every site sampled along the river. Below the Craigie Linn, trout densities were up to fifteen times higher than in the 1985 survey. Trout densities here were also up to fourfold higher than 1985 values in the 1990

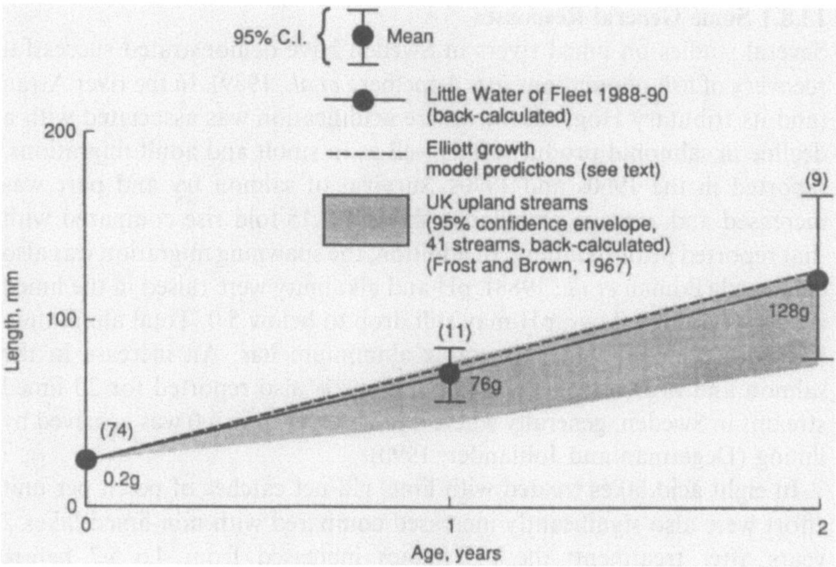

FIG. 13.10 Growth of brown trout in Little Water of Fleet (above Craigie Linn) compared with the Elliot growth model and UK upland streams (C.I. = confidence interval). The bracketed numbers and weights refer to Little Water of Fleet fish samples.

survey. It is not yet clear whether this represents natural variation or is related to upstream improvements. The fish upstream of the Craigie Linn are attributed to spawnings in Loch Fleet. Their growth too was good, comparing favourably with data from other UK upland streams and with Elliott growth model projections (Fig. 13.10). Flushing of the stomach contents revealed that the stream fish were feeding mainly on chironomid larvae, trichopteran larvae, and other aquatic fauna, with a relatively small terrestrial component (Fig. 13.8).

From the sizes of fish present in June 1990 (up to 250 mm length), a first spawning in the reaches above Craigie Linn in 1990 or 1991 is expected.

Eel populations increased post-liming in the uppermost reaches of the Little Water but remained broadly similar in other parts (Table 13.4).

13.8 COMPARISON WITH OTHER STUDIES

The biological findings at Loch Fleet are similar to those from other aquatic and watershed liming projects aimed at improving fisheries.

13.8.1 Some General Responses

Several studies on limed rivers in Sweden have demonstrated successful recovery of fish populations (eg. Appelberg *et al.*, 1989). In the river Atran (and its tributary Hogvadsan), where acidification was associated with a decline in salmonid production as well as in smolt and adult migrations, reported in the 1960s and 1970s, survival of salmon fry and parr was increased and average abundance showed a 15-fold rise compared with that reported prior to liming. In addition, the spawning migration was also enhanced (Edman *et al.*, 1988). pH and alkalinity were raised in the limed river, but at high flows, pH may still drop to below 5·0. Total aluminium has not decreased, but extractable aluminium has. An increase in the salmon and brown trout parr populations is also reported for 20 limed streams in Sweden, generally where a pH greater than 6·0 was achieved by liming (Degerman and Johlander, 1990).

In eight acid lakes treated with lime, gill-net catches of perch per unit effort were also significantly increased compared with non-limed lakes 2 years after treatment; the pH values increased from 4·6–5·7 before liming to 6·2–7·1. Wet weight per unit catch effort was also increased significantly (Eriksson and Tengelin, 1987). Liming also changed the species interactions in fish communities, with biomass of roach, pike, and cisco (white fish) falling during the 2 years after liming, with char and minnow being favoured, although a shift to greater numbers of the former species was seen in the following 3–10 years (Appelberg and Degerman, 1990).

In Norway, the river Audna was treated by liming a lake upstream, as well as by lime-dosing of the inlet stream and the main river. This has had a very positive effect, even when liming is restricted to the period of adult migration or smolt release (Rosseland and Hindar, 1988; ICES, 1989). In a small limed lake draining to the river Lygna, however, smolt production was increased, but the smolts failed to leave the lake at the normal time (Hansen *et al.*, 1984; Hansen, 1987).

Numerous hatching experiments with brown trout, Atlantic salmon, char, and brook trout have been undertaken in streambeds in Norway (Matzow *et al.*, 1975). Most of these treatments used lime mixed with stream gravel, but shell sand filtered water and sea water additions were also effective in promoting egg survival. Similar methods were used to improve water quality in the short term for fry and parr survival, as well as slurry liming in running water.

Liming of lakes in Norway has provided more stable improved water quality in the downstream waters, as well as in the lakes themselves. This has demonstrated, in addition to trout re-establishment, successful spawning, egg hatch and development, and restoration of lost perch populations, e.g. in Lake Finnetjenn (Matzow *et al.*, 1975).

13.8.2 Fish Growth and Condition

Rapid initial growth and high condition factors are a common feature of newly restored fisheries. This appears to result from high availability of food species when fish are stocked into waters previously devoid of fish and other top predators, and when the standing crop of fish is low and therefore competition limited. In Lake Store Hovvatn in southern Norway, Rosseland and Hindar (1988) found that brown trout yearlings freshly introduced into the recently limed but fishless waters doubled the growth rate of their parent stock, but with a second stocking a year later, growth returned to normal. Similar observations have been reported after stocking newly-constructed reservoirs and for restored fisheries in rivers recovering from prolonged industrial pollution (Turnpenny and Williams, 1981).

The corollary of this stituation occurs during the process of acidification, or reacidification of limed waters. As population densities of fish fall, so competition for food is reduced and improved growth may be found (Rosseland, 1986; Maitland *et al.*, 1987). This is not always the case, however. During reacidification at Lake Store Hovvatn, acid toxicity caused fish growth rates to decline (Rosseland and Hindar, 1988), and Schofield *et al.* (1989) report a similar finding in Adirondack lakes in North America.

13.8.3 Fish Diets

Responses of aquatic invertebrate communities to liming have varied in different water bodies. In general, both zooplankton and macroinvertebrate communities have been successfully restored to a structure typical of non-acid-stressed waters, but sometimes only after a considerable delay (Eriksson *et al.*, 1983; Schaffner, 1989). An initial shock response to liming is evident in some taxa, as appeared to be the case at Loch Fleet. Schaffner (1989) noted sharp reductions in rotifer communities in Adirondack lakes within a week of liming, and reductions in crustacean communities, but a longer term (over a period of 1 year) recovery of the fauna as new species became established. Reacidification has been shown to reverse this trend (Rosseland and Hindar, 1988). The depression of invertebrate communities in response to fish stocking has also been reported (Evans, 1989).

Brown trout are essentially opportunistic feeders and can adapt to a wide variety of prey types, therefore quantity rather than the type of food is more important. Observations of rapid fish growth in Loch Fleet signify that the food resource had not become limiting by 1990, but until invertebrate communities reach a stable state, estimates of potential fish production in the loch must remain speculative. Similar opportunistic patterns of predation by perch on the invertebrate community of Store Finnetjenn after liming have been reported (Matzow *et al.*, 1975).

Allochthonous sources can become especially important when the aquatic fauna is impoverished, and may be suppressed in catchments dominated by coniferous afforestation. Allochthonous inputs of prey items into streams of the Llyn Brianne catchment in west Wales were found to be nearly five times higher in moorland versus coniferous forest areas (Welsh Water Authority, 1987). Thus the development of a sufficient aquatic fauna is likely to be critical to the success of fish populations in the reaches of the Little Water of Fleet which pass through conifer plantations. The bias towards an aquatic rather than a terrestrial diet in trout taken from these reaches in the 1989 survey was evident, and growth in these fish was good. At present no data are available to assess invertebrate production in the stream and therefore the size of fish population that might be supported.

13.8.4 Spawning

Some, but not all, liming projects have been successful in restoring spawning of salmonid stocks. In southern Norway, liming of the River Audna (by continuous dosing) was successful in improving the hatch of Atlantic salmon eggs, but in Lake Store Hovvatn, a lake with no inflow stream, which had previously held a lake-spawning stock of brown trout, no natural spawning has taken place and the population has been maintained by regular restocking (Rosseland and Hindar, 1988). In a study of two limed lakes, Woods Lake and Cranberry Pond in the Adirondack Mountains (Gloss *et al.*, 1989), hatching of naturally deposited eggs of brook trout (*Salvelinus fontinalis*) was only achieved in one of the lakes (Woods Lake), and then only on prepared substrates and not natural ones. The lake water was limed directly and the inflow streams that brook trout might normally select for spawning remained toxic. Spawning levels were not sufficient to sustain the lake population.

The success of brown trout spawning at Loch Fleet is attributed to a combination of stable, high permeability gravel beds in the Altiwhat, and effective control of acid episodes in this steam achieved by liming the headwater bogs. However, the area of suitable spawning gravels in the Altiwhat is extremely restricted and it seems likely that this will prove to be the main factor, and not food supply, that eventually limits the growth of the Loch Fleet trout population.

13.8.5 Conclusions and Summary

The main feature that is evident from other studies is that ecological equilibrium is reached only slowly after liming and restocking of acidified waters. Invertebrate communities may take several years to stabilise, depending upon rates of influx of replacement species. A fish community can take some years to establish species structure, and many generations to reach a stable age structure. If the rate of reacidification following liming is rapid, and consequently the re-liming interval short, then aquatic communities cannot be expected to achieve a stable state, but will be continuously readjusting to the changing conditions. Catchment liming, by extending the longevity of treatment compared with water column liming, offers a slow rate of environmental change which should be more favourable to the creation of a stable community structure.

13.9 IMPLICATIONS FOR FISHERY MANAGEMENT

13.9.1 Loch Fleet

Loch Fleet has never been a major fishery of the Galloway region and its small size and remote location dictate that it will always remain the "connoisseur's" fishery that it was in the past, although no doubt the publicity arising out the project itself may arouse the interest of anglers. Management of the fishery will therefore principally involve first maintaining an adequate water quality for fish, and secondly restricting fishing effort.

In maintaining water quality, two separate requirements can be distinguished. Present indications from modelling studies (see Chapter 16) are that water quality in the main body of the loch will remain suitable for trout until the turn of the century. However, in the main Altiwhat spawning stream, conditions are expected to become sporadically critical for fish from the year 1993 when acid episodes may re-occur. If a self-sustaining population is to be maintained, it will be necessary to institute a higher frequency of liming targeted to this stream than is necessary for maintaining fish in the loch itself. This need could be met by in-stream dosing methods (e.g. lime-wells or by the addition of basic materials to the

TABLE 13.5
Standing crops of trout in British upland lakes (*Source*: Frost and Brown, 1967)

Lake, Location	*Standing crop* (kg ha^{-1})
Loch Kinnaird, Scotland	56–67
Wise Een Tarn, Lake District	26–38
Three Dubs Tarn, Lake District	24
Range	24–67

streambed), but reliability of these methods depends upon regular attention to dosing equipment and recharging with basic materials (see Chapter 10). Headwater bog liming has already proved itself an effective and long-lasting (predicted 7 years) treatment for the Altiwhat and would seem to be the obvious choice for this situation.

The carrying capacity of Loch Fleet for trout is not yet known, nor can the future performance of the fishery be predicted until an equilibrium state is reached. Table 13.5 lists standing crop values for comparable British upland lakes, where values range between 24 and 67 kg ha^{-1}. Taking a mid-range value of 46 kg ha^{-1}, a total standing crop of 774 kg for Loch Fleet (17 ha) might be attainable at equilibrium. Figure 13.11 illustrates a hypothetical scenario of population dynamics for the fishery, adopting typical population parameters recorded from comparable waters.

To maintain the population structure shown would require successful spawning by more than 150 adults (or 75 pairs). It seems unlikely that spawning grounds in the Altiwhat will support this number of spawners and thus these population levels are probably not presently attainable. The extension of spawning grounds by the construction of artificial spawning beds (Clay, 1961; Leitritz and Lewis, 1976) would offer one possibility; such schemes have proved successful elsewhere in Britain (Harris, 1978). This might be attempted in the lower Altiwhat; a possible alternative would be to construct spawning beds close to the loch outlet where water quality is more stable. Harriman *et al.* (1987) cite instances of other lochs in the area that have no distinct inflow, and where fish are obliged to spawn in the outflow. Fish count records for 1989 showed that, unlike previous years following fish introduction, substantial emigration from the loch outlet took place in the autumn spawning season, perhaps as a result of competition for spawning grounds in the Altiwhat with the expanded population. By constructing spawning beds at the outlet, the

Box One: Carrying Capacity of Loch Fleet

Area of loch	17ha
Range of standing crop (from Table 13.5)	408 - 1139 kg
Mid-point value	774 kg

Box Two: Age/Weight Structure

(based on data for Three Dubs Tarn: Frost & Brown, 1967)

Age	*Number of Fish*	*Mean weight, g*	*Total weight, kg*
0 (eggs)	63,000	-	-
1+	3,750	10	38
2+	1,500	60	90
3+	750	350	263
4+	246	1000	264
5+	93	1300	121
		Total:	776 kg

Box Three: Replacement Spawning Stock

Requirement (from Box Two)	63,000 eggs
Assumed egg production per ♂+♀	750 per kg
Assumed spawning efficiency	90%
Assumed first spawning age	3+
Hence, weight of 3++ spawners required	94 kg

Assuming age/weight structure from Box Two spawning requirements could be met by:

Age	*Number of Fish*	*Total weight, kg*
3+	100	35
4+	40	40
5+	15	20
Total:	155	95 kg

Box Four: Fishery Yield

After meeting replacement spawning requirements, the following numbers of age 3 ++ fish would be available for capture:

Age	*Number of Fish*	*Total weight, kg*
3+	650	228
4+	224	224
5+	77	100
Total:	951	552 kg

FIG. 13.11 Predicted population dynamics for a brown trout fishery at Loch Fleet, based on data for trout populations elsewhere in the UK (and assuming sustained water quality).

need for repeating the lime application to the inlet stream might be avoided. For this idea to work, it might be necessary to introduce an outlet-spawning stock the fry of which are genetically programmed to migrate upstream back into the loch.

13.9.2 Little Water of Fleet

The recolonisation of the Little Water of Fleet is expected to continue with further downstream losses of fish from Loch Fleet. If spawning above the Craigie Linn is successful, then the stream will develop a self-sustaining population in its own right and a rapid expansion of numbers can be expected.

An approximate idea of the standing crop that may be possible can be gained by comparison with other waters. The average width of the stream is around 5 m and length 13 km; hence the total area is 6·5 ha. For streams of circumneutral pH, a standing crop of around 4·5 $g\,m^{-2}$ is typical (see Chapter 3, Table 3.2) and is within the range recorded in the lower reaches of the river below the Craigie Linn. A standing crop potential of around 293 kg for the entire stream is therefore indicated, i.e. about one-third of the standing crop potential of the loch. In practice, afforestation effects rather than acidity may ultimately limit the standing crop to a figure below this value.

The Fleet system below the Craigie Linn has long been fished for sea trout, and catches have declined in recent years (Table 3.1). The reaches above the waterfall offer only limited scope for angling on account of marginal conifer cover for much of its length and difficult terrain elsewhere. The main value of the stream will be as a recruitment ground for fisheries below the Craigie Linn. As such, its value would be greatly enhanced by opening up access from the lower reaches by means of a fish ladder on the Craigie Linn. Clearly, this expenditure could not be recommended before a sustained improvement had been demonstrated. Additional measures may prove to be required, for example supplementary liming in stream sub-catchments, cutting back bankside conifers, and cleaning spawning gravels.

13.10 CONCLUSION AND GENERAL IMPLICATIONS

The fishery programme at Loch Fleet has demonstrated that catchment based liming treatments can be used to restore healthy and well-fed self-sustaining salmonid populations within a very short period after treatment. The catchment liming approach has significant advantages to fisheries over direct lake liming treatments in terms of

- stable water quality over a decade or more, allowing biological communities to reach a stable state, and
- restoration of suitable water quality in upstream spawning areas.

Consequential benefits are also seen in downstream waters owing to improved water quality (though dilution limits this effect) and through downstream emigration of fish.

REFERENCES

Allen, K.R. (1985). Comparison of the growth rate of brown trout (*Salmo trutta*) in a New Zealand stream with experimental fish in Britain. *J. Anim. Ecol.* **54**: 487–95.

Appelberg, M. and Degerman, E. (1990). Development and stability of fish assemblages after lime treatment. Internat. Conference Acidic Deposition: Its Effects and Impacts, Glasgow, UK, Roy. Soc. Edinburgh, Abstract p. 225.

Appelberg, M., Degerman, E., Johlander, A. and Karlsson, L. (1989). Liming increases the catches of Atlantic salmon on the west coast of Sweden. *Nord. J. Freshw. Res.* **65**: 44–53.

Clay, C.H. (1961). *Design of Fishways and Other Fish Facilities*. Dept. of Fisheries of Canada, Ottawa, 301 pp.

Degerman, E. and Johlander, A. (1990). Development of salmonid populations after reclamation of acidified streams. Internat. Conference Acidic Deposition: Its Effects and Impacts, Glasgow, UK, Roy. Soc. Edinburgh, Abstract p. 602.

Edman, G., Fleischer, S., Fritz, O. and Stibe, L. (1988). Hövadsan 1979–1986, Förorad-Kalkad-Panyttfodd: Lansstyrelsen i Hallands län, Medd. 1988: **4**, 90pp. (In Swedish, summary in English.)

Edwards, R.W., Densem, J.W. and Russell, P.A. (1979). An assessment of the importance of temperature on the growth of brown trout in streams. *J. Anim. Ecol.* **48**: 501–7.

Elliott, J.M. (1975). The growth rate of brown trout (*Salmo trutta* L.) fed on maximum rations. *J. Anim. Ecol.* **44**: 805–21.

Elliott, J.M. (1984). Growth, size, biomass and production of young migratory trout *Salmo trutta* in a Lake District stream, 1963–83. *J. Anim. Ecol.* **53**: 979–94.

Elliott, J.M. (1985). Growth, size, biomass and production for different life-stages of migratory trout *Salmo trutta* in a Lake District stream, 1966–83. *J. Anim. Ecol.* **54**: 985–1001.

Elliott, J.M. (1988). Growth, size, biomass and production in contrasting populations of trout *Salmo trutta* in two Lake District streams. *J. Anim. Ecol.* **57**: 49–60.

Eriksson, F., Hornstrom, E., Mossberg, P. and Nyberg, P. (1983). Ecological effects of lime treatment of acidified lakes and rivers in Sweden. *Hydrobiologia* **101**: 145–64.

Eriksson, M.O.G. and Tengelin, B. (1987). Short-term effects of liming on perch, *Perca fluviatilis*, populations in acidified lakes in south-west Sweden. *Hydrobiologia* **146**: 187–91.
Evans, R.A. (1989). Response of limnetic insect populations of two acidic, fishless lakes to liming and brook trout (*Salvelinus fontinalis*). *Can. J. Fish Aquat. Sci.* **46**: 342–51.
Frost, W.E. and Brown, M.E. (1967). *The Trout*. Collins, London, 316 pp.
Gloss, S.P., Schofield, C.L., Spateholts, R.L. and Plonski, B.A. (1989). Survival, growth, reproduction, and diet of brook trout (*Salvelinus fontinalis*) stocked into lakes after liming to mitigate acidity. *Can. J. Fish. Aquat. Sci.* **46**: 277–86.
Hansen, L.P. (1987). Growth, migration and survival of lake reared juvenile anadromous Atlantic salmon (*Salmo salar* L.). *Fauna Norv. Ser. A* **8**: 29–34.
Hansen, L.P., Jonsson, B. and Doving, K.B. (1984). Migration of wild and hatchery reared smolts of Atlantic salmon (*Salmo salar* L.) through lakes. *J. Fish. Biol.* **25**: 617–23.
Harriman, R., Morrison, B.R.S., Caines, L.A., Collen, P. and Watt, A.W. (1987). Long-term changes in fish populations of acid streams and lochs in Galloway, south west Scotland. *Wat. Air Soil Poll.* **32**: 89–112.
Harris, G.H. (1978). *Salmon Propagation in England and Wales*, A report by the Association of River Authorities. National Water Council Working Party, 62 pp.
ICES (1989). Report of the Study Group on Toxicological Mechanisms Involved in the Impact of Acid Rain and its Effects on Salmon, CM 1989/M:4. Copenhagen, Denmark, 67 pp.
Jensen, A.J. (1990). Growth of young migratory brown trout *Salmo trutta* correlated with water temperature in Norwegian rivers. *J. Anim. Ecol.* **59**: 603–14.
Leitritz, E. and Lewis, R.C. (1976). *Trout and Salmon Culture* (*Hatchery Methods*). Calif. Dept Fish Game, Bull. 164, 197 pp.
Maitland, P.S., Lyle, A.A. and Campbell, R.N.B. (1987). *Acidification and Fish in Scottish Lochs*. Institute of Terrestrial Ecology, Natural Environment Research Council, UK, 71 pp.
Matzow, D., Rosseland, B.O. and Skogheim, O.K. (1975). Effects of lime treatment on fish, pp. 110–30 in *Liming of Acid Water* (ed. K. Baalsrud). Dept. of Environment and Directorate for Nature Management, Trondheim, Norway, 147 pp.
Meehan, W.R. and Miller, R.A. (1978). Stomach flushing: effectiveness and influence on survival and condition of juvenile salmonids. *J. Fish. Res. Bd. Can.*, **35**(10), 1359–63.
Rosseland, B.O. (1986). Biological effects of acidification on tertiary consumers. Fish Population Responses. *Water Air Soil Poll.* **30**: 451–60.
Rosseland, B.O. and Hindar, A. (1988). Liming of lakes, rivers and catchments in Norway. *Water Air Soil Poll.*, **41**: 165–88.
Schaffner, W.R. (1989). Effects of neutralization and addition of brook trout (*Salvelinus fontinalis*) on the limnetic zooplankton communities of two acidic lakes. *Can. J. Fish. Aquat. Sci.* **46**: 295–305.
Schofield, C.L., Gloss, S.P., Plonski, B. and Spateholts, R. (1989). Production and growth efficiency of brook trout (*Salvelinus fontinalis*) in two Adirondacks

Mountain (New York) lakes following liming. *Can. J. Fish. Aquat. Sci.* **46**: 333–41.

Thorpe, J.E. (1986). Salmon enhancement: stock discreteness and choice of material for stocking, pp. 373–88 in *Atlantic Salmon: Planning for the Future* (ed. D.H. Mills and D.J. Piggins). Croom-Helm, London.

Turnpenny, A.W.H. and Williams, R. (1981). Factors affecting the recovery of fish populations in an industrial river. *Env. Poll.*, Ser. A, **26**: 39–58.

Welsh Water Authority (1987). Llyn Brianne Acid Waters Project. First Technical Summary Report, April 1987. Welsh Water Authority, 96 pp.

CHAPTER 14

Other Aquatic Biology: Flora and Fauna

R.W. Battarbee
University College London, London, UK

N.A. Logan
Glasgow College of Technology, Glasgow, UK

K.J. Murphy
University of Glasgow, Glasgow, UK

P.J. Raven*
Department of the Environment for Northern Ireland, Belfast, UK

R.J. Aston
PowerGen plc, Nottingham, UK

G.N. Foster
Scottish Agricultural College, Ayr, UK

14.1 INTRODUCTION

While the response of fish reintroduced to the Loch after liming and other treatments was a primary objective of the Loch Fleet study, the response

* *Present address*: National Rivers Authority, Bristol, UK.

of other aquatic ecosystem components was clearly important. Studies commissioned included analysis of the diatom flora (both past and present), microbiology, primary production, zooplankton communities, benthic and littoral fauna and flora, and specifically the beetle communities, potentially competing with the introduced fish as predators. For most of these studies the limited pre-liming period allowed only some sampling opportunities in the initially acid conditions, while post-liming studies have continued for more than 4 years after lime treatments.

14.2 THE DIATOM FLORA OF LOCH FLEET

14.2.1 Introduction

The diatom flora of Loch Fleet is of particular interest since it was sampled retrospectively via sediment cores to provide a history of lake acidification, and also by real-time sampling to monitor the recent response to catchment liming.

Diatom studies at Loch Fleet have been carried out at various times over the last 10 years. The two principal projects have been that of Anderson *et al.* (1986), who used diatom analysis of sediment cores from the loch to reconstruct its acidification history, and that of Cameron (1990), who has been able to compare the diatom flora of the loch before and after liming on the basis of live communities, sediment trap assemblages, and sediment core records. Because diatoms are excellently preserved in sediment cores we have a continuous record of floristic change and we can compare the post-liming flora not only with the pre-liming flora (before 1986), but also with the historic pre-acidification flora (before about 1975) and the pre-afforestation flora (pre-1961). The pre-afforestation flora was present for over 5000 years (Battarbee *et al.*, unpublished) and so can be regarded as the original natural flora of the loch.

14.2.2 Methods

In May 1985, a 3·1-m long sediment core was taken using a Livingstone corer from the north-east corner of the loch, a site with the most rapid and most uninterrupted record of sediment accumulation. Most subsequent shorter cores were located in this part of the loch. Short cores were taken using both Mackereth and Kajak devices. Current diatom periphyton samples have also been collected from various locations around the loch since 1980. Since 1986 they have been collected systematically from fixed sample locations from a range of substrates (stone, sand, and plant) and from the open water.

Cores have been measured for dry weight, loss on ignition, and wet density, and dated using ^{14}C and ^{210}Pb techniques (Battarbee *et al.*, 1988). Diatom samples from both sediment and live material were cleaned using hydrogen peroxide (H_2O_2) to remove organic matter, and the diatoms were mounted with Naphrax for examination on glass slides using high magnification ($\times$ 1000) light microscopy.

14.2.3 Results

14.2.3.1 The Preliming Flora

A full diatom diagram for the long core from Loch Fleet taken in 1985 is presented by Anderson *et al.* (1986). A summary diagram of the uppermost 2·4 m is shown in Fig. 14.1. Carbon-14 dating gives an age of 5331 $\pm$ 75 (before present) for the sample from 235 to 250 cm, and ^{210}Pb dating shows that there was an extremely rapid rise in sediment accumulation rate dating from the early 1960s, probably caused by soil (peat) erosion in the lake catchment following ploughing by the Forestry Commission in 1961.

The diagram suggests that the diatom flora of the lake was dominated for over 5000 years by the planktonic taxon *Cyclotella kützingiana* and the periphytic species, *Achnanthes minutissima* and *Brachysira vitrea*. This assemblage indicates a water pH of about 6·0 (Anderson *et al.*, 1986).

Except for a very pronounced peak of *Cyclotella comensis* at about 200–208 cm, there is virtually no change in this flora between the base of the diagram and the 100–110 cm depth. Similar long-term stability of the diatom flora has been demonstrated from other Scottish sites (Jones *et al.*, 1989; Atkinson and Haworth, 1990), a stability that has persisted despite significant climate change and the gradual accumulation of catchment peat. The *Cyclotella comensis* peak may have been the result of nutrient inwash following a fire in the catchment (Anderson *et al.*, 1986). It appears that this disturbance was temporary since the previous flora returned to the lake in almost identical proportions after a very brief period.

Strangely, unlike other upland Galloway lakes with granitic catchments, there is no indication of increasing acidity through the nineteenth and early twentieth centuries (see Flower *et al.*, 1987). The reason for this is not clear, although a study by Cook *et al.* (1987) has shown that the lake receives about 5% of its hydrological input as highly alkaline ground water, a source of neutralising capacity that may not exist at other sites and could have delayed acidification (Cook *et al.*, 1991) (see Chapter 12).

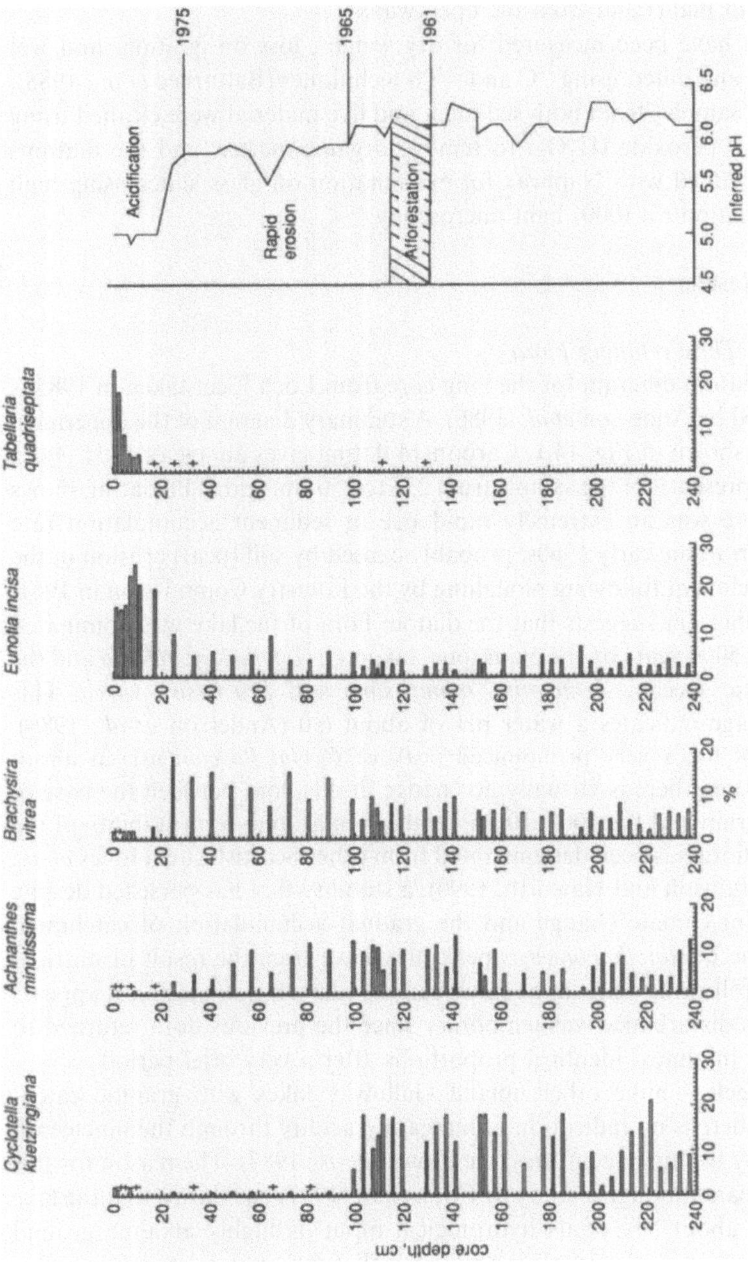

FIG. 14.1 Diatom changes and pH reconstruction in Loch Fleet sediment core (LF3) with depth taken in 1985 before liming; ages are from ^{210}Pb dating (from Anderson *et al.*, 1986). (+) indicates 'present'.

14.2.3.2 Effects of Afforestation

The first major change in the diatom flora did not occur until the 1960s when *Cyclotella kützingiana* abruptly declined, precisely coinciding with evidence of peat inwash. However, there was little change in the character of the non-planktonic flora. This differential effect strongly suggests that the plankton decline was due to physical factors, such as an increase in turbidity, rather than to acidification or other changes in water chemistry. An almost identical change occurred in the Round Loch of Glenhead (about 13 km north-west) some 300–400 years ago in response to peat erosion at that time (Stevenson *et al.*, 1990). The continued dominance of the lake diatom flora by *Achnanthes minutissima, Brachysira vitrea, Cymbella gracilis, Tabellaria flocculosa* and other taxa of a circumneutral or acidophilous character indicates that the loch had a pH after afforestation well above 5·5 and was quite capable of supporting brown trout, although the fishery was reported to have already declined (see Chapter 3).

14.2.3.3 Effects of Acidification

The uppermost 25 cm of the sediment core records the very rapid acidification of the loch when pH must have fallen at least one unit through the 1970s. By 1979, when the first reliable modern measurement was reported, the pH was already 4·4 (Wright and Henriksen, 1980). Over this period of time there had been a major change in the character of the diatom flora with the decline in taxa of circumneutral preference, such as *Achnanthes minutissima* and *Brachysira vitrea*, the expansion of *Eunotia incisa* and the appearance and expansion of *Tabellaria quadriseptata*, one of the most acid-tolerant diatoms and one that dominates the diatom flora of other acidified Galloway lakes (Flower *et al.*, 1987). At other sites in Galloway, however, this species succession took place over a period of 100 years, whereas at Loch Fleet it occurred in less than 10 years.

This change is remarkable not only for its rapidity (similarly rapid acidifications – over 10–25 years – have occurred in Scandinavian countries (Renberg and Hellberg 1982; Tolonen and Jaakkola, 1983)), but for its timing. All other acidified sites in Scotland we have studied began to change in the nineteenth century and acidified more or less in line with the notional increase in atmospheric pollution (Battarbee *et al.*, 1988). At Loch Fleet, however, acidification occurred after the downturn in national UK sulphur dioxide emissions in 1965 (Barrett *et al.*, 1988). Consequently, the reason for acidification is not clear, but one hypothesis is that the rapid accumulation of peaty sediment in the lake in the decade after ploughing isolated the lake water from direct neutralising contact with the calcium-rich pore-water of the loch sediments.

14.2.3.4 Effects of Liming

The effect of recent liming on the diatom flora of Loch Fleet has been closely monitored by Cameron (1990), both from changes in live communities (Fig. 14.1) and from the recent sediment record (Fig. 14.2). The first liming was carried out in April 1986 and a systematic survey in October 1986 showed that a major change in the composition of the diatom periphyton had already occurred. *Tabellaria quadriseptata* had declined and was being replaced by *T. flocculosa*, *Eunotia incisa*, and *Achnanthes minutissima*, species that had dominated the loch prior to acidification. However, further liming in April 1987 and a consequent continued rise in the loch water pH to approximately 6·5 resulted in the restored dominance of *Achnanthes minutissima* in the periphyton (Fig. 14.3) and the appearance of more alkaliphilous planktonic taxa, *Synedra acus* and *Asterionella formosa*. These latter taxa have never occurred before in any abundance in the loch; they are both taxa with higher pH and nutrient requirements than previous dominants and their abundance indicates that the present pH of the loch water is higher than that of the natural pre-acidification and pre-afforestation period when, as inferred from diatom reconstruction, the pH was 5·8–6·0.

The changes in the live flora are now clearly reflected in the recent sediment (Fig. 14.2). A core taken in 1988 shows a smoothing of the known record. Mixing within the sediment, possibly by bioturbation, shows that *S. acus*, not recorded in the loch until 1987, has been mixed down, while *T. quadriseptata*, which no longer occurs in the loch in any abundance, has been mixed upwards. This smoothing is reflected consequently in the reconstructed pH curve, which shows a present value of pH 6·0 for the surface sediment compared with a measured value of 6·5. The diatom-inferred pH is still influenced, and reduced, by the continued occurrence of *T. quadriseptata* in the uppermost sediment.

14.2.4 Conclusions

The diatom record of a long sediment core from Loch Fleet indicates that the diatom flora, and the inferred loch water chemistry, were virtually constant for over 5000 years until the consequences of afforestation, acidification, and liming are manifested over the last 30 years.

PLATE IV Sampling benthic invertebrates using the Freshwater Biological Association 'air-lift' sampler.

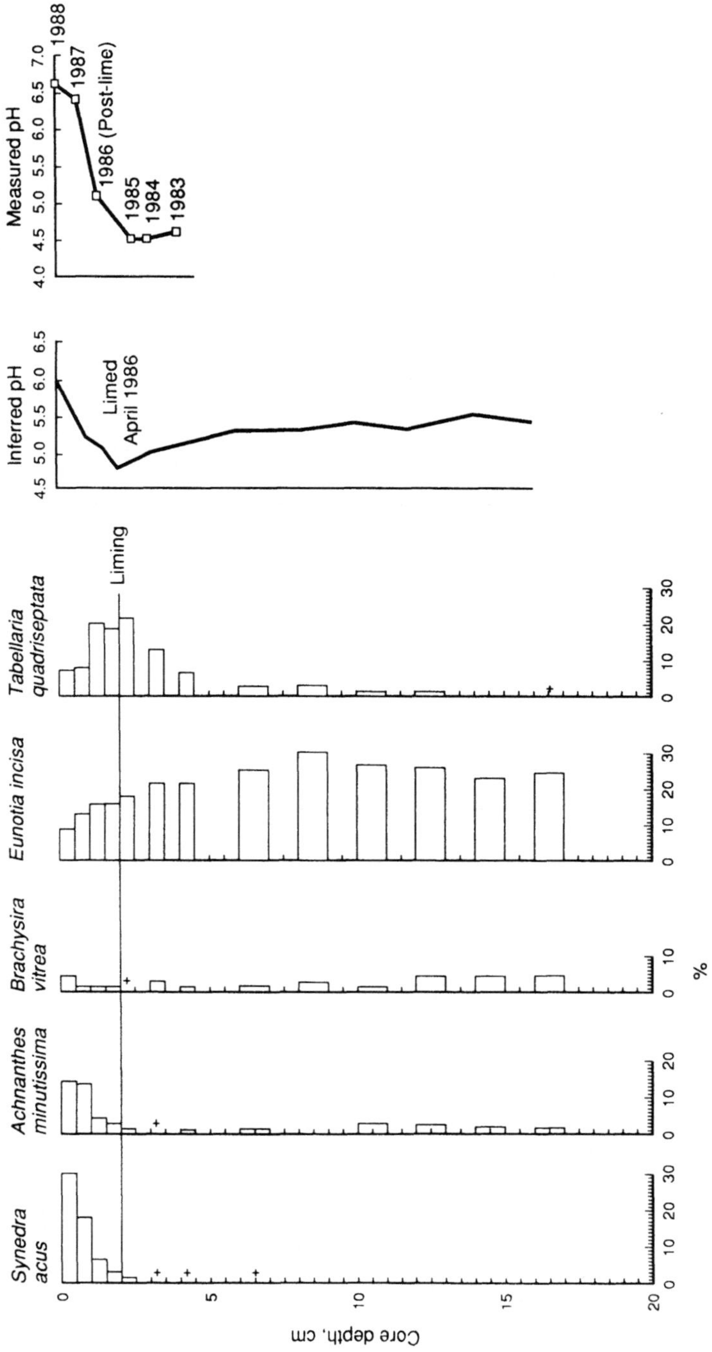

FIG. 14.2 Diatoms in a short core from Loch Fleet taken in 1988 after liming showing diatom-based pH reconstruction and measured pH (data from Cameron, 1990).

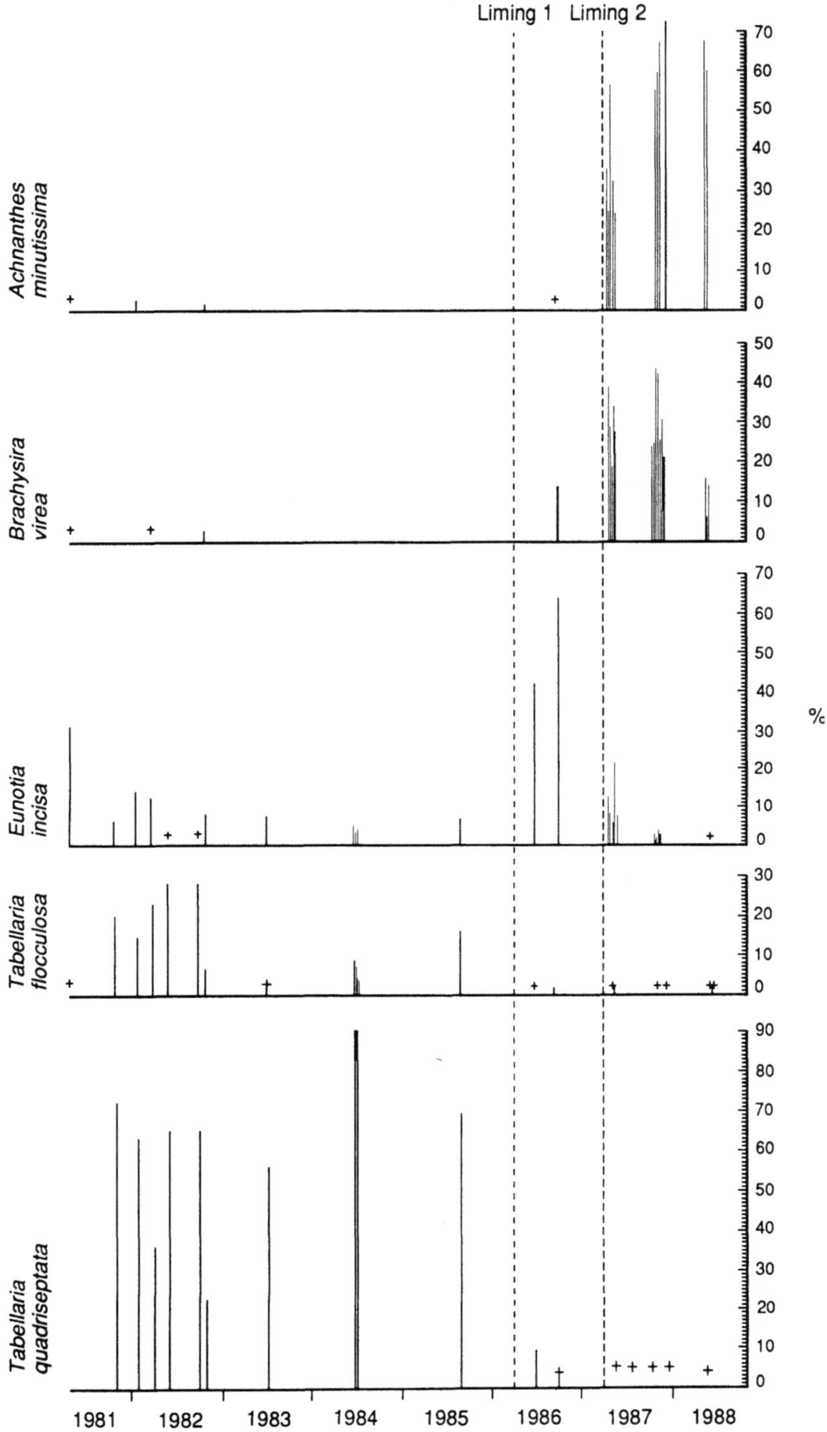

FIG. 14.3 Changes in the dominant epilithic diatom flora before and after liming (from Cameron, 1990).

Ploughing in 1961 prior to afforestation caused the loss of the diatom plankton, possibly because of the effects of soil inwash, but it did not directly cause the lake to acidify.

The acidification of Loch Fleet occurred in the mid-1970s and was associated with a major change in the composition of diatom species. The timing of this change contrasts with that of all other nearby Scottish sites studied and may be related to the indirect effects of afforestation.

Liming in 1986 and 1987 caused the re-occurrence of taxa which previously occurred in the loch, but the appearance of new alkaliphilous planktonic taxa indicates that the present pH (about 6·5) is higher than has occurred naturally in the loch before.

14.3 MICROBIOLOGICAL STUDIES

Microbiological studies to assess the activities and diversities of microbial populations were undertaken over a 2-year period in conjunction with a wider study of other lakes of different acidities within the region.

14.3.1 Methods and Programme

Little information was available on the effect of acidification on the bacterial floras of lakes in the region, so other acidified and non-acidified waters were investigated. For comparison, these waters were selected to have similar underlying geologies to Loch Fleet and were located in areas with comparable acid deposition and development of forestry in their catchments, and covered a pH range from neutral to highly acidified. Those chosen were Loch Grannoch (pH 3·3–4·2), Loch Doon (pH 4·3–5·0), Barscobe Loch (pH 5·7–6·9), and Lochinvar (pH 6·2–6·9) (Fig. 14.4).

A programme of regular sampling over two years (1986–87) was undertaken, and aerobic heterotrophic bacterial numbers, diversities, and activities were estimated for all the waters to chart temporal changes in their floras and any differences attributable to lake acidities and chemistries. These control data would then allow any effects of the treatment regimes at Loch Fleet to be recognized.

Sub-surface samples of 500 ml loch water were aseptically collected from Loch Fleet and its embayments at fortnightly intervals, weather permitting, and the other lakes were sampled at monthly intervals. Loch Fleet was also sampled daily for a fortnight each summer; water samples from different depths, and sediment samples were collected from all

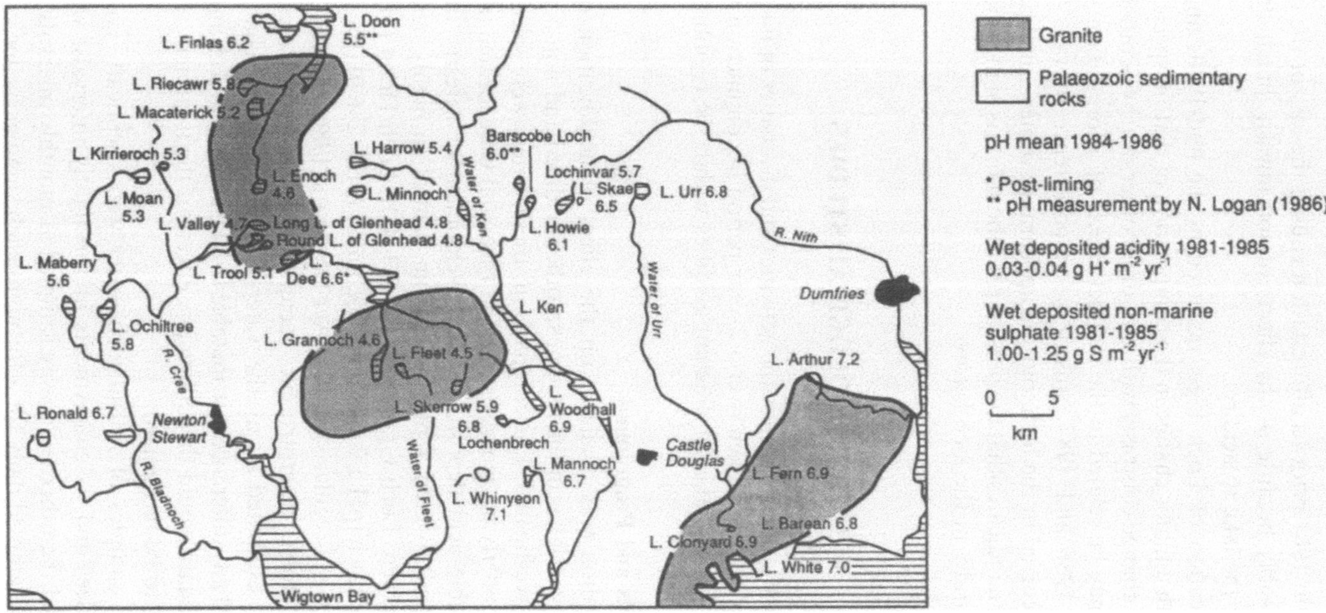

FIG. 14.4 Lakes in Galloway, showing relationship of measured pH and geology (after Battarbee *et al.*, 1988).

sampling sites. Water temperature, pH, and dissolved oxygen were measured in the field on each occasion and further samples were collected for chemical analysis. Plates of casein peptone starch agar were inoculated in triplicate with undiluted water samples or diluted sediment samples and incubated for 7 days at 25°C. Viable counts were made, colonies were classified on the basis of morphology, and the Shannon–Weaver index was calculated to give estimates of diversity, equitability, and species richness. The validities of the 31 types recognised were subsequently confirmed by computer taxonomy of representative isolates based on 32 characterisation tests. Total counts were made by a direct epifluorescent filter technique. Activity was measured as the rate of evolution of $^{14}CO_2$ from the metabolism of radioactively labelled glutamate, corrected for the viable count, and expressed as activity per bacterium.

14.3.2 Results

Other than obvious changes in water quality after liming treatment at Loch Fleet, the lakes showed similar and stable water chemistries. None of the lakes showed major differences with depth in measurements of temperature, pH, dissolved oxygen, or in the range of bacterial types present. Despite the wide range of pH values of the different lake waters the bacterial diversities were very similar, and pH showed no significant correlation with bacterial counts or activities either. The ratio of viable counts to total counts (mean 0·5) was similar for all waters. Only one type of organism, identified as a *Cytophaga* species, appeared to be affected by low pH. Counts of this organism were lower in acidified waters but increased in Loch Fleet after the catchment had been limed.

The diversities of the cultivable bacterial populations appeared to be inversely related to water temperature. As temperatures rose during the summer months some organisms became dominant whilst others became rarer or undetectable; these changes are seen as reductions in both equitability and species richness, two components of diversity (Fig. 14.5). These correlations were less distinct in the second year's data: species richness fell but equitability remained stable in the main body of Loch Fleet. The rate of temperature increase appears to be an important factor – in the first year, water temperature rose sharply to 20°C during the daily sampling period but in the second year the temperature rose only slowly to a maximum of 18°C.

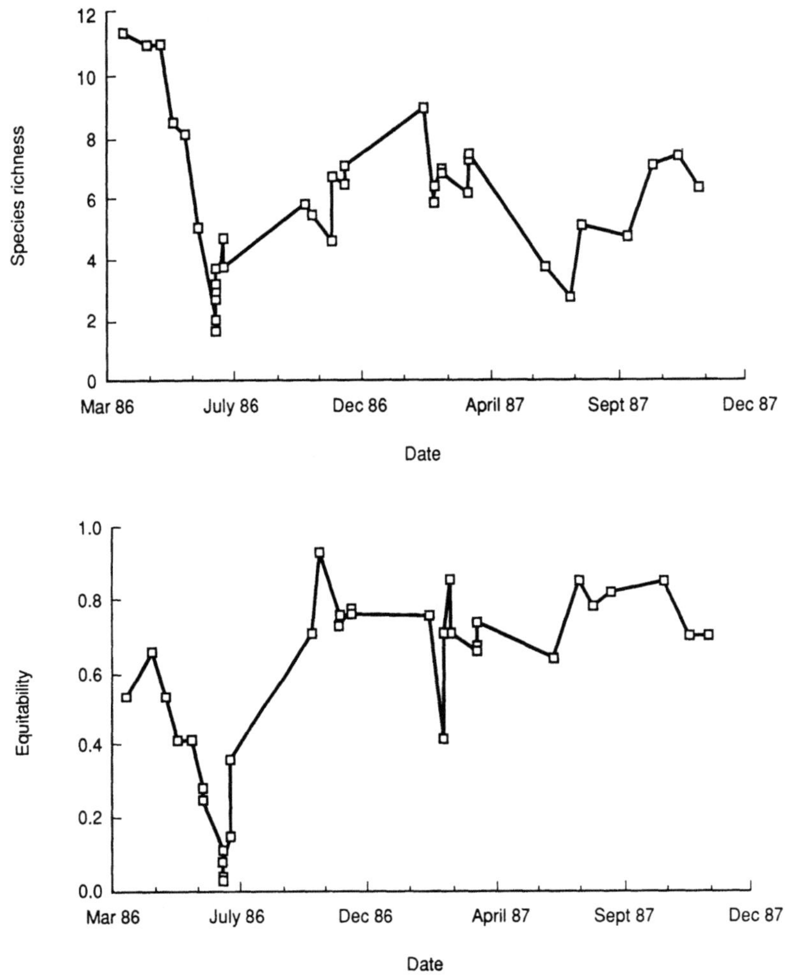

FIG. 14.5 Two indicators of microbial diversity, in relation to seasonal changes in Loch Fleet.

14.3.3 Conclusions

Lake water acidification in the range pH 3·3–6·9 appears to have no gross effect on the population densities, diversities, and activities of aerobic heterotrophic bacteria in Loch Fleet and Loch Grannoch: one organism, however, a *Cytophaga* species, was isolated in smaller numbers from acidified waters and reappeared in Loch Fleet after liming. In all of the lakes studied, diversity decreased as water temperatures rose, especially when the rise was rapid.

14.4 PRIMARY PRODUCTION

14.4.1 Introduction

Three methods were used to monitor changes in phytoplankton production: direct cell counts, chlorophyll analysis, and ^{14}C labelled uptake of sodium bicarbonate. Macrophyte production was monitored by estimating the biomass of samples collected up to depths of 70 cm. Phytoplankton production was measured by analysis of chlorophyll *a* in extracts of 10-litre water samples taken from shoreline embayments (4, 5, 8, and 9) and the loch centre and outflow area. Sampling was approximately monthly from April 1985 to October 1989. Peaks were reported for April and September, with values higher in embayments 4 and 5 than in 8 and 9, and lowest in the loch centre. The values in 1987 overall are about 10 times higher than those at the same sites in 1986; further increases were evident in 1988, but values were lower in 1989.

The amount of ^{14}C fixed per litre-hour was estimated for samples taken in July, August, and September; uptake is less than 0·1% of added sodium bicarbonate.

Additional information was obtained for water quality variables linked with productivity – dissolved oxygen, silicate, and underwater light regime.

14.4.2 Methods

Algal standing crop (phytoplankton plus resuspended epiphytic and benthic algal cells) in the water column of Loch Fleet was determined from water samples collected on each sampling occasion from open water and embayment sampling stations. The normal approach in algological studies of estimating algal standing crop in terms of chlorophyll *a* (Chl*a*) content was adopted. Immediately on return to the laboratory, water samples were filtered through Whatman 0·45 μm glass fibre filter paper, the paper ground in 10 ml acetone, then stored under refrigeration for 24 h prior to centrifuging at 2500 rev min^{-1} for 5 min. Chl*a* concentration in the decanted supernatant was determined by spectrophotometric assay, and calculated using the standard (SCOR-UNESCO) equation:

$$\mathrm{Chl}a(\mathrm{mg\,m^{-3}}) = \frac{vAK(663\mu - 663a)}{V_f l} \qquad (1)$$

where

l = light path (cm)
v = volume of acetone extractant (ml)

A = Chl*a* absorption coefficient = 11·0
K = 2·43
V_f = volume of water filtered (litres)
663μ = absorbance of unacidified sample at 663 nm
$663a$ = absorbance of sample at 663 nm after the addition of 1 drop 0·1M HCl.

The population density of algal cells present in water samples (a standard complementary approach to Chl*a* concentration used in assessing algal production) was determined by cell counts. Known volumes of water were fixed and stained with a solution of Lugol's iodine plus glycerine, and left for 21 days under refrigeration to allow the stain to be incorporated in algal cells, prior to filtering through a cellulose nitrate membrane (pore size 0·45 μm). One drop of Univert immersion oil was then added to a section of the membrane, and total numbers of cells present in 50 or 60 fields counted at × 400 magnification. Cell population density (number of cells per litre) was calculated by:

$$\text{number of cells per litre} = \frac{A_f C}{A_t V} \qquad (2)$$

where

A_f = area of filter (mm^2)
C = total cell counts in n fields
A_t = total area of n fields (mm^2)
V = volume of sample (litres)

"Snapshot" estimates of the primary productivity of the algal community present in the loch water at the time of sampling were obtained using standard ^{14}C *in situ* light and dark assay procedures.

14.4.3 Results

An overall increase in algal populations was observed in 1987 (1 year after liming). An increase in *Cyclotella* was reported for summer months, possibly reflecting higher pH levels.

Initially, low measures of production were recorded for both macrophytes and algae, comparable with those found elsewhere in nutrient-poor, low pH temperate lakes. Between 1986 and 1988, algal production, measured in terms of photosynthetic pigment standing crop, mostly attributed to diatoms and unicellular algae, at least doubled (Fig. 14.6). Over the period, submerged macrophyte production also increased substantially, by about three-fold, from a very low baseline.

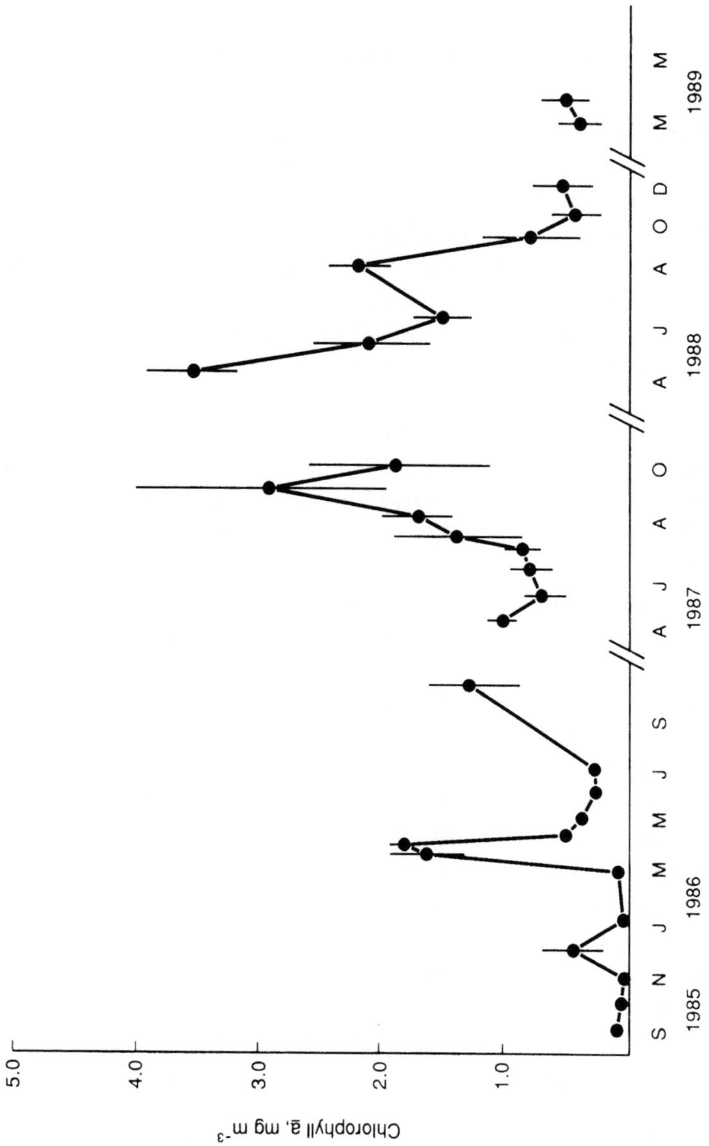

FIG. 14.6 Suspended algae measured as chlorophyll *a* (means and SE) in Loch Fleet, 1985 to 1989.

Contemporaneous with an increased calcium and alkalinity content of the water after liming, conductivity increased and some decrease in underwater light was observed in 1987–88 compared with earlier years.

14.5 MACROPHYTE STUDIES

14.5.1 Introduction

Sphagnum moss has recently invaded a number of acidified lakes, notably in Sweden and the Netherlands, frequently becoming the dominant macrophyte and displacing isoetid species such as *Lobelia dortmanna* and *Isoetes lacustris* (e.g. Grahn *et al.*, 1974; Grahn, 1977; Roelofs, 1983). It has also become established in some acid lakes in Scotland and Wales (Raven, 1988a, b, and unpublished data). Since submerged *Sphagnum* usually proliferates only in acid waters (pH $< 5{\cdot}6$), it declines rapidly when acid lakes are limed (Hultberg and Andersson, 1982). A considerable amount of *Sphagnum* had established in Loch Fleet, probably since rapid water acidification in the mid-1970s (Raven, 1986). The objective of this study (1985–88) was to assess short-term changes in the distribution and relative abundance of aquatic macrophytes after liming.

14.5.2 Methods at Loch Fleet

Aquatic macrophytes, namely vascular plants, bryophytes (mosses and liverworts), and algae discernible with the naked eye (e.g. filamentous green algae and *Batrachospermum*) were sampled by taking Ekman grab samples both randomly in the loch and at 2-m intervals along four 30-m transects aligned at right angles to the shore. Water depth, substrate, and macrophyte taxa were recorded at each sampling site and the amount of plant material recovered was estimated visually using a crude qualitative scale, comprising "present", "abundant", and "very abundant".

Pre-liming distribution data obtained in September 1985 indicated that living macrophytes were confined to water less than 5 m deep, so the three subsequent annual surveys (September 1986–September 1988) were concentrated in these peripheral areas. The transects were located on the sheltered, gently shelving south-western shore, the steep, exposed northern shore, and two sites subsequently enclosed within embayments 9 and 5, the latter representing an acid water control.

Samples of *Sphagnum* were retained to estimate mean dry weight for each of the qualitative abundance categories described above. Total *Sphagnum* dry weight biomass in the loch was estimated by multiplying the relative percentage occurrence and abundance data, the presumption being that percentage occurrence gives a good overall indication of *Sphagnum* cover. The length of the green (living) portion of *Sphagnum* shoots, which broadly indicates the amount of growth during the current growing season (Grähn, 1985), was measured to assess annual changes in condition.

Water level was measured relative to a fixed shoreline mark and all water depth data were standardised to a September 1986 reference level. Floristic changes throughout the loch were assessed by comparing the percentage occurrence data of taxa during 1985–88 and evaluated further by considering each 1 m depth interval. For simplicity, the random sublittoral and transect Ekman grab data, other than those taken within the embayments, were combined.

Annual changes in relative abundance for selected macrophytes were estimated using an index calculated by numerically weighting the abundance categories recorded in the field (Raven, 1989). For each taxon, pre-liming (1985) abundance status was represented by an index value of 100. Abundance in subsequent years was expressed as a percentage of that value. This could only be used to assess relative changes in the abundance of an individual taxon from year to year, not directly as a comparison between taxa. A more accurate assessment of change in the abundance of *Sphagnum*, based on dry weight biomass, is described above.

14.5.3 Results

Prior to liming, the species-poor aquatic flora was typical of an exposed upland oligotrophic lake with acid water (Table 14.1). Emergent vegetation was restricted to *Carex rostrata* growing in the sheltered north-west bay. The predominantly rocky, boulder strewn shoreline severely restricted macrophyte growth inshore. Both *Littorella uniflora* and *Lobelia dortmanna* were patchily distributed and abundant only in areas of sand or gravel. Vascular plants were extremely scarce on the exposed rocky eastern shoreline.

In deeper water *Isoetes lacustris* and *Sphagnum auriculatum* were locally abundant on organic sediment. Consequently, there was a distinct *Littorella–Lobelia–Isoetes–Sphagnum* depth zonation (Fig. 14.7). *Sphagnum auriculatum*, the dominant macrophyte in water 2–5 m deep, was particularly abundant on muddy sediments off the sheltered western shore, growing as a dense carpet more than 30 cm deep in places. In September 1985 the total dry weight biomass of *S. auriculatum* and *S. cuspidatum* in the loch was estimated to be 508 kg and 2·6 kg, respectively.

TABLE 14.1
The percentage occurrence of aquatic macrophytes in Loch Fleet before and after catchment liming

Macrophyte	*Before liming*	*After liming*		
	September 1985	*September 1986*	*September 1987*	*September 1988*
Sphagnum auriculatum	54·2	57·5	7·8[39·9]	7·6
Isoetes lacustris	34·5	36·8	34·6	38·2
Lobelia dortmanna	18·7	10·3	13·1	10·2
Solenostoma triste	12·1	20·7	24·8	–
Sphagnum cuspidatum	9·3	5·7	0·6	–
Littorella uniflora	8·4	2·3	11·1	10·8
Filamentous green algae	5·6[a]	4·6	11·1	58·6[a]
Juncus bulbosus var. *fluitans*	0·9	2·3	8·5	3·8
Utricularia sp.	0·9	4·6	13·7	20·4
Fontinalis sp.	–[b]	–	–	0·6
Isoetes echinospora	–[b]	–	2·0	1·9
Amblystegium serpens	–	2·3	2·6	11·5
Batrachospermum spp.	–	–	1·3	0·6
Bryum pallens	–	–	–	0·6
Calypogeia mulleriana	–	1·1	2·0	2·5
Cephalozia connivens	–	1·1	1·3	1·3
Drepanocladus fluitans	–	1·1	4·6	8·3
Lophozia sp.	–	1·1	–	–
Polytrichum commune	–	1·1	–	0·6
Scapania undulata	–	1·1	–	0·6
Number of Ekman grabs	107	87	153	157

[a] Mainly *Mougeotia* spp.
[b] Recorded during a littoral mapping survey.
Notes: Figures represent occurrence in Ekman grab samples from water less than 5 m deep. Taxa arranged in order of initial relative frequency. Nomenclature follows Holmes *et al.* (1978). [ ] includes *Sphagnum* showing signs of die-back.

The number of macrophytes increased after liming, although with few exceptions the newly established taxa (mostly bryophytes) remained minor components of the macrophyte community (Table 14.1). Twelve macrophytes have been considered in some detail, with the decline of *Sphagnum* given special attention.

Lobelia remained the most widespread vascular plant in shallow water throughout 1985–88, although there was evidence of a decline in its abundance during 1988 (Tables 14.1 and 14.2). *Littorella* became more

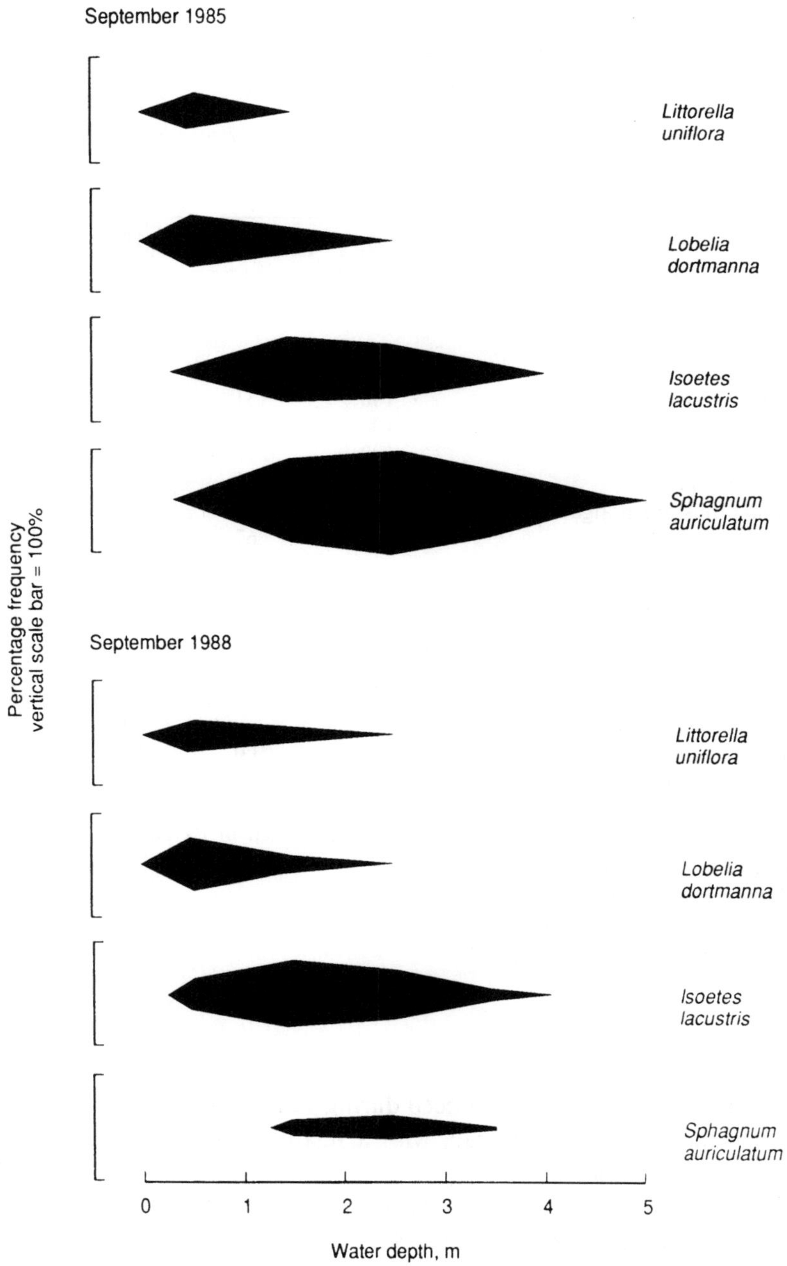

FIG. 14.7 Depth distribution of four species of macrophytes, before and after liming.

TABLE 14.2
Changes in the relative abundance index of 12 selected macrophyte taxa in Loch Fleet after catchment liming

	Abundance index[a]			
	Before		After liming	
	1985	*1986*	*1987*	*1988*
(a) Taxa with > 10% frequency in any one year				
Sphagnum auriculatum[b]	100	90	8	1
Isoetes lacustris	100	131	150	151
Lobelia dortmanna	100	118	101	50
Solenostoma triste	100	236	301	0
Littorella uniflora	100	41	240	236
Filamentous green algae	100	246	243	2217[c]
Utricularia sp.	100	767	3250	4667
(b) Taxa with > 10% frequency in at least one 1 m depth interval in any year				
Sphagnum cuspidatum[b]	100	50	20	0
Juncus fluitans	100	225	944	422
Isoetes echinospora	—	—	100	95
Amblystegium serpens	—	100	113	720
Drepanocladus fluitans	—	100	418	1328

[a] Index of 100 shown for the first year taxon recorded in Ekman grab samples.
[b] Based on dry weight biomass.
[c] Mainly *Mougeotia* spp.
Notes: Figures represent an abundance index indicating percent of the pre-liming (1985) value for each macrophyte.

widespread in water 1·1–2·0 m deep during 1987–88 (Fig. 14.7), and also became more abundant overall, with a relative abundance index more than twice that prior to liming.

The leafy liverwort, *Solenostoma triste*, increased both in distribution and abundance during 1986 and 1987, but was not recovered alive from Ekman grab samples taken in September 1988 when there was a large amount of liverwort debris in the loch. Virtually all liverworts in shallow water were completely smothered by filamentous green algae (mainly

Mougeotia spp.), which had proliferated that year (Tables 14.1 and 14.2). *Juncus fluitans* also increased in distribution and abundance after liming but suffered a reversal in 1988 when much of its foliage was thickly coated with *Mougeotia*.

Bladderwort, *Utricularia* sp., increased rapidly, particularly in water 1–3 m deep, and by 1988 it was the second most widespread vascular macrophyte (20·4% frequency), with an abundance index almost 50 times that prior to liming (Tables 14.1 and 14.2).

The distribution of *Isoetes lacustris* was similar before and after liming (Fig. 14.7), but its abundance index had increased 50% by 1987. Furthermore, many rosettes were noticeably larger and more vigorous during 1987–88. *Isoetes echinospora* increased in distribution but remained relatively scarce compared with *I. lacustris* (Table 14.1).

Of those species unrecorded before liming, two mosses, *Amblystegium serpens* and *Drepanocladus fluitans*, increased significantly in distribution and abundance during 1987–88, replacing *Sphagnum* spp. as the most frequent macrophyte taxa in deeper (3·1–4·0 m) water. *Potamogeton polygonifolius* appeared by September 1986 and within 2 years had spread to several other sites, including embayments 4 and 9.

There was little change in the percentage observed frequency of *Sphagnum auriculatum* within 6 months of liming. However, there was evidence for reduced growth during the 1986 summer. Eighteen months after liming (September 1987), there was considerable die-back and most *Sphagnum* shoots were either dead or showed less than 10 mm growth (Table 14.1, Fig. 14.8). Within 30 months of liming there had been a 99% decline in the estimated total dry weight biomass and remnant plants were confined to water 1·1–3·0 m deep (Table 14.3). The surviving *Sphagnum* was apparently in good condition, with some living shoots exceeding 40 mm in length. *Sphagnum cuspidatum* declined relatively quickly, disappearing from deeper water within 6 months and was not recorded at all from the loch 30 months after liming (Table 14.3).

In direct contrast to the open loch, *S. auriculatum* and *S. cuspidatum* continued to flourish in embayment 5 where the water remained at a pH of about 4·5–4·7 throughout. The length distribution frequency of living shoots from embayment 5 in 1988 was similar to that for the entire loch before catchment liming, the modal class length being 41–50 mm in each case (Fig. 14.8). Moreoever, the estimated dry weight biomass of *S. auriculatum* in the embayment increased from 6·0 kg to 13·4 kg during 1985–88.

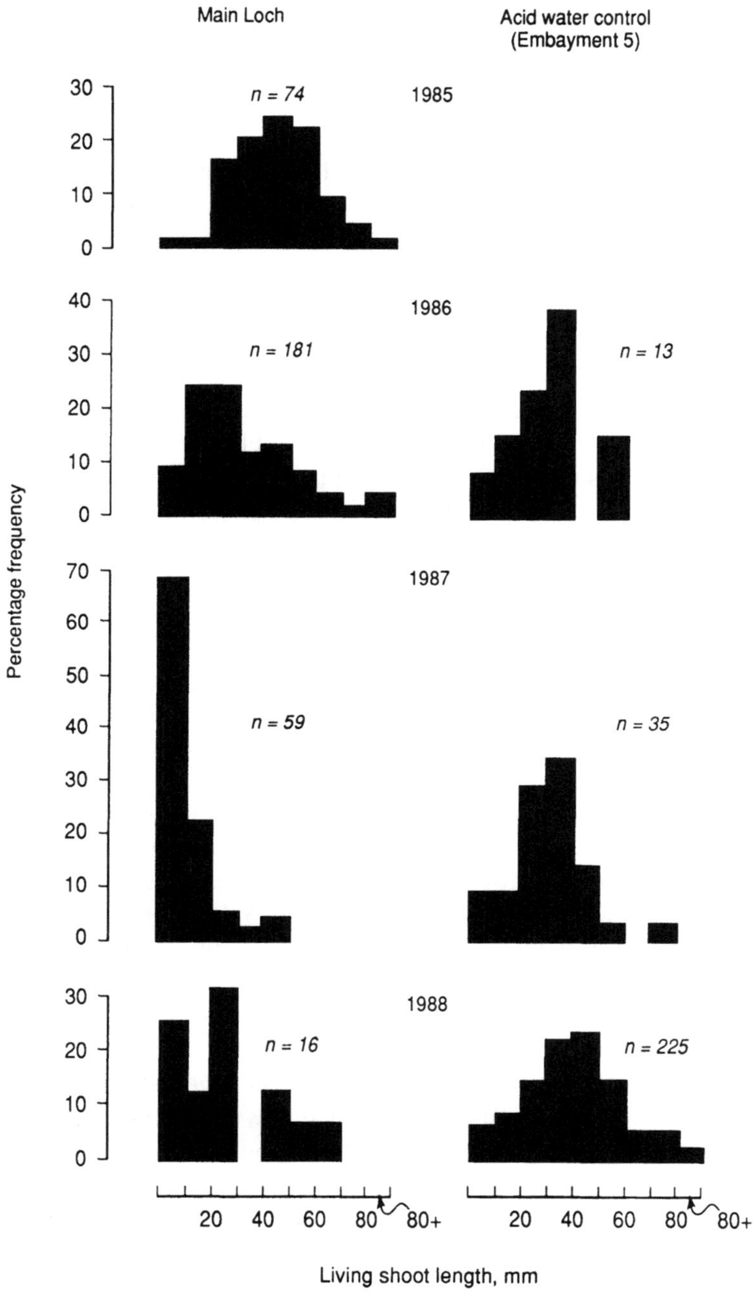

FIG. 14.8 The lengths of living *Sphagnum auriculatum* shoots before and after liming, compared with that in control enbayment 5. (n = number of observations.)

TABLE 14.3
The estimated dry weight biomass of *Sphagnum auriculatum* and *S. cuspidatum* in Loch Fleet before and after liming

(a) *Sphagnum auriculatum*

Water depth (m)	*Dry weight biomass* (kg)			
	1985	*1986*	*1987*	*1988*
≤1·0	9·4	1·5	0·9	–
1·1–2·0	143·3	139·2	11·3	1·9
2·1–3·0	205·2	182·9	24·6	3·7
3·1–4·0	138·4	136·3	1·3	–
4·1–5·0	11·9	–	1·3	–
Total	508·2	459·9	39·4	5·6

(b) *Sphagnum cuspidatum*

Water depth (m)	*Dry weight biomass* (kg)			
	1985	*1986*	*1987*	*1988*
≤1·0	0·23	–	–	–
1·1–2·0	0·59	0·52	–	–
2·1–3·0	0·37	0·79	–	–
3·1–4·0	1·42	–	–	–
4·1–5·0	–	–	0·53	–
Total	2·61	1·31	0·53	–

14.5.4 Conclusions and Summary

The decline of *Sphagnum* in Loch Fleet was broadly consistent with, but relatively slower for some surface-limed acid lakes in Sweden and upland Wales. For example, *Sphagnum* growth in Lake Trehörningen was greatly impaired very soon after liming, while in Lake V. Skälsjon, *Sphagnum* in contact with unslaked lime died almost immediately (Eriksson *et al.*, 1983). However, in Lake Lysevatten, where lime was applied to shallow areas, some depauperate *Sphagnum* survived in deeper (2–10 m) water (Hultberg and Andersson, 1982). In Llyn Berwyn, some *Sphagnum* survived in the vicinity of acid inflow streams (N. Stringer, Welsh Water Authority, pers. comm.). Because these lakes reacidified relatively quickly, *Sphagnum* was able rapidly to recolonise and proliferate. Similarly, although liming the catchment of Loch Fleet produced a more sustained increase in loch pH, the continued vigorous growth of *Sphagnum* in the acidic control embayment suggests that *Sphagnum* would probably recolonise if the water were to reacidify in future.

Assessing changes of macrophytes other than *Sphagnum* is difficult because there are very few other studies for comparison and, since each catchment is unique and the method of lime application not standardised, directly comparable results are unlikely. In Lake Bredvatten, for instance, both *Isoetes* and *Lobelia* recolonised areas vacated by *Sphagnum* after liming (Hultberg and Andersson, 1982), but in three other limed lakes in Sweden, no such response occurred (Eriksson *et al.*, 1983). Despite evidence that *Littorella* was extending into deeper water and that *Isoetes* growth was more vigorous after liming, isoetid species had clearly not recolonised areas vacated by *Sphagnum* in Loch Fleet within 30 months of liming. However, given suitable conditions, this might occur in the longer term because evidence from the sedimentary record indicates that *Isoetes* was more abundant in Loch Fleet prior to acidification (Anderson *et al.*, 1986).

In Loch Fleet, *Utricularia* sp., *Amblystegium serpens*, and *Drepanocladus fluitans* responded quickly to liming by colonising deeper water (2–4 m) as *Sphagnum* died away. *Utricularia* also increased in Lake Trehörningen after liming (Eriksson *et al.*, 1983), but the appearance and subsequent increase of *D. fluitans* in Loch Fleet contrasts with its rapid decline in Llyn Berwyn after liming (N. Stringer, Welsh Water Authority, pers. comm.).

The relative scarcity of *Juncus fluitans* in Loch Fleet before liming and its increase after liming contrasts with evidence suggesting that this species usually thrives in acid waters (e.g. Roelofs, 1983) and declines rapidly when water is deacidified (Aulio, 1987). However, the appearance of *Potamogeton polygonifolius* in Loch Fleet can be compared with the colonisation of L. Gårdsjön (Lazarek, 1986), L. Trehörningen, and L. Langsjön (Eriksson *et al.*, 1983) by *P. natans* soon after liming.

Algae can have a significant impact on the depth distribution of macrophytes through shading (Sand-Jensen and Søndergaard, 1981). Shading by filamentous green algae, notably *Mougeotia* spp., may therefore have contributed to the reduced water transparency of Loch Fleet (2·0 m Secchi disc depth in 1988 compared with 2·7 m prior to liming) and consequently caused a reduction in the maximum depth of macrophyte growth from 5 m to 4 m. Furthermore, direct smothering by filamentous algae probably caused the demise of *Solenostoma triste* in Loch Fleet during 1988. The apparent decline in *Lobelia* abundance that year (Table 14.2) may also have been due to the shading effect of the algae, because *Lobelia* is known to be adversely affected by the presence of dense epiphytic growth (Lazarek, 1986).

The sudden proliferation of filamentous green algae in 1988 suggests that the Loch Fleet ecosystem was still unstable 30 months after catchment liming. Longer-term floristic changes will depend upon numerous factors, including water chemistry, nutrient availability, water transparency, competition, and grazing pressure. This will depend to a large extent on the long-term suitability of the loch to support a sustainable trout fishery and consequently the catchment management required to maintain suitable water quality.

14.6 MACRO-INVERTEBRATES OF LITTORAL, BENTHIC, AND STREAM COMMUNITIES

14.6.1 Introduction

Invertebrates are an important part of the aquatic ecosystem with a specific role as items of fish diet, providing an important link by assimilating energy and nutrients derived from detritus and primary production. Furthermore, selected species are indicators or monitors of the degree of acidification, or its reversal. Studies at Loch Fleet included regular sampling from 1984 to 1990 of benthic communities of streams and the loch littoral, as well as aquatic beetles, significant as predators in the pre-limed fishless loch. The invertebrate community studies included a preliming period fom July 1984 to April 1986 when no fish were present; a post-liming period from summer 1986, prior to the reintroduction of fish in summer 1987; and a period after fish stocking from May 1987 to October 1989. Evidence was sought for changes in the invertebrate community that might be associated with water quality changes and fish predation.

14.6.2 Methods

14.6.2.1 Site and Sampling Points

The bottom substrates colonised by aquatic invertebrates varied from fine black peaty mud in the deep section at the northern end of the loch, to bare rock substrates along parts of the eastern shore. In deep water in the southern section of the loch the substrate was white clay mixed with stones, gravels, and syntic sand. The main inlet and outlet streams are fast-flowing with beds made up of gravel and stones and, in one section of the outlet stream, bare rock.

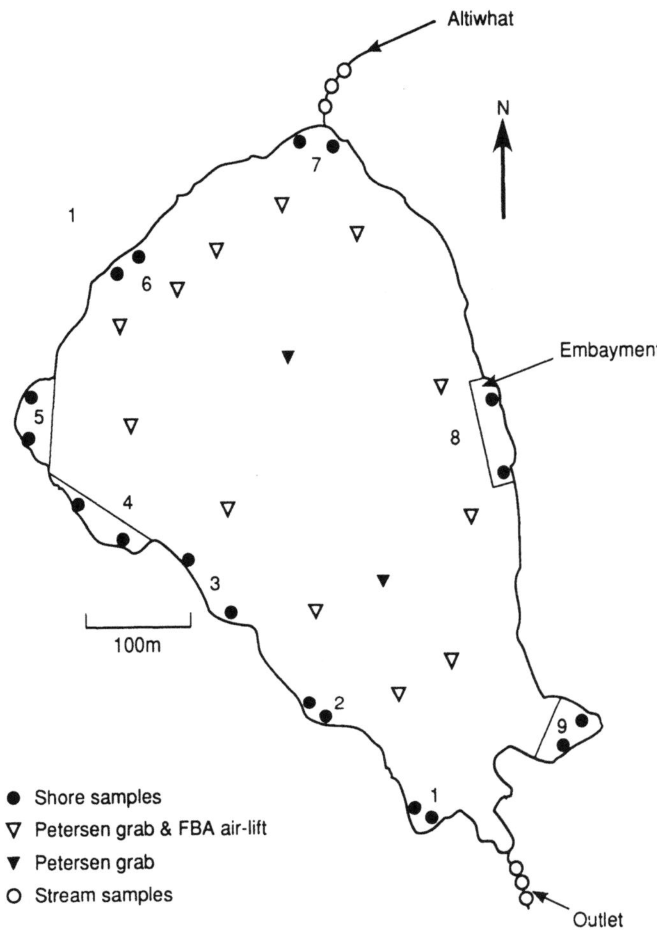

FIG. 14.9 Loch Fleet: invertebrate sampling points and methods used (numbers refer to bays).

Figure 14.9 shows the sampling points used for invertebrates and an indication of the sampling methods used. The samples were taken in spring, summer, and autumn in a sampling period extending from summer 1984 to autumn 1990. The shore samples were arranged in pairs in nine bays, four of which were separated from the loch by polyethvlene curtains for variable periods. In design, the enclosed embayments were to receive water only from the up-slope section of catchment (Fig. 14.9) (see Chapter 4), although some leakage and mixing with water from the main water body occurred during windy/stormy weather.

14.6.2.2 Sampling Methods

Invertebrates near the loch shore were sampled with a hollow metal cylinder which was worked into the substrate to enclose an area of approximately $0{\cdot}16\,m^2$. The enclosed substrate was agitated with a fork and large stones and aquatic weeds removed from the sampling area on to a sieve (mesh size 250 μm). Any invertebrates suspended in the water inside the cylinder were then pumped on to the sieve for collection.

Samples from substrates in deep water were taken both with a Petersen grab and an F.B.A. airlift sampler (Drake *et al.*, 1983) fitted with an extension tube to work at 4 m depth (Plate IV). Abundance is calculated as numbers per square metre.

The main inlet and outlet streams were sampled with a square framed handnet (mesh size 120 μm) fitted with side deflection plates to channel the stream flow into the net. The net was held downstream of a quadrat covering an area of $0{\cdot}09\,m^2$.

The samples were preserved in formalin and the invertebrates sorted into major groups and counted. Animal groups present in very small numbers were grouped together as "others" and included the Platyhelminthes, Hirudinea, Mollusca, Odonata, Megaloptera, and Diptera (excluding the Chironomidae and Simulidae). Crustacea were distinguished only as copepods and cladocerans.

For statistical treatment of the data the total numbers and the numbers of invertebrates per square metre of substrate area in each group were transformed to $\log_{10}$ to render the variances independent of the means (Sokal and Rohlf, 1981). The data were then analysed using a general linear model procedure recommended in SAS (1985) for models where there are unequal numbers of observations for different variables. This was followed by comparison between the numbers of invertebrates per unit area before liming, after liming, and after the introduction of trout, using Duncan's multiple range test (Duncan, 1975).

14.6.3 Results

Figures 14.10 and 14.11, and Tables 14.4, 14.5, and 14.6 show changes in the invertebrate populations after the application of lime and after the introduction of fish.

14.6.3.1 Loch Observations

Liming was followed by a reduction in numbers (all groups combined) during the first year, and recovery to preliming numbers in subsequent years in deep water samples but not in shore samples (Figs 14.10 and

14.11). In the embayments, numbers before and after the introduction of trout were not significantly different from other shore samples ($p > 0{\cdot}05$).

Figures 14.10 and 14.11 show the invertebrates of various groups (numbers m^{-2}) before liming, after liming, and after the introduction of fish. Statistically significant differences were found in numbers of some groups between these three periods. In the year after liming, for example, there were significant reductions in the numbers of Chironomidae, Hydracarina, and Oligochaeta (Fig. 14.11 and Table 14.4). Comparison of the numbers before liming with those after liming, together with the presence of fish (Table 14.4), indicate reductions in the numbers of Chironomidae, Coleoptera, Hydracarina, and Oligochaeta in shore samples and of Ephemeroptera (leptophlebids) in deep water. The Trichoptera, however, were more abundant than before liming in the shore samples.

Table 14.4 (column B), which compares numbers after liming with and without fish, shows that the Chironomidae and Trichoptera were more abundant after the introduction of fish than before, while numbers of Hydracarina were further decreased in shore samples after the introduction of fish.

14.6.3.2 Stream Observations

An increase in the total numbers of invertebrates in the inlet stream occurred after liming and before the introduction of trout (Table 14.5) but there was no significant change in numbers in the outlet stream (Table 14.6). Following the introduction of trout there was a decrease in the total number of invertebrates in the inlet but not in the outlet stream.

In contrast to the loch, the numbers of Chironomidae, Coleoptera, and Ephemeroptera increased in the inlet stream in the year after liming (Table 14.5). The numbers of Ephemeroptera in this stream continued to increase after the introduction of fish but there were significant reductions in Coleoptera and Trichoptera. No representatives of the Ephemeroptera were present in the inlet stream before liming, but after liming it was colonised by *Baetis rhodani* (family Baetidae). This species had not been recorded previously in any part of the loch catchment and until this time all Ephemeroptera in both the loch and the outlet stream were from the more acid-tolerant family Leptophlebidae, able to colonise deep water (Savage, 1986).

In the outlet stream none of the changes in numbers in any of the groups after liming, but before the introduction of fish, was significant at the 95%

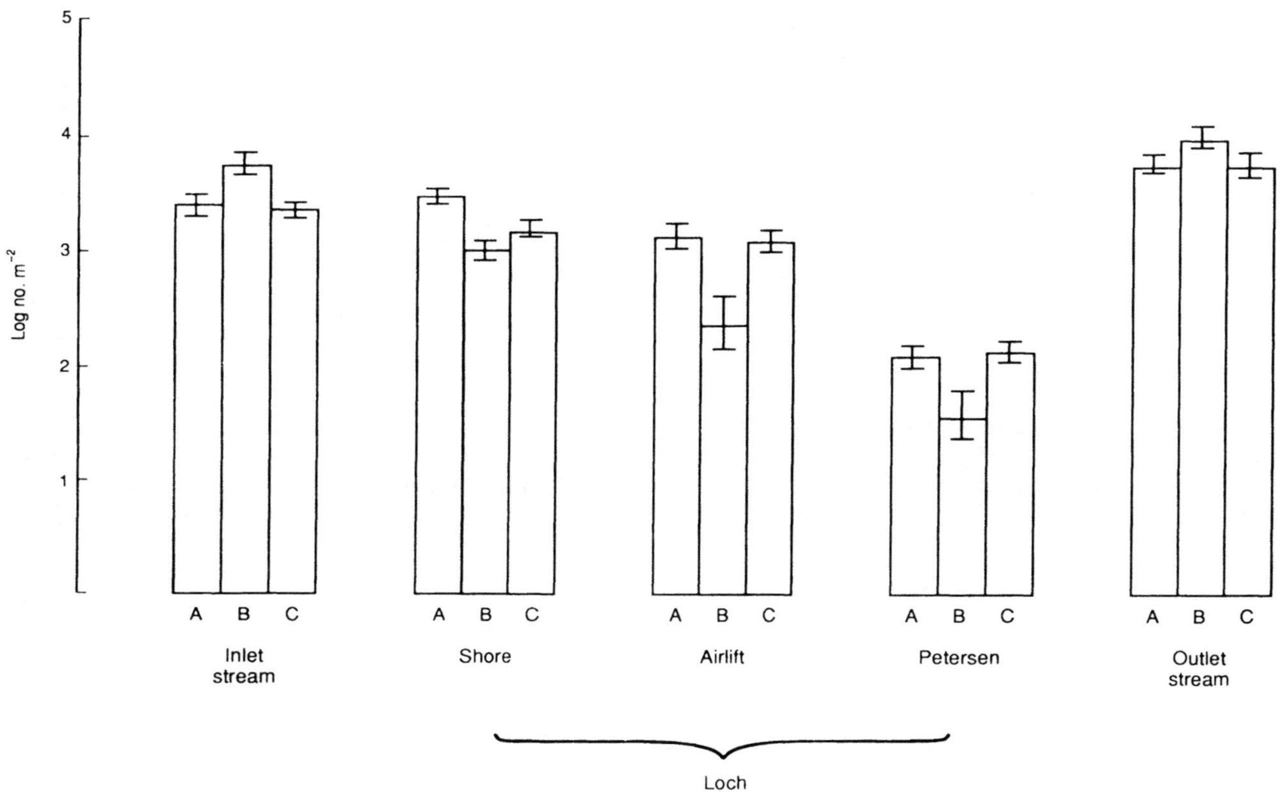

FIG. 14.10 Numbers (means and SE) of invertebrates (all groups combined) in Loch Fleet, Altiwhat stream, and loch outlet. A = pre-liming; B = post-liming, before fish restocking; C = post-liming after fish restocking.

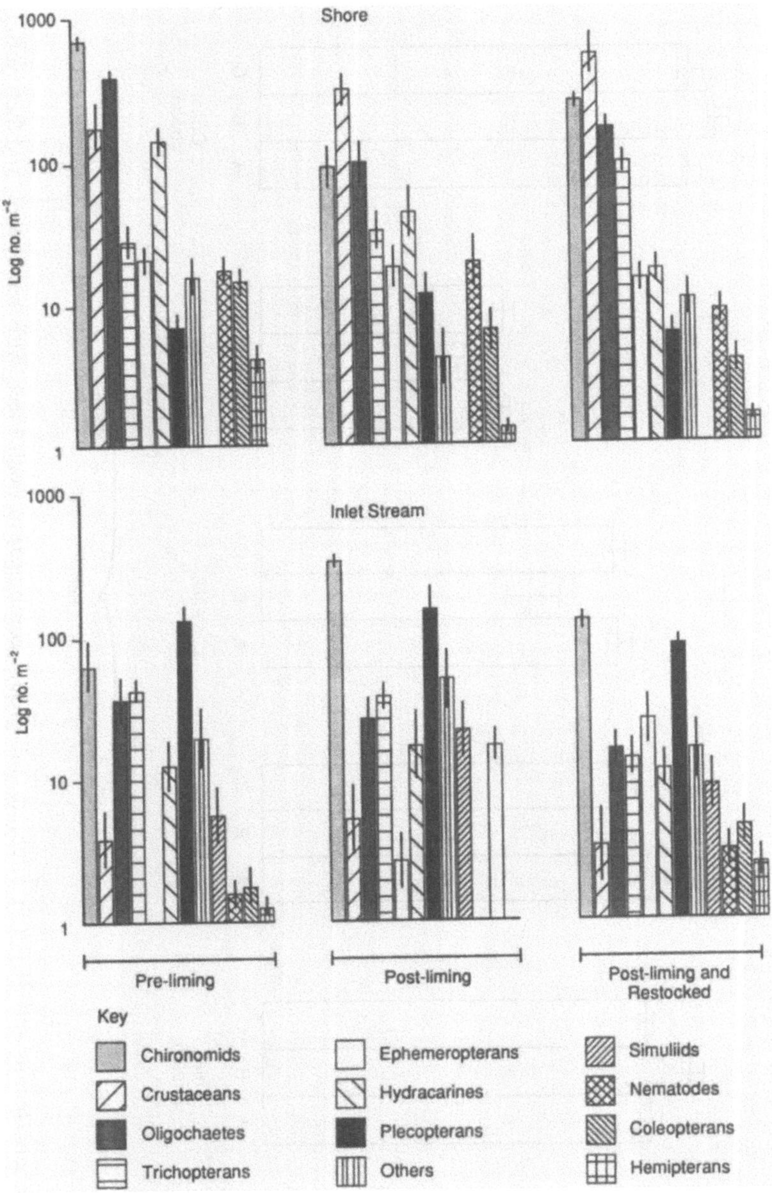

FIG. 14.11 Numbers of invertebrates of various groups in Loch shore and Altiwhat stream samples (log means and SE).

TABLE 14.4
Loch benthos and shore: significant differences in catch of macroinvertebrates, before liming, after liming, and after trout introduction

	Column A			*Column B*			*Column C*		
	Shore	*Loch*		*Shore*	*Loch*		*Shore*	*Loch*	
		AL	*PE*		*AL*	*PE*		*AL*	*PE*
Chironomidae	− − −	−	− − −	+ + +	+	+ + +	− − −	ns	ns
Coleoptera	ns	ns	ns	ns	ns	ns	− − −	ns	ns
Crustacea	ns	ns	ns	ns	ns	+	ns	ns	ns
Ephemeroptera	ns	−	ns	ns	ns	+	ns	−	ns
Hemiptera	ns	ns	ns	ns	ns	ns	ns	ns	ns
Hydracarina	− − −	ns	− − −	− − −	ns	ns	− − −	ns	− − −
Nematoda	ns	ns	ns	ns	ns	ns	ns	ns	ns
Oligochaeta	− − −	ns	ns	ns	ns	ns	− − −	ns	ns
Plecoptera	ns	ns	ns	ns	ns	ns	ns	ns	ns
Trichoptera	ns	− − −	ns	+ + +	+ + +	ns	+ + +	ns	ns
Others	ns	ns	ns	ns	ns	ns	ns	ns	ns
All groups	− − −	− −	− −	+ + +	+ + +	+ +	− − −	ns	ns

Notes: Column A: before liming compared with after liming (and trout absent).
Column B: after liming with trout absent compared with after liming, trout present.
Column C: before liming and with trout absent compared with after liming with trout present.

Symbols: + indicates increase; − decrease; *p* (probability) is shown as follows:
+/−, $p < 0{\cdot}5$; ++/−−, $p < 0{\cdot}01$; +++/−−−, $p < 0{\cdot}001$;
ns = not significant.
AL = airlift samples.
PE = Petersen grab samples.

level or above (Table 14.6). After the introduction of fish, however, Ephemeroptera, Plecoptera, and Trichoptera were significantly increased compared with the preliming numbers, the Trichoptera made up mainly of *Hydropsyche* and *Ryacophila* spp. In contrast, the numbers of Chironomidae were significantly less after the introduction of fish than before.

The changes in the invertebrate populations following catchment liming in 1986 and before the introduction of fish (Tables 14.4–14.6) give some measure of the short-term effects of liming, although the period of study was limited and valid conclusions limited by long periods between samplings. Interpretation of any long-term effects is further complicated by the effects of predation by the trout introduced in 1987 and 1988. With these

TABLE 14.5
Altiwhat: significant differences in catch of macroinvertebrates

	Column A	*Column B*	*Column C*
Chironomidae	++	ns	ns
Coleoptera	+++	−−−	+++
Crustacea	ns	ns	ns
Ephemeroptera	+++	+++	+++
Hemiptera	ns	+	ns
Hydracarina	ns	ns	ns
Nematoda	ns	+	ns
Oligochaeta	ns	ns	ns
Plecoptera	ns	ns	ns
Trichoptera	ns	−−	−−
Others	ns	ns	ns
All groups	+	−	ns

Note: See footnotes for Table 14.4.

reservations, the following is considered to be the most likely explanation for the statistically significant changes recorded.

14.6.3.3 Post Liming, Without Fish

The decrease in number of invertebrates in the loch (but not in tributary streams) in the year following liming in spring 1986 may have resulted from liming, possibly attributable either to the direct effect of increased

TABLE 14.6
Loch outflow: significant differences in catch of macroinvertebrates

	Column A	*Column B*	*Column C*
Chironomidae	ns	−	ns
Coleoptera	ns	ns	ns
Crustacea	ns	ns	ns
Ephemeroptera	ns	ns	+
Hemiptera	ns	ns	ns
Hydracarina	ns	ns	ns
Nematoda	ns	ns	ns
Oligochaeta	ns	ns	ns
Plecoptera	ns	ns	+
Trichoptera	ns	+++	+++
Others	ns	ns	ns
All groups	ns	ns	ns

Note: See footnotes for Table 14.4.

pH on an invertebrate community hitherto adapted over a long period to low pH, or to precipitation of toxic forms of aluminium on to the loch bottom. However, it should be appreciated that the changes in loch water quality post-liming were gradual, and not excessive (i.e. calcium from 1 mgl^{-1} (50 μeq litre^{-1}) to less than 4 mgl^{-1} (200 μeq litre^{-1})). Schneider and Dillon (1976), Broberg (1978), and Raddum *et al.* (1984) also recorded an initial decrease in numbers of invertebrates after direct liming to lakes.

This reduction in the numbers of some invertebrate groups in the loch in the year after liming, and the concurrent increase in their numbers in the inlet stream and absence of any significant change in the outlet stream, calls for some explanation. If toxic forms of aluminium precipitated out after liming, this would apply to both the loch and its inlet streams when the catchment was limed; possibly, aluminium precipitates may have been flushed out of the streams and deposited in lake sediments.

14.6.3.4 Post Liming, after the Introduction of Fish

Where the numbers of invertebrates were significantly less after the introduction of fish than in the period before liming (i.e. total numbers of Chironomidae, Coleoptera, Hydracarina, Oligochaeta, and Ephemeroptera in the loch, and Trichoptera in the inlet stream; Tables 14.4 and 14.5), there are three possible explanations apart from random annual variation, which cannot be discounted. These are: (1) that the reductions were caused by the direct or indirect effects of liming similar to those occurring in the first year; (2) that they were caused by fish predation; and (3) that they were caused by a combination of these factors.

Where the numbers were higher after the introduction of fish than before liming (Crustacea and Trichoptera in the loch, Ephemeroptera and Plecoptera in the outlet stream, Ephemeroptera in the inlet stream; Tables 14.4 and 14.5) we can propose that the increases were probably brought about by the direct or indirect effects of liming, since they occurred in spite of predation by trout.

An increase in Ephemeroptera has also been reported in other limed waters (Eriksson *et al.*, 1983; Hasselrot *et al.*, 1984). Weatherly and Ormerod (1987) also report higher numbers of Ephemeroptera and some species of Trichoptera in circumneutral (but unlimed) streams than in more acid streams.

The increase in abundance of Trichoptera in the outlet stream after the introduction of trout, but their concurrent decrease in the inlet stream (Tables 14.5 and 14.6), requires some further explanation. Here, it seems possible that predation by trout is greater in the inlet stream since this is narrower and smaller than the outlet. It is also possible that suspended food, suitable for the trichopteran species found, is more abundant in the outlet stream.

The question arises whether the eventual increase in these groups after liming was a response to increased pH or increased calcium concentration. In the case of *Baetis rhodani* there is evidence that high calcium is more important than high pH, since this species has been recorded at pH values as low as 4·2, provided that calcium levels have been around $3\,mgl^{-1}$ ($150\,\mu eq\,litre^{-1}$) or above (Aston *et al.*, 1985).

Preliminary observations on the stomach contents of the trout introduced into the loch indicate that Trichoptera form the major part by weight in summer, although there were also higher numbers of small terrestrial insects. In other seasons, terrestrial insects are unlikely to provide a significant dietary component. The increase in Trichoptera in the loch after liming has probably played an important part in sustaining the restored trout population.

14.6.4 Summary and Conclusions

Results of changes in invertebrate communities reported for other liming studies are similarly equivocal. Some studies (e.g. Schneider and Dillon, 1976; Hultberg, 1983; Raddum *et al.*, 1984) also show a decrease in invertebrate numbers after liming, at least over the short term. Others (e.g. Bengtsson *et al.*, 1980; Hasselrot *et al.*, 1984; Herricks and Cairns, 1977; Weatherley and Ormerod, 1988) report an increase, at least for some groups. Undoubtedly, year-by-year and seasonal shifts in invertebrate populations make interpretation difficult, especially when sampling periods are short and sampling intervals long, as was the case at Loch Fleet. It is also clear that the invertebrate communities in different locations (inlet, loch, and outlet) were significantly different at any one time, so that sampling could not have been consolidated. Evidently, sufficient invertebrate material was produced in the system to sustain the growing trout population, as shown by stomach content analysis and trout growth. More subtle changes in the assemblage of invertebrate communities would require a much more intensive programme of sampling and analysis.

14.7 CHANGES IN THE BEETLE FAUNA OF LOCH FLEET

14.7.1 Introduction

It has been observed in a number of fishless (and acid) lakes that the absence of fish is often associated with a wealth of other predatory organisms such as beetles, dipteran predators (e.g. Corixids), and predatory insect larvae (e.g. dragon-flies). Corixids and beetles are at an advantage in acid fishless lochs since their sensititivity (as air-breathers) to water quality conditions is low, while as predators they can profit by the absence of fish.

A limited study was undertaken to monitor changes in the beetle population at Loch Fleet following the introduction of fish by comparing populations in open water and in the impounded fishless embayments.

14.7.2 Methods

The adult populations of Coleoptera were sampled in Loch Fleet in June and September from 1985 to 1988. Each sampling unit comprised a 1-min sweep along the shore with a D-framed net fitted with a 1-mm mesh bag, beetles being identified in the field and returned alive to the loch. Sweeps were taken from 18 fixed points, two in each of the nine sectors, plus two further sweeps chosen either at random or to compensate for loss of one of the impounded areas, making a total sample of 20 units.

14.7.3 Results

Thirty-six species were found in the loch and in adjacent wetland areas during the course of the survey (see Appendix 14). Most of the fauna are species associated either with running water or with acid mires. Three (*Potamonectes griseostriatus*, *Agabus arcticus*, and *Gyrinus minutus*) are restricted in Britain to acid, open, standing water, usually on high ground. A few species (*Laccophilus minutus*, *Hyphydrus ovatus*) are on the northern edge of their British distribution in the Southern Uplands and are usually associated with lowland ponds and slow running water. Species typically associated with running water (*Oreodytes sanmarkii*, *Oulimnius tuberculatus*, *Elmis aenea*) occurred in the loch, mainly around the inflow of the Altiwhat and in embayment 9 near to the outflow. Some species, *Hydroporus gyllenhalii*, *H. melanarius*, *H. nigrita*, *H. obscurus*, *H. pubescens*, *H. tristis*, *Agabus guttatus*, and *Laccobius atratus*, were largely confined to neighbouring bog pools and flushes.

The number of species recorded in the loch fell from June to September 1985 and remained low thereafter. Species numbers were generally lower

TABLE 14.7
Numbers of adult water beetles (catch per minute) in Loch Fleet

Year:	*1985*		*1986*		*1987*		*1988*	
Month:	*vi*	*ix*	*vi*	*ix*	*vi*	*ix*	*vi*	*ix*
Hydroporus palustris	12·3	47·7	4·6	15·8	2·6	4·9	6·1	2·6
Potamonectes depressus elegans	0·1	0·5	0·1	0·1	0·1	1·9	0·4	0·2
Individuals per sample	16·6	66·3	7·0	19·8	5·1	12·2	10·2	3·9
Total no. of species	23	14	7	8	9	12	10	9

in embayment 5 (the control area still receiving acid water) than elsewhere, this being partly due to the relatively small size of the embayment restricting the number of sweeps that could be made.

With the exception of *Potamonectes griseostriatus*, the numbers of which fell within each year, and of 1988, in which all of the commoner species were in decline, beetle numbers usually increased from the early to the late season sample. The total catch of beetles fell from 1985 to 1988; most of this was associated with a decline in the commonest species, *Hydroporus palustris* (Table 14.7). The numbers of *P. depressus elegans* increased in 1987 but then declined in 1988.

The decline in numbers of *H. palustris* was greater in the open parts of the loch than in the impounded areas, such that there were significant differences between densities of open water and impounded areas in 1987 and 1988 (Table 14.8). Densities changed least in embayment 5, but the differences from other impounded areas could not be tested for significance owing to the small number of samples involved.

H. palustris remained dominant in impounded areas in 1987 but was displaced as the dominant species of open water by *P. elegans*, the difference in distribution of *H. palustris* and *P. elegans* between the open water and the impounded areas being significant ($p < 0{\cdot}01$).

14.7.4 Conclusions and Summary

The general decline in beetle densities and occurrences began before the introduction of fish. An overall decline in invertebrate numbers was also noted (see Section 14.6.3, above). Either there was a response to the liming that began in April 1986, or climatic differences between 1985 and 1986 resulted in lower breeding successes in 1986. The slower rate of decline of beetle populations in the acid water embayment 5 suggested that the decline began because of liming.

TABLE 14.8
Changes in numbers of adult *Hydroporus palustris* and *Potamonectes elegans* in impounded areas and in open water, September 1985–88

	Beetles per minute (arithmetic mean)		*Statistical difference by ANOVA*
	Open	*Impounded*	
Hydroporus palustris			
1985	55·8	18·8	N.S.
1986	11·0	19·5	N.S.
1987	0·46	10·2	$p < 0{\cdot}001$
1988	0·69	4·8	$p < 0{\cdot}01$
Potamonectes elegans			
1985	0·57	0·25	N.S.
1986	0·10	0	nos. too low
1987	2·27	1·44	N.S.
1988	0·08	0·22	nos. too low

Note: Analysis of variance was carried out using data (n) transformed to $\log_{10}(n + 1)$.

14.8 OVERALL CONCLUSIONS

The varied studies of the fresh water communities of the open loch and tributary streams reported here have included all principal biological components – heterotrophs, diatoms, algae, macrophytes, planktonic and benthic invertebrates, and predators (beetles and fish). In spite of this wide canvas, findings are diverse and not all can be incorporated into a comprehensive "story" or model to represent the community response to liming. The reasons for this may be:

- that year-by-year variations reflecting abiotic and biotic factors and their interactions confound the data;
- that measurement of changes in the community is confounded by the relatively rapid sequence of events, from liming to fish reintroduction.
- that sampling has been insufficient in frequency and detailed analysis to provide enough representative data.

These problems relate more to the invertebrate community than to the plant community, reflecting its greater diversity both of species and life-styles.

Recent studies of invertebrate communities of acid waters (e.g. Sutcliffe and Hildrew, 1989; Ormerod *et al.*, 1990, Rutt *et al.*, 1990) suggest that, at least in "stable" fresh waters categorised as acid, moderately acid, or circumneutral, water chemistry and biology can be specified in terms of pH, calcium concentration (for hardness), and aluminium concentration. This allows the prediction, with reasonable certainty, of the invertebrate community expected for various sulphur/acid deposition scenarios, and even for changes in land use and lime applications to replenish soil base reserve.

The lack of such a clear indication from the Loch Fleet biological data suggests that sampling may have been inadequate, or that changing conditions through the post-liming period have been too rapid to achieve any community stability. It is certainly more challenging to establish community response to such changes than to categorise differences between classes of streams, or to predict changes at some future time, without the critical test or validation. Similarly, studies following direct lake or stream liming, which follow short-term changes in selected invertebrate species (e.g. Brettum and Hindar, 1985, Dickson, 1988), do not yet provide for an integrative community perspective, nor for prediction or achievement of a stable community.

Nonetheless, the catchment liming at Loch Fleet, with the primary objective of establishing a fishery, has shown evidence of the successful reappearance of hitherto absent species, as well as a sufficiency of prey to meet the needs of the fish population. No doubt further studies will be needed to document the future structure and stability of this ecosystem.

REFERENCES

Anderson, N.J., Battarbee, R.W., Appleby, P.G., Stevenson, A.C., Oldfield, F., Darley, J. and Glover, G. (1986). Palaeolimnological Evidence for the Recent Acidification of Loch Fleet, Galloway. Working Papers, No. 17, Palaeoecology Research Unit, University College London, 71 pp.

Atkinson, K.M. and Haworth, E.Y. (1990). Devoke Water and Loch Sionascaig: recent environmental changes and the post-glacial overview. *Phil. Trans. Roy. Soc., Lond.* **B327**: 349–55.

Aulio, K. (1987). Rapid decline of mass occurrence of *Juncus bulbosus* in a deacidified freshwater reservoir. *Memoranda, Soc. Fauna Flora Fennica* **63**, 41–44.

Aston, R.J., Sadler, K., Milner, A.G.P. and Lynam, S. (1985). The Effect of pH and Related Factors on Stream Invertebrates. CERL Report TPRD/L/2792/N84, 29 pp.

Barrett, C.F., Atkins, D.H.F., Cape, J.N., Crabtree, T., Davies, T.D., Derwent,

R.G., Fisher, B.E.A., Fowler, D., Kallend, A.S., Martin, A., Scriven, R.A. and Irwin, J.G. (1988). *Acid Deposition in the United Kingdom*. Review of UK Acid Rain Review Group (2). Warren Spring Laboratory, Stevenage, 104 pp.

Battarbee, R.W., Anderson, N.J., Appleby, P.G., Flower, R.J., Fritz, S.C., Haworth,. E.Y., Higgitt, S., Jones, V.J., Kreiser, A., Munro, M.A.R., Natkanski, J., Oldfield, F., Patrick, S.T., Richardson, N.G., Rippey, B. and Stevenson, A.C. (1988). *Lake Acidification in the United Kingdom 1800–1986*. Ensis Publishing, London, 68 pp.

Bengtsson, B., Dickson, W. and Nyberg, P. (1980). Liming lakes in Sweden. *Ambio* **9**: 34–36.

Brettum, P. and Hindar, A. (1985). Effects of lime treatment on the biological system, pp. 79–109 in *Liming of Acid Water* (ed. K. Baalsrad). Dept. of Environment and Directorate of Nature Management, Trondheim, Norway, 147 pp.

Broberg, O. (1978). Kalningsmedel, kalkning och effekter av kalming. *Vatten* **2**: 104–16 (cited in Weatherley, 1988).

Cameron, N.G. (1990). Unpublished Ph.D. Thesis, University of London.

Cook, J.M., Edmunds, W.M. and Robins, N.S. (1987). Groundwater Contribution to Loch Fleet, Galloway. Hydrogeological Report 87/4, British Geological Survey.

Cook, J.M., Edmunds, W.M. and Robins, N.S. (1991). Groundwater contribution to an acid upland lake (Loch Fleet, Scotland) and the possibilities for amelioration. *J. Hydrol.* **125**: 111–128.

Dickson, W. (editor) (1988). *Liming of Lake Gårdsjön*. National Swedish Environmental Protection Board, 327 pp.

Drake, C.M., Winstanley, B., Ohnstad, F.R. and Elliot, J.M. (1983). *The F.B.A. Air-Lift Sampler User Manual*. F.B.A. Occasional Publication No. 22, 20 pp.

Duncan, D.B. (1975). "*t*-tests" and intervals for comparisons suggested by the data. *Biometrics* **31**: 339–59.

Eriksson, F., Hornström, E., Mossberg, P. and Nyberg, P. (1983). Ecological effects of lime treatment of acidified lakes and rivers in Sweden. *Hydrobiologia* **101**: 145–64.

Flower, R.J., Battarbee, R.W. and Appleby, P.G. (1987). The recent palaeolimnology of acid lakes in Galloway, southwest Scotland: diatom analysis, pH trends and the role of afforestation. *J. Ecol.* **75**: 797–824.

Grahn, O. (1977). Macrophyte succession in Swedish lakes caused by deposition of airborne substances. *Water Air Soil Poll.* **7**: 295–306.

Grahn, O. (1985). Macrophyte biomass and production in Lake Gårdsjön – an acidified clearwater lake in S.W. Sweden. *Ecol. Bull.* (*Stockholm*) **37**: 203–12.

Grahn, O., Hultberg, H. and Landner, L. (1974). Oligotrophication – a self accelerating process in lakes subjected to excessive supply of acid substances. *Ambio* **3**: 93–94.

Hasselrot, B., Andersson, I.B. and Hultberg, H. (1984). Ecosystem Shifts and Reintroduction of Arctic Char (*Salvelinus salvelinus* (L.)), Report SVN No. 61: 78–92, Inst. Freshwater Research, Drottningholm, Sweden.

Herricks, E.E. and Cairns, J. (1977). The effects of lime neutralisation on stream ecology, pp. 477–86 in *32nd Annual Purdue Industrial Waste Conference*.

Holmes, N.T.H., Whitton, B.A. and Hargreaves, J.W. (1978). *A coded List of*

Freshwater Macrophytes of the British Isles. Water Archive Manual Series, No. 4, Department of the Environment, Water Data Unit, Reading.

Hultberg, H. (1983). Liming of acid-stressed ecosystems: induced chemical and biological effects. *VDI-Berichte* **500**: 409–14.

Hultberg, H. and Andersson, I.B. (1982). Liming of acidified lakes: induced long-term changes. *Water Air Soil Poll.* **18**: 311–31.

Jones, V.J., Stevenson, A.C. and Battarbee, R.W. (1989). Acidification of lakes in Galloway, Southwest Scotland: a diatom and pollen study of the post-glacial history of the Round Loch of Glenhead. *J. Ecol.* **77**: 1–23.

Lazarek, S. (1986). Responses of the *Lobelia* – epiphyte complex to liming of an acidified lake. *Aquatic Botany* **25**: 73–81.

Milner, A.G.P. (1989). Diets of Brown Trout from Loch Fleet. CERL Memorandum ESTD/L/ELS/0009/M89, 11 pp.

Ormerod, S.J., Weatherley, N.S. and Gee, A.S. (1990). Modelling the ecological impact of changing acidity in Welsh streams, pp. 279–310 in *Acid Waters in Wales* (eds. R.W. Edwards, A.S. Gee and J.H. Stoner). Kluwer, Dordrecht, Netherlands, 337 pp.

Raddum, G.G., Hagenlund, G. and Halvorsen, G.A. (1984). Effects of Lime Treatment on the Benthos of Lake Søndre Boksjö, SVN Report No. 61: 167–176. Inst. Freshwater Research, Drottningholm, Sweden, 202 pp.

Raven, P.J. (1986). Occurrence of Sphagnum Moss in the Sublittoral of Several Galloway Lochs, with Particular Reference to Loch Fleet. Working Papers, No. 13, Palaeoecology Research Unit, University College London, 45 pp.

Raven, P.J. (1988a). Occurrence of *Sphagnum* moss in the sublittoral of several small oligotrophic lakes in Galloway, southwest Scotland. *Aquatic Botany* **30**: 223–30.

Raven, P.J. (1988b). Changes in the distribution of aquatic macrophytes in Loch Fleet, with particular reference to sublittoral *Sphagnum*, in *The Loch Fleet Project. Report of the Intervention Phase (2) 1986–1987* (eds. G. Howells and T.R.K. Dalziel). CEGB, SSEB, NSHB, British Coal 102 pp (+annexes).

Raven, P.J. (1989). Short-term Changes in the Aquatic Macrophyte Flora of Loch Fleet, S.W. Scotland, Following Catchment Liming, with Particular Reference to Sublittoral Sphagnum. Research Papers, No. 39, Palaeoecology Research Unit, University College, London, 42 pp.

Renberg, I. and Hellberg, T. (1982). The pH history of lakes in south-western Sweden, as calculated from the subfossil diatom flora of the sediments. *Ambio* **11**: 30–33.

Roelofs, J.G.M. (1983). Impact of acidification and eutrophication on macrophyte communities in soft waters in the Netherlands. I. Field observations. *Aquatic Botany* **17**: 139–55.

Rutt, G.P., Weatherley, N.S. and Ormerod, S.J. (1990). Relationships between physicochemistry and macroinvertebrates of British upland streams: the development of modelling and indicator systems for predicting fauna and detecting acidity. *Freshw. Biol.* **24**: 463–80.

Sand-Jensen, K. and Søndergaard, M. (1981). Phytoplankton and epiphyte development and their shading effect on submerged macrophytes in lakes of different nutrient status. *Int. Rev. Ges. Hydriobiol.* **66**: 529–52.

SAS (1985). *User's Guide: Statistics*. Version 5 Edition. SAS, Cary NC USA. 956 pp.

Savage, A.A. (1986). The distribution, life-cycle and production of *Leptophlebia vespertina* (L.) (Ephemeroptera) in a lowland lake. *Hydrobiologia* **133**: 3–19.

Schneider, W.A. and Dillon, P.J. (1976). Neutralisation and fertilisation of acidified lakes near Sudbury Ontario, pp. 93–100 in *Proc. 11th Canadian Symposium 1976*. Water Pollution Research Canada.

Sokal, R.R. and Rohlf, F.G. (1981). *Biometry. The Principles and Practice of Statistics in Biological Research*. 2nd edn. W.H. Freeman & Co., San Francisco. 859 pp.

Stevenson, A.C., Jones, V.J. and Battarbee, R.W. (1990). The cause of peat erosions: a palaeolimnological approach. *New Phytologist* **114**: 727–35.

Sutcliffe, D.W. and Hildrew, A.G. (1989). Invertebrate communities in acid streams, pp. 131–30 in *Acid Toxicity and Aquatic Animals* (eds. R. Morris, E.W. Taylor, D.J.A. Brown and J.A. Brown). Cambridge University Press, Cambridge, UK, 282 pp.

Tolonen, K. and Jaakkola, T. (1983). History of lake acidification and air pollution studied on sediments in South Finland. *Ann. Bot. Fenn.* **20**: 57–78.

Weatherley, N.S. (1988). Liming to mitigate acidification of freshwater ecosystems: a review of the biological consequences. *Water Air Soil Poll.*, **39**: 421–437.

Weatherley, N.S. and Ormerod, S.J. (1987). The impact of acidification on macroinvertebrate assemblages in Welsh streams: towards an empirical model. *Env. Poll.* **46**: 223–40.

Wright, R.F. and Henriksen, A. (1980). Regional survey of lakes and streams in southwestern Scotland, April 1979. SNSF Project IR 72/80, 63 pp.

APPENDIX 14: CHECKLIST OF AQUATIC BEETLES RECORDED IN LOCH FLEET FROM 1985 TO 1988

Haliplidae

Haliplus fulvus (Fab.)
H. lineatocollis (Marsham)
H. ruficollis (DeGeer)

Dytiscidae

Laccophilus minutus L.
Hyphydrus ovatus L.
Hygrotus inaequalis (Fab.)
Hydroporus erythrocephalus L.
H. gyllenhalii Schiödte
H. melanarius Sturm
H. nigrita (Fab.)

H. obscurus Sturm
H. palustris L.
H. planus (Fab.)
H. pubescens (Gyllenhal)
H. tristis (Paykull)
Potamonectes assimilis (Paykull)
P. depressus elegans (Panzer)
P. griseostriatus (DeGeer)
Strictotarsus duodecimpustulatus (Fab.)
Oreodytes sanmarkii (Sahlberg)
Agabus arcticus (Paykull)
A. bipustulatus L.
A. guttatus (Paykull)
Rhantus exsoletus (Forster)
Dytiscus marginalis L.

Gyrinidae
Gyrinus minutus Fab.
G. substriatus Stephens

Hydrophilidae
Helophorus aequalis Thomson
H. brevipalpis Bedel
H. flavipes Fab.
Cercyon impressus (Sturm)
Laccobius atratus (Rottenburg)
Anacaena globulus (Paykull)

Hydraenidae
Limnebius truncatellus (Thunberg)

Elmidae
Elmis aenea (Müller)
Oulimnius tuberculatus (Müller)

CHAPTER 15

Terrestrial Biology in Limed Catchments

R.S. Clymo
Queen Mary and Westfield College, University of London, London, UK

G.N. Foster
Scottish Agricultural College, Ayr, UK

J. MacKay
Glenlee Power Station, New Galloway, UK

and contributions from

J. Robertson
The Macaulay Land Use Research Institute, Aberdeen, UK

R. Shore
Institute of Terrestrial Ecology, Huntingdon, UK

D.I. Skidmore
CERC Research Consultants, Cambridge, UK

15.1 INTRODUCTION

Although the focus of many restoration/liming studies lies with the improvement of surface water quality and its associated fauna and flora, a variety of studies on the associated terrestrial ecosystems have been undertaken at Loch Fleet, as well as at other restoration sites. The objectives are partly to deduce soil and hydrological conditions as revealed by vegetation, and to assess the contribution of terrestrial components, for example for fish diets, and furthermore to provide a basis for assessing any changes in the associated terrestrial environment as a consequence of liming or other restoration techniques.

The concern that catchment liming would alter soil and vegetation characteristics of upland terrain, as well as possible direct effects on components of the terrestrial fauna, has led to speculation that adverse effects might ensue. At Loch Fleet, several investigations have been undertaken during the post-liming period to identify and quantify any unwelcome responses. These studies include surveillance of wetland, moorland and forest vegetation, some terrestrial invertebrate communities, small mammals, and birds. Few studies were made prior to liming in 1986, but observations have been assessed by reference to "control" areas within the catchment, or to the current state of knowledge of terrestrial fauna and flora of the Galloway uplands.

In addition, as explained in Chapter 11, lime applications are expected to result in soil changes (pH rise, increase in cation exchange capacity, and percentage base saturation) which might have some longer-term consequences on the upland vegetation.

15.2 VEGETATION SURVEY AT LOCH FLEET

15.2.1 Conditions Prior to Liming

The vegetation of the Loch Fleet catchment prior to liming was recorded in October 1984 and May 1985 at marked points on a regular 50-m grid. A small subsample of these points was resurveyed in October 1987 after treatments had been applied. The whole catchment will be resurveyed in 1992.

At the start of the project, the moorland vegetation was dominated by flying bent (*Molinia caerulea*) and heather (*Calluna vulgaris*) (Fig. 15.1). A total of more than 70 species was recorded (see Appendix 15.1). A number of other upland plants and communities characterise the environmental conditions at each site, i.e. soil type, degree of flushing, base status, landform, and exposure. The planted forest (1963) on the south-west of the catchment occupies only about 10% of the total area. It comprises a mixture of Sitka spruce (*Picea sitchensis*) and lodgepole pine (*Pinus*

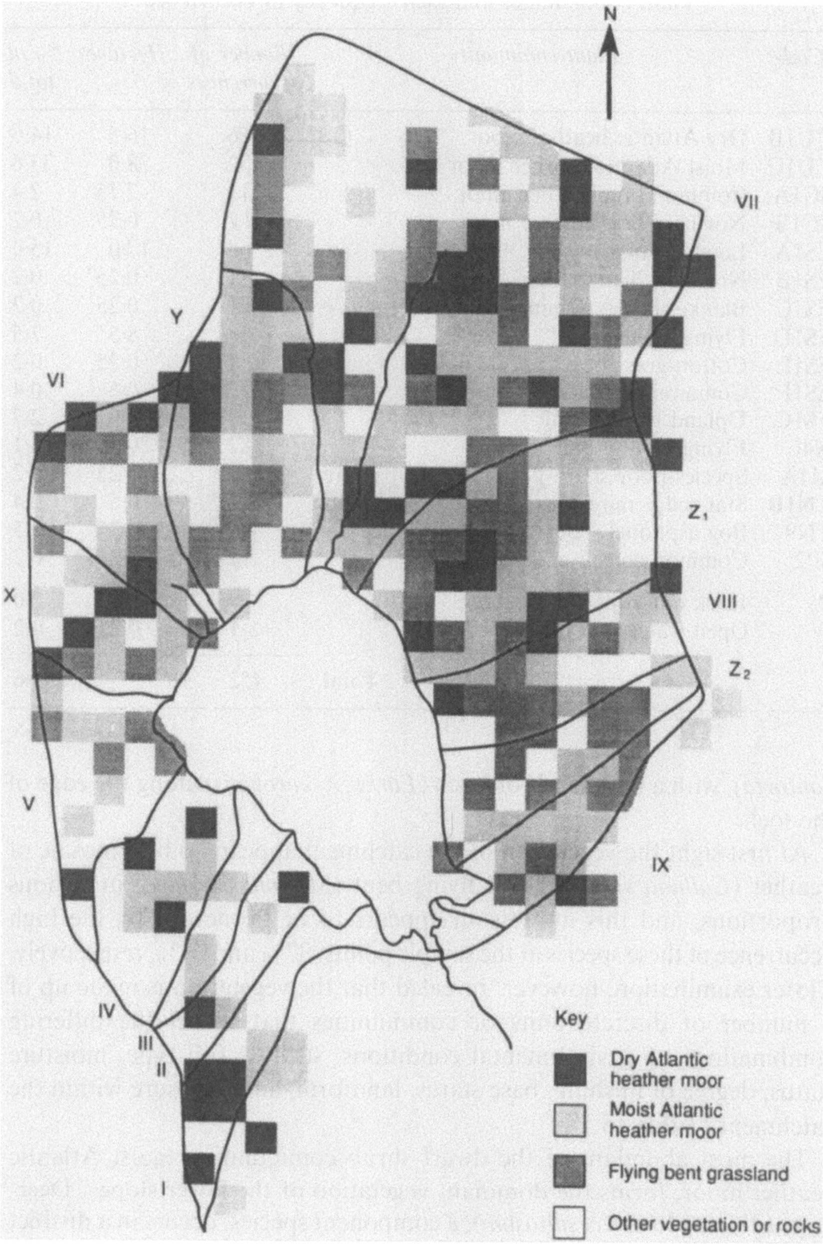

FIG. 15.1 Loch Fleet and its catchment: distribution of major plant communities.

TABLE 15.1
Plant communities and their frequency of occurrence

Code	*Plant community*	*Number of occurrences*	*Hectares*	*% of total*
CU1B	Dry Atlantic heather moor	66	16·5	14·6
CU1C	Moist Atlantic heather moor	152	38·0	33·6
SC1A	Common bog heather moor	11	2·75	2·4
SC1B	Northern bog heather moor	1	0·25	0·2
ES1A	Lowland blanket bog	68	17·0	15·0
ES1B	Northern blanket bog	1	0·25	0·2
ES1C	Blanket bog – terminal phase	1	0·25	0·2
ES1D	Flying bent bog	34	8·5	7·5
ES1E	Cotton-grass bog	1	0·25	0·2
ES1F	Common cotton-grass mire	2	0·5	0·4
ES1G	Upland blanket bog	12	3·0	2·7
N4C	Flying bent grassland	82	20·5	18·1
M1A	Species-poor sharp-flowered rush pasture	1	0·25	0·2
CN1B	Star sedge mire with sharp-flowered rush	2	0·5	0·4
CN9	Bog asphodel – bog moss water track	7	1·75	1·5
SP2	Common cotton-grass pool	1	0·25	0·2
R	Rock outcrops and boulders	9	2·25	2·0
W	Open water	1	0·25	0·2
	Total	452		99·6

contorta), with a few stands of larch (*Larix* × *eurolepis*) along the edge of the loch.

At first sight the vegetation of the catchment appears to be a mosaic of heather (*Calluna vulgaris*) and flying bent (*Molinia caerulea*) in various proportions, and this impression appears to be borne out by the high occurrence of these species at the sample points, 87% and 75%, respectively. Closer examination, however, revealed that the vegetation is made up of a number of discrete units or communities that reflect the differing combinations of environmental conditions, such as soil type, moisture status, degree of flushing, base status, landform, and exposure within the catchment (Table 15.1).

The most abundant of the dwarf shrub communities, moist Atlantic heather moor, forms the dominant vegetation of the lower slopes. Deer-grass (*Trichophorum cespitosum*), a component species, occurs in a distinct zone across the upper, eastern margin of the catchment where there is a marked break in slope. This species is found also on the exposed, rocky ridges on either side of the deep gully of the Altiwhat in sector VII. Dry Atlantic heather moor is confined to the steep slopes of mounds and gullies and is most extensive north of the loch on the steep upper slopes where it is grazed by red deer. Lowland blanket bog occurs mainly on lower slopes. On the high ground west of the forestry and on the high peat plateau of sector VII it is replaced by its upland counterpart. Two flush communities are widespread, each dominated by dense tussocks of flying bent (*Molinia caerulea*). Flying bent bog is differentiated by the presence of cotton-grass (*Eriophorum vaginatum*) and is most common on the northern and eastern flanks. Flying bent grassland, extensive in the northern part of the catchment, is also heavily grazed by deer in the more sheltered sites.

There are clear, definable relationships between vegetation and the environment, with various environmental properties being reflected in the presence of specific communities or species (Table 15.2). The dominant environmental influence is that of flushing, which to some degree affects virtually every site in the catchment. Areas least flushed are the steep slopes at the highest point of the northern watershed and the peat "flats" under the coniferous plantations in the south-west. The middle and lower slopes of the northern and eastern sectors are the most flushed. The percentage cover abundance of flying bent grass (*Molinia caerulea*) provides a crude estimate of the degree of flushing.

The link between vegetation and soil within the catchment is primarily one associated with the structure of the organic soil horizons and soil wetness. There is a gradient from dry heather moor on well-structured, peaty soils which are occasionally or frequently waterlogged, to bog heather moor on soils subject to prolonged waterlogging, then to blanket bog on unstructured peat where the waterlogging is prolonged or very prolonged.

15.2.2 Conditions Following Liming

The vegetation resurvey 2 years after liming (but comprising only 12 limed sites) revealed only slight changes following liming, with a few additional species which were thought unlikely to be new colonisers, but rather to have been undetected during the initial survey. There was some indication from one resurveyed site on peat in sector VI, however, that there was a slight trend to greater dominance of the grass *Molinia caerulea* over *Calluna* which had died at that sample site. In sector VIII, *Calluna* is regenerating after 1987 muirburn. With the exception of *Calluna* damage at the single site, the changes in the short term appear to be slight or

TABLE 15.2
Diagnostic plant species and their frequency of occurrence

Plant species	Diagnostic for	Occurrence	
		Sites	%
Calluna vulgaris	moorland	394	87·2
Empetrum nigrum	exposure	16	3·5
Erica cinerea	dry heath	203	44·9
E. tetralix	moist and wet heath	259	57·3
Vaccinium myrtillus	dry heath; exposure	56	12·4
Molinia caerulea	acid flushing	341	75·4
Nardus stricta	acid peaty soils	64	14·2
Carex echinata	flushing	1	0·2
C. panicea	flushing	43	9·5
Juncus acutiflorus	better base status flushes	9	2·0
Eriophorum vaginatum	blanket bog	90	19·9
Trichophorum cespitosum	moist and wet heath, exposure	180	39·8
Anemone nemorosa	better base status	1	0·2
Narthecium ossifragum	wet heath and bog: flush channels	28	6·2
Potentilla erecta	in wet communities, flushing	112	24·8
Viola palustris	better base status flushes	2	0·4
Hypnum jutlandicum	acid moorland	338	74·8
Racomitrium lanuginosum	exposure	25	5·5
Rhytidiadelphus loreus	exposure	29	6·4
Sphagnum auriculatum	flush channels	7	1·5
S. capillifolium	moorland	61	13·5
S. papillosum	blanket bog	45	10·0
Thuidium tamariscinum	better base status	1	0·2
Larix × eurolepis	coniferous plantations	8	1·8
Pinus contorta	coniferous plantations	47	10·4
Picea sitchensis	coniferous plantations	37	8·2

insignificant. Four years after liming the catchment presents the same picture overall, only two sites indicating changed communities.

Although the soil conditions at limed sites are changing, the indications were that aerial shoots of *Sphagnum* were killed (3 sites, sectors VI and VII) and *Calluna* cover was reduced by liming (1 site, sector VI). The dominance of *Molinia* is favoured, in the short term, by both muirburn and liming of the moist Atlantic heather moor community.

15.3 WETLAND/MOSS STUDIES

15.3.1 Introduction

Mosses have no roots. Their leaves are, in most cases, only one cell thick with no waxy cuticle (or no more than a very thin one). Most mosses take in water and solutes directly through their leaves. Some species, including the bog moss *Sphagnum*, move solutes around internally. Mosses are therefore particularly susceptible to chemicals in the environment. *Sphagnum* has disappeared from large parts of the southern Pennines, probably as a consequence of high concentrations of sulphur dioxide and bisulphite aerosol in the atmosphere there (Ferguson *et al.*, 1978).

Sphagnum is, quantitatively, much the most important moss at Loch Fleet. It is a component of the heather moorland and of the *Molinia* moorland that covers much of the catchment; it is co-dominant in some of the upper catchment blanket bog (particularly in sector VII); and submerged species were prominent in the loch itself prior to liming, at 2–5 m deep (extending to greater depths than any other macrophyte) (see Chapter 14). Although the large-scale vegetation survey undertaken at Loch Fleet (Section 15.2) categorises only a limited area of the catchment (3.1% of 110 ha) as blanket bog, in many areas *Sphagnum* species are commonly found, although not dominant. Over the catchment as a whole, the biomass of *Sphagnum* is clearly less than that of *Molinia* or *Calluna* on the moorland terrain, or of the planted forest. Methods of lime application concentrated effects within limited areas; in sector IV, runoff from the forest is routed through deep ploughed furrows, and in sector VII about 2.7 ha of the treated area (in total about 10 ha) is upland blanket bog dominated by *Sphagnum* species.

Sphagnum produces up to 30% of its dry mass as polymeric uronic acids. These are formed in the –COOH form, but the acid can dissociate, so the H^+ can be exchanged by other cations in rain or in groundwater seeping among the plants (Clymo, 1963). As a consequence the surrounding water is made more acid. A second important property of *Sphagnum* is that it is sensitive to the *combination* of high pH and high calcium concentration produced by liming (Clymo, 1973). The species differ in their susceptibility. *Sphagnum papillosum* is among the more sensitive species, and on sector VII is exposed to permanent or intermittent limed flows, while *S. capillifolium* growing at slightly higher levels above the standing water can avoid direct exposure to the limed water. Two other species, *S. inundatum* and *S. recurvum*, are less sensitive. A further species,

S. subnitens, found in the forested sector IV, is another more resistant species.

Lime was added to the Loch Fleet catchment in different ways in the different sectors (see Chapter 9). In the upper wetland area of sector VII, where most of the upland blanket bog of the catchment is found, limestone was distributed in the standing water and wetter areas to provide a source of limed water to the Altiwhat, specifically to counter acid pulses in this spawning tributary during periods of heavy rainfall. It was thus intentionally focused in these areas and inevitably has produced a mosaic of concentrations of calcium and of H^+ in the standing water of the bog. As a consequence in sector VII, the pattern of addition and the consequent pattern of exposure concentrations is highly heterogeneous. In effect, a large number of undesigned and uncontrolled small-scale exposures has resulted from the original large-scale designed experiment. In these circumstances much may be learned from observation, but it is difficult to produce numerical results.

15.3.2 Observations

In Loch Fleet itself, *Sphagnum auriculatum* (one of the Subsecunda) was abundant prior to liming in the depth zone from 2 to 5 m. The responses of this species to the changed water quality within the lake itself are reported in Chapter 14 (Section 14.5.3); there is reduced growth and a substantial reduction in biomass. The closely related species, *S. inundatum*, found in the upland bog of sector VII, is surviving in the intermittently submerged areas, but not spreading much.

In sector IV mosses are particularly abundant in the ploughed furrows within the forest. Lime was applied below the canopy as a slurry, and runoff concentrated effects into the furrows. It may be noted that the conditions within the forest reflect the land use and management of that area over the past three decades and shading has restricted the understory. Six months after treatment some mosses were thriving; these included *Hypnum jutlandicum*, *Dicranum scoparium*, and *Polytrichum commune*. Two species of *Sphagnum* – *S. subnitens* = *S. plumulosum* and *S. recurvum* – both survive and are apparently healthy. But two other species – *S. capillifolium* and *S. papillosum* – were completely dead. The different response of these two pairs is consistent with their general ecology. The pair that died are species characteristic of ombrotrophic (rain-water-dependent) bog, but the two that survived are often found in flushed situations, subject to a greater supply of solutes. These responses do, of course, depend on *local* dose effects.

In preliminary trials of limestone and chalk applications at 10 and 20 t ha^{-1} on a series of small plots on open moorland just outside the catchment near sector IX, vegetation survey at 12 months after treatment showed rather equivocal results, with *Sphagnum* (no species given) present in some plots only, and the moss *Hypnum jutlandicum* reduced in some, but increased in others. There was no consistent difference related to the dose applied.

In the upper part of sector VII (Fig. 15.2), in the source area of the Altiwhat Burn, where lime was spread at an average rate of 10 t ha^{-1} over about a third of the sector, more conspicuous effects are seen. In this difficult terrain, distribution was uneven and in the wetter areas there are still (3 years later) in numerous places patches of limestone congealed to a semi-solid mass of several centimetres (even up to 100 cm) across. The hydrology of the area is complex and varies with flow, with at least six sub-catchments (Fig. 15.2), one much larger than the rest and with four distinct parts. The complex pattern of water flow has interacted on the original distribution of lime to give a bewildering pattern of doses in the water seeping past the plants.

The results of differential survival to this varied exposure are similar to those seen in sector IV. Wherever lime is still visible, *Sphagnum papillosum* is dead, and *S. capillifolium* is often dead but survives on low hummocks above the reach of limed water. In contrast, both *S. auriculatum* and *S. recurvum* have survived in places immediately adjacent to, or intermixed with, the remains of these two, more sensitive, species. The effects are clearly related to water flow patterns: in some cases *S. papillosum* has been killed at least 30 m down-slope from the nearest visible lime deposit. In other places patches of *S. papillosum* survive in backwaters, presumably growing aside from the flow of limed water. The plants are able to reacidify small flows of water by the cation exchange process already described. In the 3 ha of this limed sub-catchment to the north-west of the Altiwhat, about 35% of the area was dominated by *Sphagnum*. Twelve months after liming, in 3% of the area, *Sphagnum* was still alive but in 32% (almost all *S. papillosum*) was dead. Probably most plants died within a few weeks of liming.

Three years later the sheets of dead *Sphagnum* remain a whitish grey colour. The apices of the plants have disintegrated, but much of the rest remains, although it is compacting. There is a possibility that erosion may occur. If it were to do so, it might be rapid, but there is as yet no obvious sign, although there is evidence of past erosion (prior to liming) outside the limed areas of sector VII. Nor is there any sign of recovery of these two

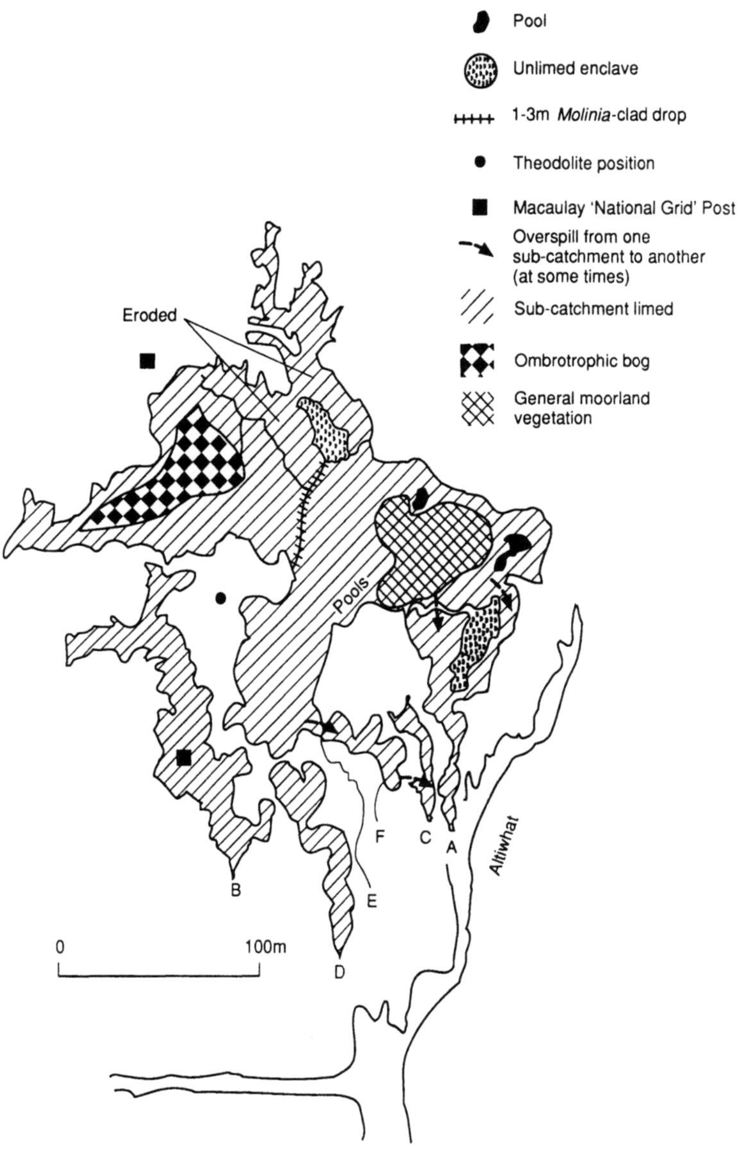

FIG. 15.2 Loch Fleet limited wetland study area in sector VII. Subcatchments (A to F) as in text.

species of *Sphagnum*, although the less sensitive species (*S. inundatum* and *S. recurvum*) were growing actively 4 years after liming.

15.3.3 Conclusions

At 6 months and 12 months after liming, the two most sensitive *Sphagnum* species were affected where they were directly exposed to limed water in sectors IV and VII. Other species are less damaged, even *S. capillifolium*, where they may grow above the usual level of standing water. While *Sphagnum* damage is still evident 3 years after liming in the upland area of sector VII, many other plants characteristic of bogs (the moss *Racomitrium lanuginosum*, bog heather, *Erica tetralix*, bog cotton, *Eriophorum vaginatum*, sundew *Drosera* sp., bog asphodel, *Narthecium ossifragum* are evidently healthy.

The possibility of increased erosion post-liming has also been considered, and permanent plots established to assess it. So far there is no evidence of erosion. It may be noted that there are indeed areas of long-standing and substantial erosion in sector VII and elsewhere in the upper parts of the catchment where liming did not occur.

15.4 TREE STUDIES

15.4.1 Introduction

Although afforestation in the Loch Fleet catchment covers only 10% of the area, some studies of tree response to liming were undertaken because of greater interest in more densely forested sites elsewhere.

Calcium is known to affect plant growth directly through its action on hormone transport. It also has an indirect action on plant productivity through its interaction with soil ions, reducing or increasing their availability to the plant. Of significant importance is likely to be the increased availability of bicarbonate ions, which may affect cytokinin transport (Marschner, 1986). These indirect effects may not be immediately obvious but are likely to be evident in long, multiple rotations (Gessll, 1987). At excessive rates of application, lime-induced chlorosis may occur and this may either be a direct effect of calcium or an indirect effect due to iron deficiency or that of some other element such as phosphorus or potassium (Rafahi, 1982; Marschner, 1986).

The effects of nutrients on tree productivity are most easily assessed through the measurement of morphological characters. The most commonly used characters, and those that are most straightforward to assess, are overall tree height and diameter at breast height (dbh). However, these characters are unlikely to show significant responses within the short term. Needle length has also been used widely to monitor the vigour of conifers (e.g. Binns *et al.*, 1980).

Conifers are outbreeding trees and are, therefore, genetically and morphologically heterogeneous. When planted in close stands they are also subject to substantial site differences between trees and even between branches within a tree owing to differential light penetration of the canopy. It is therefore necessary to sample a large number of sites and trees to obtain a reliable assessment of mean tree performance.

15.4.2 Observations on Limed and Unlimed Areas

Lodgepole pine and spruce, 25 years old, were present at Loch Fleet in mixed stands, of which one area had been limed (see Chapter 9, Table 9.4) in 1986. Observations made after the cessation of growth in 1988, when 15 trees of each species were sampled from the limed sector IV and from the adjacent, unlimed sector III. Growth cycles, one of which is produced in each year on each lateral branch, were identified from the pre-liming year 1985 and from the post-liming years 1986, 1987, and 1988. Appropriate growth cycles were sampled from two branches from each of an unshaded, upper whorl and a shaded, lower whorl. Characteristics of the growth cycles, i.e. length, needle density, and needle length, which are considered to be indicative of tree vigour, were measured.

Soil pH below the trees showed considerable heterogeneity, with unexpectedly high pH values occurring even within the unlimed subcatchment. This heterogeneity, together with the inter-tree variability noted in the assessments of all the characters, made the analysis and interpretation of results subject to confounding effects. These preliminary results suggest the possibility of a slight growth depressant effect on the vigour of lodgepole pine but, in contrast, a beneficial effect on Sitka spruce. These results must be considered as short-term effects giving no indication of the effects of calcium on long-term rotations.

Assessment of the effect of liming on coniferous trees has also been made by foliar analyses (Wilson *et al.*, 1988). Needle samples collected from pine and spruce in 1986, 1987, and 1988 showed an increase in foliar nitrogen and phosphorus concentrations, possibly as a result of a flush of soil microbial activity after liming increased the mineralisation of organic matter. Thus, for upland sites such as Loch Fleet, characteristically nitrogen and phosphorus deficient, liming appears to have had a beneficial effect on tree mineral nutrition, at least in the short term. Longer-term effects during the expected 50-yr rotation of forest plantations would require longer-term monitoring.

15.5 GROUND BEETLE AND SPIDER COMMUNITIES

15.5.1 Introduction

Recent studies of small limed ombrogenous bog plots (e.g. Mackenzie, 1989) have shown some differences in invertebrate communities between limed and control plots, specifically in mites, Collembola, and ants, 18 months after liming. In the same trials, predatory species such as beetles increased non-significantly, but spiders were 30% fewer. It was considered valuable to discover if this change had occurred with the larger scale trials at Loch Fleet and work was undertaken in 1988 to detect if any effects had occurred. Terrestrial invertebrates selected for study were ground beetles (*Carabidae*) and spiders (*Aranea*), predatory species common on moorland; 26 beetle species and 68 spider species were recorded. The view was taken that such predators would be better indicators of any change than lower trophic groups. The ecology of these two groups is well known and each includes species that are sensitive to environmental changes as well as opportunist species that colonise disturbed habitats.

Traps were set out in all sectors (limed and unlimed) and each supplied with about 50 ml of proprietary antifreeze solution (ethylene glycol), a killing agent and preservative that is neither attractive nor repellent to invertebrates. The traps were covered with 2 cm chicken wire to prevent access by mammals and birds. Trap catches were collected monthly, the material being bulked into catches for each line of traps, giving two replicates per sectoral treatment. Catches were identified to species level. Data were subject to multivariate analysis.

15.5.2 Findings (Appendices 15.2 and 15.3)

Of all the species of spiders identified (Merrett *et al.*, 1985) most catches were dominated by *Pardosa pullata*. The groupings identified by TWINSPAN were differentiated by reference to the abundance of the commonest species, *P. pullata*, or to the presence of some less common species.

The soils of the "wetter" moorland of sectors VII and IX were characterised paradoxically by the lowest percentage of water and organic matter and the highest dry bulk densities. This was because these areas are flat and received inorganic material from beneath the blanket bog up-slope. Thus the soil of the sloping moorland comprises almost pure peat, whereas more inorganic material is incorporated into the peat of the flat moorland. Despite the apparent paradox of low soil moisture values, the flatter parts of catchment were the only ones with open pools of water and thus able to support moisture-loving species.

The DECORANA ordination (Fig. 15.3) confirmed the division into flat moorland with standing water (sectors VII and IX), afforested areas (FC and IV), and the rest of the samples on sloping, unafforested moorland. The moorland treated by muirburn formed an outlying group. There was no tendency for limed sectors (IV, VI, VII, IX, and Z_1) to be associated in ways that could not be explained by other features.

Diversity (Williams, 1964) was generally lowest in the afforested area (FC), most variable in the watershed mire with spot applications of lime (VII), and highest on the fringe of an afforested area (IX).

Of all the species of carabid identified (Pope, 1977), the species most often caught was *Pterostichus diligens*, an upland wetland species. *Carabus* species were common in many catches. TWINSPAN of abundance classes split the samples into two distinct groups: from afforested areas and from moorland (Fig. 15.4). The moorland group could not be subdivided in a meaningful way.

The lowest values for diversity were associated with open moorland receiving the highest rate of lime treatment (sector VI, site 1) and with untreated moorland (sector X, site 1). The second highest diversity value for ground beetles was for the same trap transect (sector VII, site 2) which had the highest diversity of spiders. The diversities of the ground beetle fauna were generally low.

15.5.3 Conclusions

The principal predator assemblages of moorland invertebrates, ground beetles and spiders, could thus be divided into those of open moorland and those of afforested areas. It was not possible to detect any effects on assemblages associated with liming.

More spider species were found than ground beetle species and it was possible to subdivide the spider assemblages of moorland into those of the sloping area and the flat area with open pools, and that associated with muirburn. The two flat areas of moorland, on sectors VII and IX, were both subject to low rates of lime, but on the basis of lack of discernible effect on the other moorland sectors, it is reasonable to conclude that liming had no effect in these sectors.

TWINSPAN indicated the possibility of using the abundance of *Pardosa pullata* as an indicator of assemblage type. This could lead to false conclu-

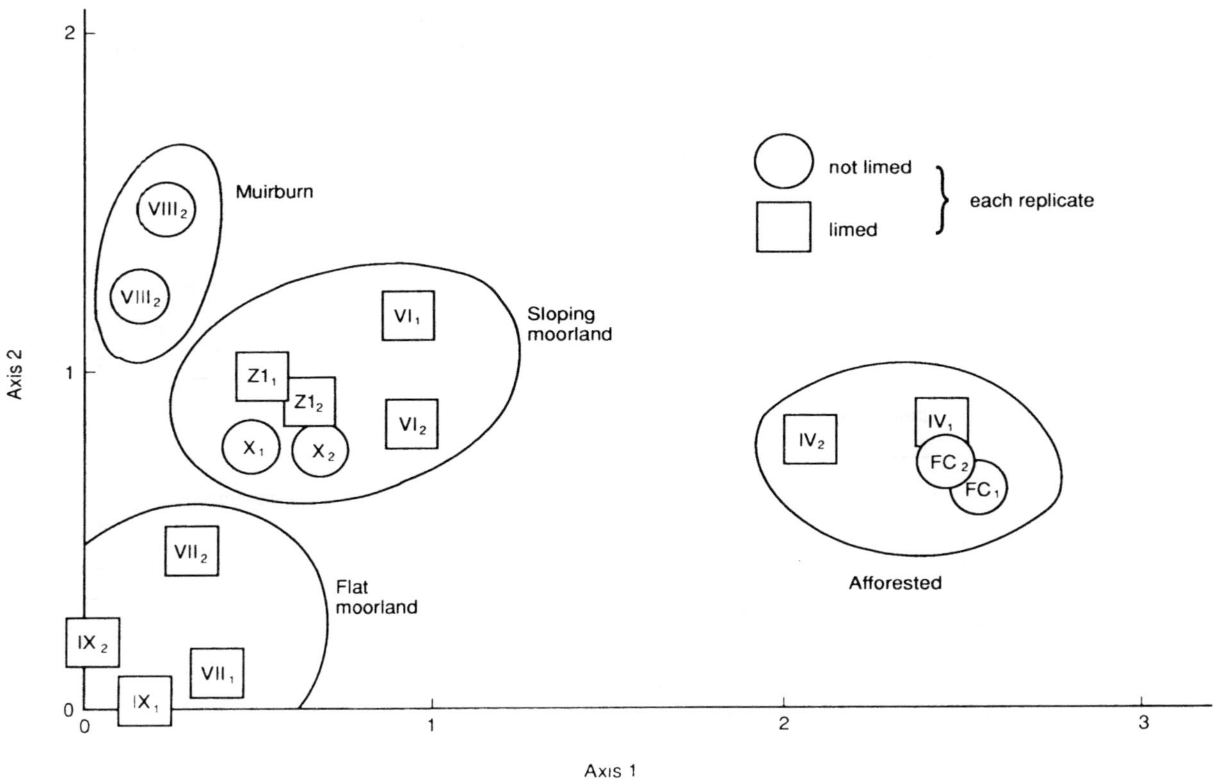

FIG. 15.3 DECORANA ordination of spider assemblages for limed as unlimed sectors of Loch Fleet catchment. FC is a forest control site. outside the Fleet catchment.)

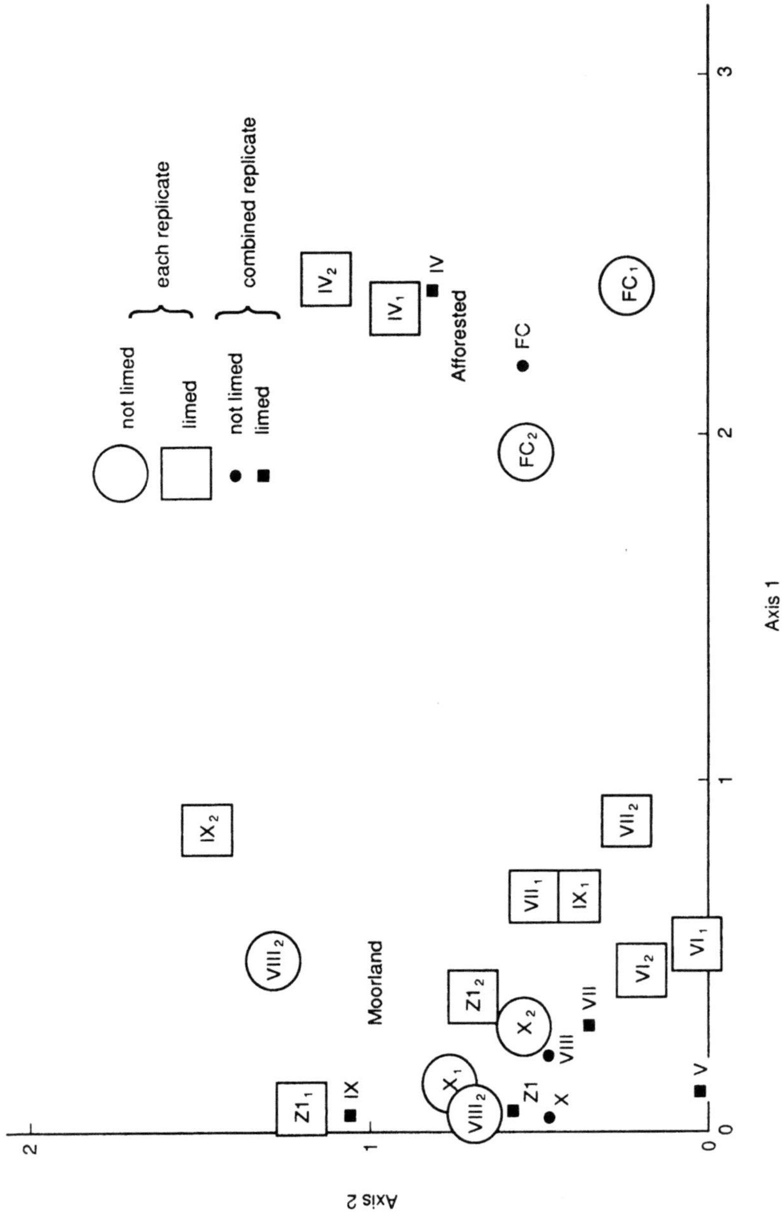

FIG. 15.4 DECORANA ordination of ground beetle assemblages for limed and unlimed sectors of Loch Fleet catchment (FC is a forest control site outside the Loch Fleet catchment.)

sions owing to trapping efficiency varying with the terrain. The presence or absence of less abundant species would be more appropriate even though TWINSPAN of presence/absence data was less effective than TWINSPAN of abundance classes.

The presence and abundance of *Carabus* species did not appear to be affected by liming. Most species of this genus are particularly sensitive to habitat disturbance in lowland grassland (Tietze, 1985) and upland grassland (Luff and Rushton, 1988).

Diversity was higher for spiders than for ground beetles. Diversity did not appear to be affected by liming.

15.6. OTHER TERRESTRIAL FAUNA

15.6.1 Small Mammals (Shore, 1990)

The small mammal fauna of upland moors is dominated by common shrews, *Sorex araneus*, and pygmy shrews, *S. minutus* (Butterfield *et al.*, 1981; Yalden, 1981), both of which are protected under the Wildlife and Countryside Act (1981). Moorland and upland sheepwalk is also marginal habitat for field voles, *Microtus agrestis* (Corbett and Southern, 1977), which are a major prey of protected birds of prey on moorland (Davis and Davis, 1981; Newton *et al.*, 1982). Mackenzie (1989) found that the number of shrews incidentally captured in her pitfall traps (set to catch invertebrates) was greater in control plots than in small (less than 25 m^2), experimentally limed plots (4·05 vs 2·9 shrews per trap, respectively; $\chi^2_{(1)} = 4{\cdot}7$; $p = 0{\cdot}02$). There was no significant effect of liming on vole captures. These results suggested that shrew activity was lower on limed plots than on controls, possibly reflecting differences in the abundances of prey organisms.

It was judged useful to study whether catchment liming at Loch Fleet has affected small mammal populations. Small mammal numbers on sectors VI and Z_1 (limed in 1986 and 1987, respectively) and sectors X and Y (non-limed controls) at Loch Fleet were estimated by pitfall trapping in August 1989. Results from Loch Fleet indicate that, at least 2 years after application, there are no significant differences between limed and control areas in the numbers of common shrews and field voles captured (Table 15.3). However, significantly fewer female pygmy shrews were captured on limed than on control sites overall (16 vs 31; $\chi^2_{(1)} = 4{\cdot}79$; $p < 0{\cdot}05$), although there was no difference in the number of male pygmy shrews captured in limed and control areas.

TABLE 15.3
Small mammal catches at Loch Fleet, 1989

Sites	*Common shrew*		*Pygmy shrew*		*Field vole*	
	Male	*Female*	*Male*	*Female*	*Male*	*Female*
Y (control)	12	15	28	23	1	1
VI (lime)	17	16	24	11	4	0
X (control)	8	10	12	8	0	1
Z_1 (lime)	17	10	15	5	2	1
Total:						
Control	20	25	40	31	1	2
Lime	34	26	39	16	6	1

Interpretation of these findings at Loch Fleet calls for analysis of the results obtained from other sites. Catchment liming may potentially have both short- and long-term effects on shrews. At Llyn Conwy, N. Wales, fewer common and pygmy shrews were pitfall trapped on the small (5 × 5 m) limed plots than on control areas up to 2 months and 5 months after application, respectively, and live-trapping suggested that the number of pygmy shrews on limed sites was reduced up to 4–5 months after application (Shore, 1990). Liming may initially reduce shrew activity by decreasing prey availability; fewer small invertebrates were captured in pitfall traps on limed than on control plots (Mackenzie, 1989). However, the reduction in shrew activity appears to be short lived. At Llyn Conwy, common and pygmy shrews were live-trapped in equal numbers on limed and control areas 2 months and 5 months after application, and 1 year after liming, pitfall trap capture success was not significantly different for limed and control areas at both Llyn Conwy and at Llyn Brianne, mid-Wales.

Longer-term effects of liming on shrews were indicated by results from Loch Fleet and Llyn Brianne where pitfall trap captures of common shrews were always slightly greater on limed than on control areas (Table 15.3; Shore, 1990). These data suggest that, at least 1 year after application, there is an increase in common shrew activity on limed sites. Although results from Loch Fleet also suggest that capture success of female pygmy shrews is lower on limed than on control areas, this may not have been due to liming because the number of female pygmy shrews captured was not greater on control than on limed areas at any other site.

Live-trapping studies suggest that there may be an initial decrease in field vole numbers following liming (Shore, 1990). However, there is no evidence from pitfall trap studies that liming has any consistent, long-term effect on field vole activity.

Overall it is clear from pitfall trap studies that shrews and voles do occur on limed areas. However, the possible effects of liming on small mammal populations are still uncertain because pitfall traps monitor activity, not abundance. Differences in animal activity between sites may not reflect differences in population numbers; for example, animals may be more active if prey is scarce. Liming may also alter forage quantity and quality without significantly altering population numbers; areas with poor food availability may support the same number of individuals as better areas but contain fewer adult, dominant animals and many individuals in poor condition. Live-trapping studies are currently being carried out to determine if liming affects population dynamics. Age, body size, and condition of shrews captured at Loch Fleet and Llyn Brianne will be examined to study any effects of liming on nutrition and population structure.

15.6.2 Other Observations (MacKay, 1989; Appendix 15.4)

The bleak upland moorland surrounding Loch Fleet imposes severe limitations on the range of wildlife to be found there. The only British mammal that is exclusively montane is the mountain hare (*Lepus timidus*). Other mammals, such as the red deer (*Cervus elaphus*), spend the summer on the upland moors but migrate to lower ground in winter, especially into the conifer plantations. Plantations also provide shelter for roe deer (*Capreolus capreolus*), foxes (*Vulpes vulpes*), stoats (*Mustela erminea*), rabbits (*Oryctolagus cuniculus*), and wood mice (*Apodemus sylvaticus*). The dense rough upland grassland is the habitat for small mammals, especially the short-tailed vole (*Microtus agrestis*), common shrews (*Sorex araneus*), and pigmy shrews (*S. minutus*).

Adders (*Vipera berus*) are fairly common around Loch Fleet, feeding on the small mammals. They also eat lizards (*Lacerta vivipara*), toads (*Bufo bufo*), and frogs (*Rana temporaria*), which are all common and were so even prior to liming around the loch. The palmate newt (*Triturus helveticus*) is also found.

Several species of birds found at Loch Fleet are characteristic of upland and moorland areas. These include the red grouse (*Lagopus lagopus*), black grouse (*Lyrurus tetrix*), ravens (*Corvus corax*), meadow pipits (*Anthus pratensis*), stonechats (*Saxicola torquata*), and wheatears (*Oenanthe oenanthe*). The meadow pipits occur in large numbers during the summer, although they are dependent on insect populations. In turn, they are the main food for the hen harriers (*Circus cyaneus*) and the peregrine falcon (*Falco peregrinus*), which can often be seen to the north and east of the loch. The short-eared owl (*Asio flammeus*) feeds mostly on small mammals. Buzzards (*Buteo buteo*) occasionally soar over the loch, and very rarely a golden eagle (*Aquila chrysaetos*) can be seen.

Many of the birds that visit Loch Fleet are typical of freshwater habitats, such as the ducks, mallard (*Anas platyrhynchos*), and teal (*A. crecca*), which visit the loch in spring and early summer, often making several unsuccessful nesting attempts. Tufted duck (*Aythya fuligula*), pochard (*A. ferina*), and goldeneye (*Bucephala clangula*) are all regular winter visitors. The common sandpiper (*Actitis hypoleucos*) breeds at Loch Fleet in early summer. Grey wagtails (*Montacilla cinerea*), pied wagtails (*M. alba yarrellii*), and dippers (*Cinclus cinclus*) are also fairly common, the latter present in the Altiwhat even before liming. Almost all species present probably occurred before liming, except possibly the fish eaters.

Since fish were reintroduced to the loch in 1987, several fish eaters have found the loch. In 1989 a single cormorant (*Phalacrocorax carbo*) was seen almost daily throughout the summer. Mink (*Mustela vison*) and otters (*Lutra lutra*) have also been recorded. The first sighting of a heron (*Ardea cinerea*) was in September 1989 near the outlet. Gulls are often seen in flight overhead but seldom land on the loch.

The forested area also gives shelter to a number of species, including the carrion crow (*Corvus corone corone*), which is mainly a scavenger, but also preys on frogs and toads in spring when they are spawning. The crow has learnt how to remove the toad's distasteful skin. Other birds, found mainly in the woodland, include chaffinches (*Fringilla coelebs*), robins (*Erithacus rubecula*), willow warblers (*Phylloscopus trochilus*), siskins (*Carduelis spinus*), wrens (*Troglodytes troglodytes*), and blackbirds (*Turdus merula*), although the latter two can also be found elsewhere around the loch, especially in the Altiwhat gully.

Other birds seen or heard in the forest, especially along the forest roads approaching Loch Fleet, include the cuckoo (*Cuculus canorus*), wood pigeons (*Columba palumbus*), jays (*Garrulus glandarius*), crossbills (*Loxia curvirostra*), bullfinches (*Pyrrhula pyrrhula*), and woodcocks (*Scolopax rusticola*).

15.7 CONCLUSIONS

Although it has sometimes been claimed (eg. Persson, 1982) that catchment liming will change terrestrial communities, this has seldom been documented, perhaps not surprising in view of the relatively few projects followed through for sufficient time. The opportunity to record and measure such effects has been provided at Loch Fleet with studies extending through four post-liming years. While direct contact with applied lime or high pH/high calcium drainage water has damaged two *Sphagnum* species in a limed upland wetland, other vegetational changes of significance are not evident, nor are invertebrate or vertebrate communities apparently affected. Thus concern relates only to possibly adverse changes to ombrotrophic *Sphagnum* wetlands (MacKenzie, 1989). Reference to historic changes in the Cairnsmore-of-Fleet area (NCC, 1983) suggests that major husbandry changes (the introduction of rabbits in 1830, sheep monoculture early in the nineteenth century, the reintroduction of red deer in 1900) and changing land-use practices (forestry since 1945) have had evident and extensive effects on the area. The limited area and extent of *Sphagnum* effects at Loch Fleet is notable and should be set in the context of benefits to the fishery and other aquatic components.

In the case of direct lake or stream liming, there is little justification for considering terrestrial effects, although changes in the aquatic invertebrate community will no doubt increase the population or density of riparian birds such as the dipper.

REFERENCES

Binns, W.O., Mayhead, G.J. and MacKenzie, J.M. (1980). *Nutrient Deficiencies of Conifers in British Forests*. British Forestry Comm. Leaflet No. 76.

Butterfield, J., Coulson, J.C. and Wanless, S. (1981). Studies on the distribution, food, breeding biology and relative abundance of the pygmy and common shrews (*Sorex minutus* and *S. araneus*) in upland areas of northern England. *J. Zool. Lond.* **195**: 169–80.

Clymo, R.S. (1963). Ion exchange in *Sphagnum* and its relation to bog ecology. *Ann. Bot.* (*Lond.*) NS **27**: 309–24.

Clymo, R.S. (1973). The growth of *Sphagnum*: some effects of environment. *J. Ecol.* **61**: 849–69.

Corbett, G.B. and Southern, H.N. (1977). *The Handbook of British Mammals*, 2nd Edn. Blackwell Scientific Publishers, Oxford, 520 pp.

Davis, P.E. and Davis, J.E. (1981). The food of the red kite in Wales. *Bird Study* **28**: 33–40.

Ferguson, P., Lee, J.A. and Bell, J.N.B. (1978). Effects of sulphur pollutants on the growth of *Sphagnum* species. *Env. Poll.* **16**: 151–62.
Gessll, M. (1987). *Plantation Silviculture.* Springer-Verlag, Berlin, 322 pp.
Luff, M.L. and Rushton, S.P. (1988). The effects of pasture improvement on the ground beetle and spider fauna of upland grasslands. *Env. Aspects Appl. Biol.* **17**(1): 67–74.
MacKay, J. (1989). Unpublished report to Loch Fleet Project.
Mackenzie, S. (1989). The ecological effects of liming on ombrogenous bog communities, Unpublished report to the Nature Conservancy Council. University of Manchester.
Marschner, N. (1986). *Mineral Nutrition in Higher Plants.* Academic Press, New York, 674 pp.
Merrett, P., Locket, G.H. and Millidge, A.F. (1985). A check list of British spiders. *Bull. Brit. Arachnol. Soc.* **6**: 381–403.
NCC (Nature Conservancy Council) (1983). Land-use History of Cairnsmore of Fleet NNR. Nature Conservancy Council, Southwest Region, Scotland, 51 pp. + annexes.
Newton, I., Davis, P.E. and Davis, P.E. (1982). Ravens and buzzards in relation to sheep-farming and forestry in Wales. *J. Appl. Ecol.* **19**: 681–706.
Persson, G. (1982). *Acidification Today and Tomorrow.* Ministry of Agriculture, Sweden, pp. 232.
Pope, R.D. (editor) (1977). A check list of British Insects. Part 3: Coleoptera and Strepsiptera. *Handbooks for the Identification of British Insects* **11**(3). Royal Entomological Society.
Rafahi, H. (1982). Proprietes physico-chimiques et mineralogues des sols des forets de chenes de l'ouest de l'Iran. *Can. J. Soil Sci.* **62**: 39–48.
Shore, R.F. (1990). The effect of liming moorland catchments on small mammal populations. Unpublished report to National Power. Institute of Terrestrial Ecology, Huntingdon, 27 pp.
Tietze, F. (1985). Veranderungen der Arten- und Dominanzstruktur in Laufkäfer-taxozonosen (Coleoptera–Carabidae) bewirtschafteter Graslandökosysteme durch Intensivierungsfaktoren. *Zool. Jaarb. (Systematik)* **112**: 367–82.
Williams, C.B., (1964). *Patterns in the Balance of Nature.* Academic Press, London, 324 pp.
Wilson, E.J., Roberts, D.J. and Skeffington, R.A. (1988). The Effect of Lime Application to the Loch Fleet Catchment on Tree and Soil Chemical Properties – An Initial Report. CERL Report TPRD/L/3283/R88, 25 pp.
Yalden, D.W. (1981). The occurrence of the pygmy shrew *Sorex minutus* on moorland and the implications for its presence in Ireland. *J. Zool. Lond.* **195**: 147–56.

APPENDIX 15.1: PLANT SPECIES RECORDED WITHIN THE CATCHMENT

Trees (planted)

Larix × eurolepis	Larch

Picea sitchensis	Sitka spruce
Pinus contorta	Lodgepole pine
Other trees/shrubs	
Salix aurita	Eared willow
Betula pendula	Silver birch
Sorbus aucuparia	Mountain ash
Dwarf shrubs	
Calluna vulgaris	Common heather
Empetrum nigrum	Crowberry
Erica cinerea	Purple heather
E. tetralix	Cross-leaved heath
Myrica gale	Bog myrtle
Vaccinium myrtillus	Bilberry
Ferns	
Blechnum spicant	Hard fern
Dryopteris dilatata	Broad buckler fern
Oreopteris limbosperma	Lemon scented fern
Polypodium vulgare	Polypody
Grasses	
Agrostis canina montana	Velvet bent
A. capillaris	Brown bent
A. stolonifera	Creeping bent
Anthoxanthum odoratum	Scented vernal grass
Deschampsia flexuosa	Wavy hair grass
Festuca ovina	Sheep's fescue
F. vivipara	Sheep's fescue (variety)
Holcus mollis	Creeping soft grass
Molinia caerulea	Purple moor grass, flying bent
Nardus stricta	Mat grass
Sieglingia decumbens	Heath grass
Sedges and rushes	
Carex binervis	Green-ribbed sedge
C. echinata	Star sedge
C. nigra	Common sedge
C. panicea	Carnation grass

C. pilulifera	Pill sedge
Juncus acutiflorus	Sharp-flowered rush
J. articulatus	Jointed rush
J. bulbosus	Bulbous rush
J. conglomeratus	Conglomerate rush
J. effusus	Soft rush
J. squarrosus	Heath rush
Luzula multiflora	Many flowered wood rush
Eriophorum angustifolium	Common cotton grass
E. vaginatum	Hare's tail grass
Trichophorum cespitosum	Deer grass
Mosses	
Andreaea alpina	
Aulacomnium palustre	
Campylopus atrovirens	
C. flexuosus	
Dicranella heteromalla	
Dicranum scoparium	
Hylocomium splendens	
Hypnum jutlandicum	
Leucobryum glaucum	
Plagiothecium undulatum	
Pleurozium schreberi	
Pholia nutans	
Polytrichum commune	Hair moss
Racomitrium lanuginosum	Woolly fringe moss
Phytidiadelphus loreus	Hooked moss
Sphagnum auriculatum	Bog moss
S. capillifolium	Bog moss
S. cuspidatum	Bog moss
S. papillosum	Bog moss
S. recurvum	Bog moss
Thuidium tamarischinum	
Liverworts	
Calypogeia muellerana	
Cephaloziella sp.	
Diplophyllum albicans	
Lophocolea bidentata	

Odontoschisma sphagni
Pleurozia purpurea

Lichens

Cladonia coccifera
C. impexa
C. furcata
C. floerkeana
C. squamosa
C. uncialis
Cornicularia aculeata
Parmelia physodes

Forbs

Anemone nemorosa	Wood anemone
Cerastium fontanum	Common mouse ear
Cirsium arvense	Creeping thistle
C. palustre	Marsh thistle
C. vulgare	Spear thistle
Crepis paludosa	Marsh hawk's beard
Dactylorhiza ericetorum	Heath spotted orchid
Digitalis purpurea	Foxglove
Drosera rotundifolia	Sundew
Epilobium angustifolium	Rose bay willow herb
E. palustre	Bog willow herb
E. montanum	Broad-leaved willow herb
Euphrasia officinalis	Eyebright
Galium saxatile	Heath bedstraw
Listera ovata	Twayblade
Narthecium ossifragum	Bog asphodel
Pedicularis sylvatica	Lousewort
Pinguicula vulgaris	Common butterwort
Polygala serpyllifolia	Heath milkwort
Potentilla erecta	Common tormentil
Ranunculus flammula	Lesser spearwort
Rumex acetosa	Common sorrel
R. obtusifolius	Broad-leaved dock
Senecio jacobaea	Ragwort
Solidago virgaurea	Golden rod
Stellaria alsine	Bog stitchwort

Succisa pratensis	Devil's bit scabious
Trifolium repens	White clover
Veronica serpyllifolia	Thyme-leaved speedwell
Viola palustris	Bog violet/marsh violet
Aquatic/marginals	
Carex rostrata	Beaked sedge
Isoetes echinospora	Spring quillwort
I. lacustris	Quillwort
Littorella uniflora	Shore weed
Lobelia dortmanna	Water lobelia
Potamogeton polygonifolius	Bog pond weed
Utricularia intermedia	Bladderwort

APPENDIX 15.2: SPIDERS FROM PITFALL TRAPS AROUND LOCH FLEET IN 1988

Identified by Dr. S.P. Rushton.
Names follow Merrett, Locket & Millidge (1985).

Gnaphosidae	
Drassodes cupreus	(Blackwall)
Haplodrassus signifer	(C.L. Koch)
Micaria pulicaria	(Sundevall)
Clubionidae	
Clubiona diversa	Pickard-Cambridge
C. reclusa	Pickard-Cambridge
C. trivialis	C.L. Koch
Liocranidae	
Agroeca proxima	(Pickard-Cambridge)
Thomisidae	
Xysticus cristatus	(Clerck)
Oxyptila trux	(Blackwall)
Salticidae	
Neon reticulatus	(Blackwall)

Lycosidae	
Pardosa nigriceps	(Thorell)
P. palustris	(L.)
P. Pullata	(Clerck)
Alopecosa pulverulenta	(Clerck)
Trochosa terricola	Thorell
Pirata piraticus	(Clerck)
Theridiidae	
Robertus lividus	(Blackwall)
Pholcomma gibbum	(Westring)
Linyphiidae	
Ceratinella brevipes	(Westring)
Walckenaeria acuminata	Blackwall
W. antica	(Wider)
W. atrotibialis	(Pickard-Cambridge)
W. cuspidata	(Blackwall)
W. nudipalpis	(Westring)
Dicymbium tibiale	(Blackwall)
Dismodicus bifrons	(Blackwall)
Gonatium rubellum	(Blackwall)
G. rubens	(Blackwall)
Pepocranium ludicrum	(Pickard-Cambridge)
Pocadicnemis pumila	(Blackwall)
Oedothorax gibbosus	(Blackwall)
O. retusus	(Westring)
Pelecopsis mengei	(Simon)
Silometopus elegans	(Pickard-Cambridge)
Tapinocyba pallens	(Pickard-Cambridge)
Monocephalus fuscipes	(Blackwall)
Gongyllidiellum vivum	(Pickard-Cambridge)
Micrargus herbigradus	(Blackwall)
Erigonella hiemalis	(Blackwall)
Savignya frontata	(Blackwall)
Diplocephalus cristatus	(Blackwall)
Erigone atra	(Blackwall)
E. dentipalpis	(Wider)
Hilaira excisa	(Pickard-Cambridge)
Aphileta misera	(Pickard-Cambridge)

Porrhoma pallidum	Jackson
Agyneta decora	(Pickard-Cambridge)
A. olivacea	(Emerton)
A. ramosa	Jackson
Meioneta saxatlis	(Blackwall)
Centromerus dilutus	(Pickard-Cambridge)
C. prudens	(Pickard-Cambridge)
C. sylvaticus	(Blackwall)
Centromerita concinna	(Thorell)
Saaristoa abnormis	(Blackwell)
S. firma	(Pickard-Cambridge)
Bathyphantes gracilis	(Blackwall)
B. parvulus	(Westring)
Tapinopa longidens	(Wider)
Stemonyphantes lineatus	(L.)
Bolyphantes luteolus	(Blackwall)
Lepthyphantes alacris	(Blackwall)
L. Ericaeus	(Blackwall)
L. mengei	Kulczynski
L. pallidus	(Pickard-Cambridge)
L. zimmermanni	Bertkau
Linyphia clathrata	Sundevall
Microlinyphia pusilla	(Pickard-Cambridge)

APPENDIX 15.3: GROUND BEETLES CAUGHT IN PITFALL TRAPS 1988.

Names follow Pope (1977) with exception of *Pterostichus rhaeticus*.

Cychrus caraboides	(L.)
Carabus glabratus	Paykull
C. granulatus	L.
C. problematicus	Herbst
C. violaceus	L.
Leistus rufescens	(Fab.)
Notiophilus biguttatus	(Fab.)
N. germinyi	Fauvel
Loricera pilicornis	(Fab.)
Dyschirius globosus	(Herbst)

Patrobus assimilis	Chaudoir
Trechus obtusus	Erichson
Pterostichus diligens	(Sturm)
P. niger	(Schaller)
P. rhaeticus (originally confused with *P. nigrita* (Paykull))	Heer
P. versicolor	(Sturm)
Olisthopus rotundatus	(Paykull)
Agonum fuliginosum	(Panzer)
Amara lunicollis	Schiödte
Harpalus latus	(L.)

APPENDIX 15.4: WILDLIFE OF LOCH FLEET

Mammals	
Cervus elaphus	Red deer
Capreolus capreolus	Roe deer
Lepus timidus	Blue hare
Oryctolagus cuniculus	Rabbit
Vulpes vulpes	Fox
Lutra lutra	Otter
Mustela vison	Mink
M. erminea	Stoat
Sorex araneus	Common shrew
S. minutus	Pigmy shrew
Microtus agrestis	Short-tailed vole
Apodemus sylvaticus	Wood mouse
Talpa europaea	Mole
Reptiles	
Lacerta vivipara	Common lizard
Vipera berus	Adder
Amphibians	
Triturus helveticus	Palmate newt
Bufo bufo	Common toad
Rana temporaria	Common frog

Birds

Anas platyrhynchos	Mallard
A. crecca	Teal
Aythya fuligula	Tufted duck
A. ferina	Pochard
Bucephala clangula	Goldeneye
Corvus corax	Raven
C. corone corone	Carrion crow
Larus argentatus	Herring gull
L. fuscus	Lesser black-backed gull
L. canus	Common gull
Phalacrocorax carbo	Cormorant
Lyagopus lagopus	Red grouse
Larurus tetrix	Black grouse
Numenius arquata	Curlew
Actitis hypoleucos	Common sandpiper
Haematopus ostralegus	Oyster catcher
Gallinago gallinago	Snipe
Aquila chrysaetos	Golden eagle
Buteo buteo	Buzzard
Circus cyaneus	Hen harrier
Falco peregrinus	Peregrine falcon
Asio flammeus	Short-eared owl
Ardea cinerea	Heron
Phylloscopus trochilus	Willow warbler
Anthus pratensis	Meadow pipit
Montacilla alba yarrellii	Pied wagtail
M. cinerea	Grey wagtail
Cinclus cinclus	Dipper
Troglodytes troglodytes	Wren
Saxicola torquata	Stonechat
Oenanthe oenanthe	Wheatear
Erithacus rubecula	Robin
Turdus merula	Blackbird
Fringilla coelebs	Chaffinch
Carduelis spinus	Siskin

Not seen at Loch Fleet but common in surrounding forestry:

Cuculus canorus	Cuckoo
Columba palumbus	Wood pigeon

Garrulus glandarius	Jay
Scolopax rusticola	Woodcock
Loxia curvirostra	Crossbill
Pyrrhula pyrrhula	Bullfinch
Butterflies	
Pieris napi	Green-veined white – occasional in flight June/July
Vanessa atalanta	Red admiral– feeding on heather flowers, Sector V, Sept. 1987
Aglais urticae	Small tortoiseshell – near loch outlet, Sept. 1989
Clossiana selene	Small pearl-bordered fritillary – In flight esp. around May–July
Coenonympha pamphilus	Small heath – common at base of Altiwhat June/July
Moths	
Lasiocampa quercus callunae	Northern eggar – males day flying, June/July
Ematurga atomaria	Common heath – common around loch, May–August
Ceramica pisi	Broom moth – green and yellow striped caterpillar often found on various foodplants around loch, July–September
Dragonflies	
Pyrrhosoma nymphula	Large red damselfly – common around outlet and pools in July
Coenagrion pulchellum	Variable damselfly – common over loch and embayments, especially 4, June–August
Cordulegaster boltonii	Golden ringed dragonfly – Common especially around outlet, July/August
Aeshna juncea	Common hawker – pair in sector V, August 1989

PART V

GENERAL ISSUES

CHAPTER 16

Targets and Time-scales of Liming Treatments

T.R.K. Dalziel
PowerGen plc, Nottingham, UK

A. Dickson
National Power plc, London, UK

P. Warfvinge
University of Lund, Lund, Sweden

M.V. Proctor
National Power plc, Leatherhead, Surrey, UK

16.1 INTRODUCTION

In this chapter the targets for acceptable drainage water quality of liming treatments will be explored, and the ways in which predictions can be made of the length of time over which liming treatments may be expected to produce drainage water quality above the established target value will be discussed.

Predictions of treatment effectiveness times at Loch Fleet are being made in a number of ways. One method uses extrapolations of calcium flux data collected since liming to predict future calcium release and stream water calcium concentrations and acidity. In parallel, a process-oriented model of the catchment known as "SLiM" (Soil Liming Model)

(Warfvinge *et al.*, 1990) has been developed. This model calculates the change in soil solution and runoff chemistry as affected by the limestone application. It incorporates rate equations for key soil processes such as dissolution of limestone and cation exchange, as well as equilibrium reactions for soil and stream water. The chemistry of the stream water is calculated as the volume-weighted average of modelled runoff from treated and untreated parts of the catchment. The programme of soil sampling and analysis for pH, exchangeable acidity, exchangeable calcium, base saturation, and free lime described in Chapter 11 should also provide information on amounts of limestone remaining on the treated sectors and its partitioning into free lime and exchangeable calcium pools. This chapter concentrates on the extrapolation and modelling methods to predict treatment effectiveness times.

16.2 TARGETS AND TIME-SCALES FOR LIMING TREATMENTS

16.2.1 Target Water Quality

The effectiveness of any liming treatment must be assessed in terms of both the quality of the drainage water produced as a result of the treatment and how long an acceptable quality of water will be maintained. The minimum water quality that permits survival of a fish population depends on the sensitivity of the most vulnerable stage in the life history. For brown trout, the freshly fertilised egg is most sensitive to low pH and low calcium concentrations but is relatively insensitive to aluminium (Brown, 1983). As the eyed egg develops it becomes less sensitive to water quality, but on emergence from the egg membrane the newly hatched larva becomes sensitive to acidity and aluminium concentrations, particularly at low calcium concentrations. The yearling fish develops more tolerance to acidity and low calcium concentrations but remains sensitive to aluminium. There are interactions between the toxicity of these chemical components so that the toxicity of acid conditions is offset by high calcium concentrations.

High concentrations of aluminium exhibit their peak toxicity from pH 5·0 to pH 5·5 in conjunction with low calcium concentrations, possibly because a greater proportion of the total aluminium is in the toxic form. The fraction of total aluminium that is toxic will depend on pH, inorganic materials such as fluoride or silicate, and organic materials present. A threshold concentration of toxic aluminium that would not harm trout is around 30 μg litre^{-1} (Howells *et al.*, 1990). In terms of the main water quality factors that are important for brown trout, a minimum defined threshold for calcium and maximum defined thresholds for acidity and aluminium might be set as follows: calcium, greater than 100 μeq litre^{-1}; pH, greater than 6; total aluminium, less than 100 μg litre^{-1}. This water quality should protect all life stages from acute effects (Howells *et al.*, 1983). It is acknowledged, however, that for a self-sustaining population more rigorous criteria may be required. It should also be noted that sensitivity to these conditions varies with species, and also between strains of a single species.

16.2.2 Effectiveness

Target water quality should never be set close to the level of acute toxicity, and to ensure safety both during acid episodes characteristic of high flows and exposure over the longer term, average water quality should be defined at some level above the observed acute effect threshold to ensure conditions suitable for growth and reproduction. It is important to note that since different life stages have different sensitivities and inhabit different parts of a river system, a liming programme should identify the critical life stage and location to be protected. Since adult fish are generally less acid sensitive, it follows that protection of a put-and-take fishery is less demanding than for a self-sustaining population.

Most research has involved salmonid species which are more sensitive to acidity than other freshwater fish species (Howells *et al.*, 1983) and limiting conditions for salmonids seem likely to be more rigorous than those for other species. In northern Europe, including the UK, upland waters sensitive to acidification are usually "salmonid" rather than "coarse fish" waters. In Sweden, however, where many acid lakes have mixed species populations, roach and perch are also reported to be affected (Howells, 1990). In Canada and the USA, both salmonid and other lake species are affected (Magnuson *et al.*, 1984). Much less is known about the specific water quality requirements of these species, but they are likely (for basic physiological reasons) to be similarly restricted to a minimum calcium level and on maximum levels of acidity and aluminium, although possibly with different thresholds.

So far as time-scales are concerned, catchment liming, properly targeted, should offer the prospect of a more economically viable option than direct lake liming if it can be demonstrated that the treatment effectiveness times are sufficiently greater than those of direct water applications, so as to offset the additional costs likely to be incurred with catchment liming. With this in mind, it is clear that the longer a catchment liming treatment can remain effective in meeting the desired target water quality, the better.

16.3 PREDICTIONS FROM CALCIUM FLUX DATA

Hydrochemical budget calculations are described in Chapter 12. Using calcium flux data since liming, calcium release from each sector was calculated for seven 6-month periods following liming. Predictions were then made of:

- release of applied calcium from each of the sectors;
- stream water (or sector drainage water) calcium concentrations; and
- stream water acidity

Regressions of log [calcium remaining] vs either cumulative stream water flow (i.e. volume of water) or log [cumulative flow] were used to predict future amounts of calcium remaining on each sector and hence amounts released. It is implied, therefore, that the regression relationships can adequately describe the gradual decline with time of the amount of calcium initially applied to each sector. Future stream water calcium concentrations were predicted employing a similar technique but using log [net calcium concentration] vs either cumulative flow or log [cumulative flow]. Net calcium concentration is defined as gross (or total measured) calcium concentration minus the "background" calcium, i.e. that deriving from rainfall and from the catchment through groundwater or by natural processes such as weathering. Rainfall calcium is measured in samples collected in a standard bulk rainfall collector situated within the catchment. The contribution to "background" calcium made by groundwater or by natural processes can be estimated from data for the period before liming.

Future stream water acidity can be predicted if the future alkalinity is estimated. Alkalinity can be estimated from the difference between the sum of major base cations (C_B) and the sum of strong acid anions (C_A), which at Loch Fleet can be adequately represented as

$$C_B = [Ca^{2+}] + [Mg^{2+}] + [Na^+] + [K^+]$$

and

$$C_A = [SO_4^{2-}] + [NO_3^-] + [Cl^-]$$

Using Loch Fleet data for verification it can be demonstrated that

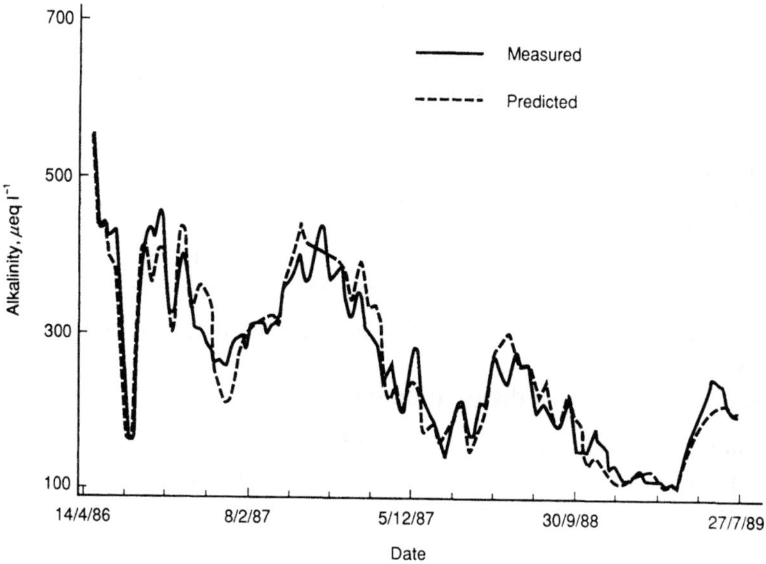

FIG. 16.1 Measured and predicted alkalinity, Altiwhat stream, April 1986 (at liming) to July 1989.

stream water alkalinity following liming can reliably be predicted using measured calcium concentration alone, since calcium has become by far the most dominant cation. This implies that the variations of C_B can be adequately defined by differences in $[Ca^{2+}]$ alone and that other cations can be represented by constants. Similarly, the value for C_A can be deduced either from measured data and assigned as a constant, or varied according to expected trends in sulphate and nitrate deposition. A modified definition of alkalinity on which pH predictions are based is therefore

$$[\text{Alk}] = [Ca^{2+}] + C_{\text{bal}} - C_A$$

where C_{bal} = average seasonal or annual difference between calcium and the total acid equivalent (or average C_B − average $[Ca^{2+}]$).

Plotting alkalinity derived in this way against measured Loch Fleet values (see Chapter 12) gives a very good fit (Fig. 16.1). Predictions of pH were then evaluated using a solution of the carbonate equilibrium derived earlier (Stumm and Morgan, 1983), modified to take into account the influence of organic acids. Two assumptions have to be made in asssessing the organic acid effect on pH prediction:

TABLE 16.1
Values of Oliver equation constants used for each sector to achieve satisfactory pH model verification

Sector	C_1	C_2	C_3
L. Fleet IV	0·96	0·95	0·024
L. Fleet VI	0·96	0·95	0·009
L. Fleet VII	0·96	1·00	0·001
Oliver *et al.* (1983)	0·96	0·90	0·039

- the organic matter is represented by a single characteristic organic acid; and
- the acid has a known dissociation constant.

A suitable method available for describing an organic acid for inclusion in a model is that of Oliver *et al.* (1983), which is based on statistical analyses of samples from streams in North America. Oliver *et al.* (1983) concluded from their data that there was so little variation in the dissociation characteristics for a variety of water samples that there was justification in using this approach. If the quantity of organic anion is defined as

$$[A^-] = \frac{K_{OA}[X_T]}{K_{OA} + [H^+]}$$

where

$[A^-]$ = concentration of organic anions,
K_{OA} = the organic acid equilibrium constant,
$[X_T]$ = concentration of carboxyl groups,

then Oliver *et al.* (1983) deduced that

$$pK_{OA} = C_1 + C_2(\text{pH}) - C_3(\text{pH})^2$$

where C_1, C_2, and C_3 are constants.

Oliver *et al.* (1983) found that, for their samples, a reasonable mean carboxyl content could be estimated from the total organic carbon (TOC) assuming 10 μeq carboxyl per mg TOC. This factor was also used to provide carboxyl estimates from TOC measurements made at Loch Fleet. The Oliver equation was modified by calibration of the values of the constants during pH model verification. The best fit calibrations for each sector were achieved using the constants given in Table 16.1.

Figure 16.2 shows amounts of water (stream water volume) and total calcium export from each of the treated sectors in each of the 6-monthly periods following liming, on which predictions are based. Three main

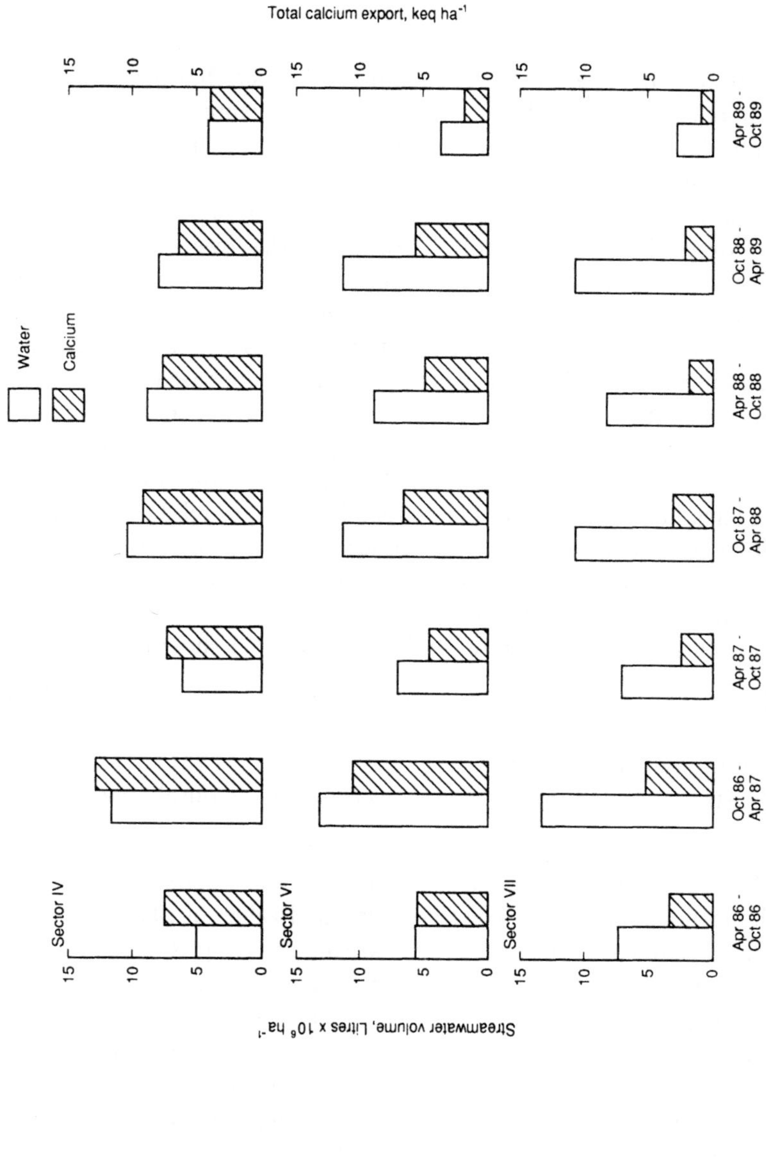

FIG. 16.2 Stream water discharge volume and total calcium export from sectors IV, VI, VII calculated for sequential six-month periods following liming.

points emerge. First, stream water volume, not surprisingly, has had an influence on the calcium release from each sector; the greater the stream water volume (i.e. cumulative flow over 6 months), the more calcium was released. Secondly, as time has elapsed, the influence of stream water volume on calcium release has reduced, i.e. with the same volume of stream water less calcium was released during later periods than in the first few months following liming. Finally, it can be seen that the amounts of calcium released from each sector at any time were approximately proportional to the limestone application rates expressed per unit of total sector area (see Chapter 9), i.e. the highest application rate (23·9 t ha^{-1} on sector IV) produced the greatest release of calcium since liming, while an application of only 3·3 t ha^{-1} on sector VII produced the lowest release.

Since stream water volume is crucial, amounts of applied calcium remaining and stream water calcium concentrations for each sector were related to cumulative stream water volume (cumulative flow), rather than related to time *per se*. Water quality predictions were based on 6-monthly periods using average stream water volumes for the four April–October (summer) periods and the three October–April (winter) periods which had elapsed since liming. The results of these computations were then converted back into amounts of calcium remaining on, or released from, each sector, or to stream water calcium concentration in the flow from each sector vs time. Predictions of acidity were modelled from the recorded and predicted calcium concentrations.

Using this approach, Figures 16.3, 16.4 and 16.5 show predictions of

(a) the amount of applied calcium remaining on each sector with time;
(b) the calcium concentrations in stream water from each sector over 6-monthly periods; and
(c) the stream water acidity on each sector over 6 monthly periods.

In each case the accuracy of the predictions can be judged by the goodness of fit between modelled and measured values (also shown) for the period over which the model was based. The target threshold values of 100 μeq Ca litre^{-1} and pH > 6 to ensure fishery survival are also given. It is recognised that these targets give a $Ca^{2+} : H^{+}$ (eq) ratio far greater than the critical ratio (>4) associated with healthy salmonid fisheries in lakes in southern Norway (Chester, 1984). It is desirable, however, to have such a safety margin, which should protect against the possibility of episodic changes in water quality that are more likely in streams than in lake water, and to protect sensitive life stages. Summer and winter average calcium concentrations measured in the year prior to liming are also shown.

The flux calculations, upon which the predictions made are based, show that considerable quantities of applied limestone still remained on the three sectors 3·5 years after liming. The greatest proportion of applied calcium released has been from sector VII, i.e. 24·1% (15·9 keq Ca ha^{-1}), whereas 10·3% (49·3 keq ha^{-1}) was released from sector IV and 8·5% (36·2 keq ha^{-1}) from sector VI. The curve of the amount of calcium remaining on sector VII vs time (Fig. 16.3) shows a distinct "levelling off". This pattern is not yet so discernible on sectors IV and VI (Figs 16.4 and 16.5), although it seems likely that with time it will become so. There are several possible reasons for this pattern. First, since application, the limestone may have percolated downwards into the soil profile and become less available to the runoff water. Consequently, less calcium and associated neutralising capacity can be released by overland flow to stream water and liming effectiveness diminishes. This theory is confirmed from analyses of Loch Fleet catchment soils (see Chapter 11), which show a progressive increase of calcium with depth in the soil profile and a greater cation exchange capacity, as well as an increase in soil pH between October 1986 and December 1987. A second factor could be an organic or inorganic coating of limestone particles, which may render the limestone less readily soluble in soil or runoff water. Investigations using energy dispersive X-ray analysis (EDAX) (Allison, 1989), however, failed to provide evidence of such particle coating. Another explanation may be found in the buffer mechanisms in the soil system. When limestone is added to an acid soil, the dissolution of this material affects not only the soil water, but also the exchange complexes. When undissolved limestone particles are present the rate at which acid soil water is neutralised is determined by that of limestone dissolution. When no more limestone is present, the buffering is taken over by the soil exchange complex. This process may have a different rate of calcium release. The base saturation of the soil will gradually decline as the soil reacidifies, as will the ability of the soil to neutralise acid discharge. Since the limestone application rate on sector VII only slightly exceeded the exchangeable soil acidity in the upper 5 cm of soil, this effect might be expected to show on this sector within a few years of treatment.

Stream water calcium concentration on each of the sectors declined over the 3·5 years of observation. On sectors IV and VI, however, the predictions indicate a settling down to a more steady calcium concentration over the next several years. On sector IV this is predicted to be at about 10–15 mg l^{-1} (500–750 μeq Ca $litre^{-1}$) and on sector VI between about 5–10 mg l^{-1} (250–500 μeq $litre^{-1}$). The decline in calcium concentrations

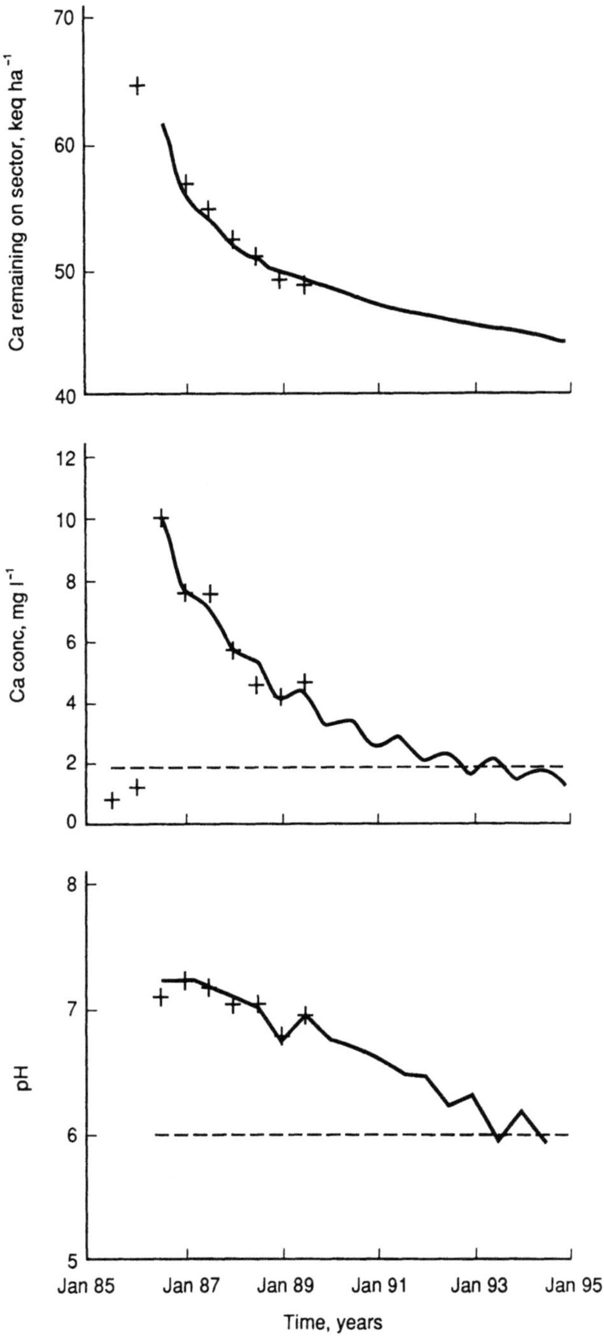

FIG. 16.3 Predicted amount of calcium retained on sector VII, predicted runoff calcium concentration and pH. Measured values are also shown (+). Target thresholds for calcium and pH are indicated (– – –).

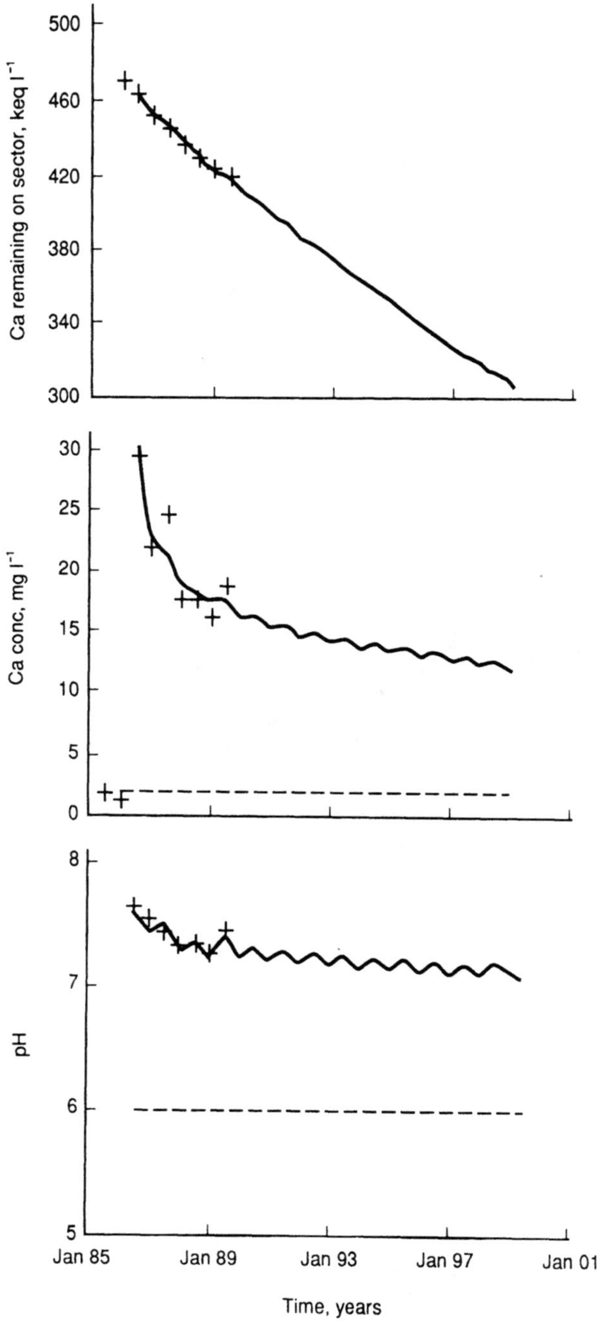

FIG. 16.4 Predicted amount of calcium retained on sector IV, predicted runoff calcium concentration and pH. Measured values are also shown (+). Target thresholds for calcium and pH are indicated (– – –).

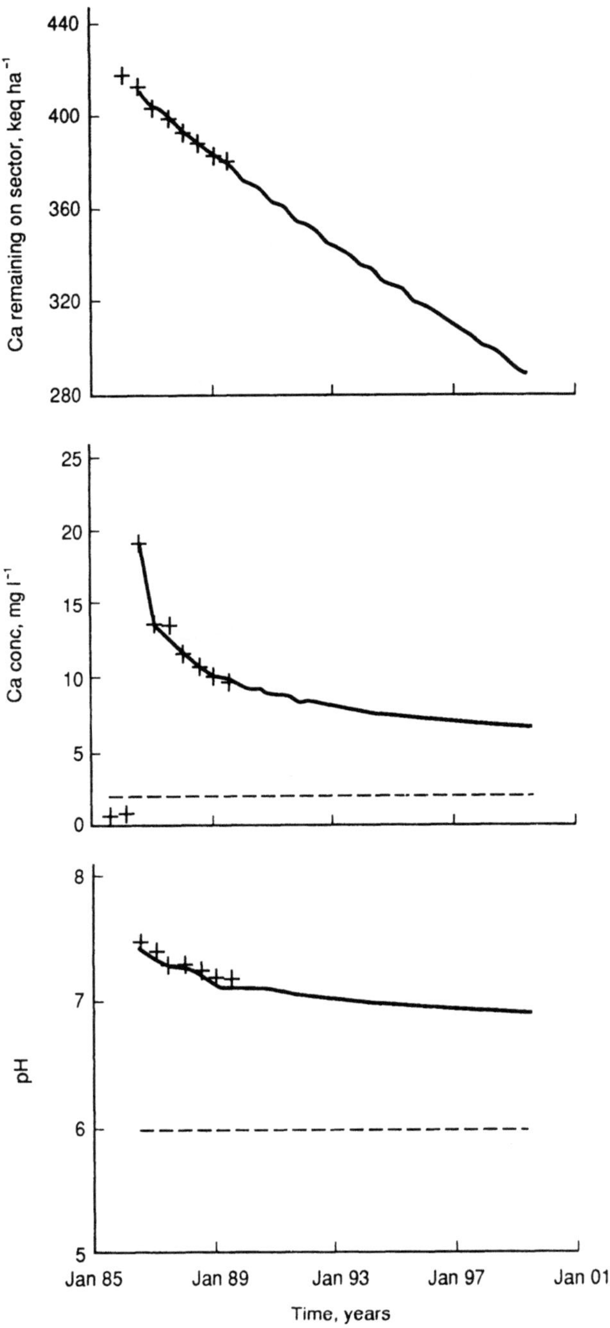

FIG. 16.5 Predicted amount of calcium retained on sector VI, predicted runoff calcium concentrations and pH. Measured values are also shown (+). Target thresholds for calcium and pH are indicated (– – –).

since the maxima recorded immediately following liming of these two sectors has had little impact on present pH values, which are predicted to remain at values around pH 7 on both sectors until the end of the century. On sector VII the situation is very different. Since the rate of calcium release has levelled off, calcium concentrations below 100 μeq litre^{-1} and pH below 6 are predicted in 1993. Given the importance of the Altiwhat, which drains sector VII, as the principal or only spawning and nursery stream for the trout population of the loch (see Chapter 13), and that this stream will be subject to episodic adverse water quality once neutralising capacity is reduced, suggests that careful monitoring of water quality and fish response in this stream is needed.

So far as the loch as a whole is concerned, assuming only the predicted contributions of calcium from sectors IV and VI, it seems likely that a loch water calcium concentration greater than 100 μeq litre^{-1} will be sustained until at least the end of the century. This calculation is based on the assumptions that the rest of the catchment will continue to contribute calcium to the loch through natural processes at the same rate as before liming, and that any calcium sequestration or release by loch sediments is minimal.

16.4 THE SOIL LIMING MODEL

16.4.1 Introduction

The soil liming model (SLiM) considers the chemistry of the four principal components of a limed catchment:

- the baseline runoff from unaffected parts of the watershed;
- runoff from soil layers influenced by the limestone addition;
- stream water chemistry resulting from mixing from the above sources; and
- lake water chemistry resulting from many different sources.

While the baseline chemistry measurements serve as input, the chemistry of the other environments are calculated by the model. SLiM calculates the soil and soil solution chemistry, and hence the contribution of the limed areas. The chemistry of stream and lake water are products of the mixing of water from different sources. SLiM depends, therefore, not only on adequate dissolution chemistry, but also on a good description of flow pathways in the catchment.

The soil chemistry component is dynamic, and is mathematically described by four chemical processes occurring simultaneously in the soil. The processes considered are:

- dissolution and deactivation of limestone;
- cation exchange involving calcium and exchangeable acidity;
- leaching and accumulation of dissolved species in the soil profile; and
- aqueous equilibria for carbonic acid, dissolved organic carbon, and aluminium.

Among processes not considered are silicate weathering, sulphate adsorption, and nitrogen transformations. This is justified since their influence is rather small compared with the large changes caused by liming. Detailed descriptions of the model's structure and processes can be found in the literature (Warfvinge, 1988; Warfvinge and Sverdrup, 1988a, b; Warfvinge and Sverdrup, 1989).

Although the different processes are described individually by unique differential equations, they interact in two respects. First, they communicate and interact by changes in the composition of the soil solution. For instance, an increase in soil pH caused by the dissolution of limestone will alter the speciation of components in the carbon dioxide–aluminium equilibria and the aluminium. This will in turn affect the transport of ions within and from the soil profile. Secondly, a number of input conditions affect more than one subsystem, for example temperature, which affects equilibrium constants as well as limestone dissolution rate constants.

16.4.2 Assumptions and Model Input

The above reactions and processes refer to chemical transformations in the soil environment. SLiM also contains a description of the mixing of water discharged from treated and untreated parts of a catchment. As for all mathematical models describing chemical interactions in a catchment, it was necessary to develop a conceptual model of the environment. This is expressed in terms of equations and conditions. The conceptual model constituting the foundation for SLiM is shown in Fig. 16.6.

It is assumed in the model that the limestone dissolution is confined to the upper few centimetres of the soil, and that this soil layer is homogeneous. Only water flowing through this layer into the stream is neutralised. This implies that only that fraction of the catchment adjacent to streams and saturated areas provides water with increased alkalinity. This is consistent with the experience from soil liming in Sweden and elsewhere where water with elevated ANC draining to lower, acid horizons becomes reacidified.

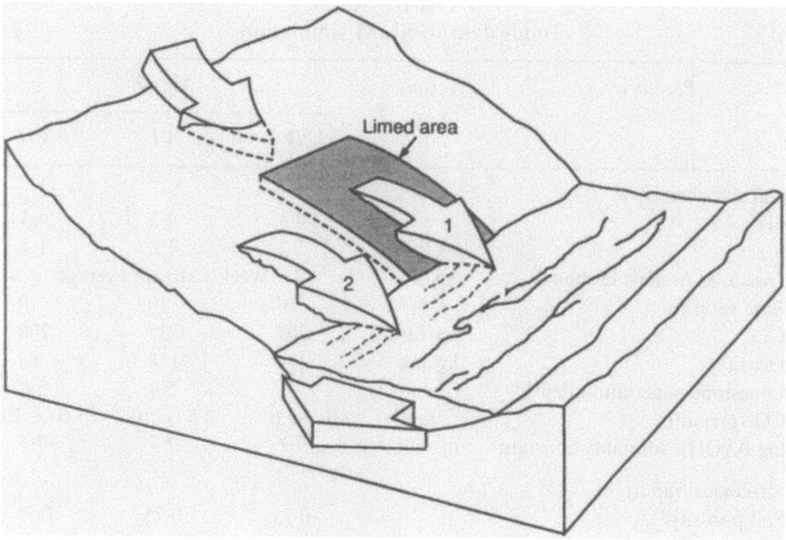

FIG. 16.6 Concepts employed in the SLiM model; inflow comes from two sources; (1) water percolating vertically through a limed soil layer, and (2) water unaffected by liming.

Preliminary model runs assuming that all water was routed through the uppermost, limed, soil layer could not reproduce short-term variations in water chemistry. During the first 3 years following liming at Loch Fleet, pH, calcium, and alkalinity were positively correlated with high flow. The hypothesis was that during low flow, runoff is routed through lower soil layers and so not neutralised by soil liming. This would increase the influence of the baseflow on the stream chemistry in these conditions. On an empirical basis the discharge from the limed layer was calculated as a function of the change in flow rate:

$$X_{\text{flow}} = const \cdot X_{\text{flow}} + (1 - const) \cdot \frac{Q_{\text{total}}}{Q_{\text{average}}}$$

where

X_{flow} = ratio of surficial to baseflow in limed areas,
$const$ = a calibrated value,
$Q_{\text{total}}, Q_{\text{average}}$ = present and average flow rates ($m^3 s^{-1}$).

Relationships like this can be modified and calibrated in a quite sophisticated manner to yield maximum agreement between data and model

TABLE 16.2
Input data to SLiM simulations

Parameter	*Unit*	*Sector*		
		IV	*VI*	*VII*
Initial soil chemistry:				
pH		4·3	4·3	4·3
$[Ca^{2+}]$	mg litre^{-1}	1·2	1·2	1·4
Dissolved organic carbon	mg litre^{-1}	Weekly stream average		
Base saturation	%	10	10	10
CEC	meq litre^{-1}	207	232	200
Density	kg litre^{-1}	134	118	83
Limestone penetration depth[a]	cm	7·5	5·0	5·0
CO_2-pressure[a]	atm	10×10^{-3}	$2{\cdot}5 \times 10^{-3}$	$1{\cdot}0 \times 10^{-3}$
log $Al(OH)_3$ solubility constant[a]	$m^6 mol^{-2}$	7·7	7·7	7·7
Hydrological input:				
Soil porosity[a]		0·75	0·75	0·75
Flow rate	ms^{-1}	Weekly average		
Stream chemistry:				
Baseline $[Ca^{2+}]$	mg litre^{-1}	Weekly average		
Baseline ANC	μeq litre^{-1}	Weekly average		
Dissolved organic carbon	mg litre^{-1}	Weekly average		
CO_2-pressure	atm	$1{\cdot}6 \times 10^{-3}$	$0{\cdot}9 \times 10^{-3}$	$0{\cdot}9 \times 10^{-3}$
log $Al(OH)_3$ solubility constant	$m^6 mol^{-2}$	7·7	7·7	7·7
Basin characteristics				
Discharge area to basin ratio		0·30[a]	0·30[a]	0·54[a]
Limed area to total area		1·0	1·0	0·35
Limestone characteristics:				
Dose	t ha^{-1}	23·4	20·8	9·42
Density	kg m^3	2750	2750	2750

[a]Refers to an estimated value.

calculations. However, there is no independent way to estimate X_{flow}, and in any case would only serve to make the integrated soil liming model appear more complete and perfected than it actually is. The same values (*const*) were chosen for sectors IV and VI, resulting in a baseflow component ranging from 0% to 30% of total flow during periods of drought. Sector VII seems more dynamic in this respect, where the constant was chosen to allow the baseflow component to amount to a maximum of 75% with a mean of 30%.

With these modifications, stream ANC_s was calculated as:

$$ANC_s = X_{flow} \cdot X_{limed} \cdot ANC_{limed} + ((1 - X_{flow}) \cdot (1 - X_{limed})) \cdot ANC_0$$

where

ANC_{limed} = ANC of discharge from limed soil layer,
ANC_0 = ANC of discharge from unlimed soil layer, and
X_{limed} = ratio of discharge area to total subcatchment area.

The baseline runoff, ANC_0, is calculated from two independent properties: the flow rate and measured concentration of strong acid anions, C_A. The regression models were developed utilising pre-liming data for each sector. The capabilities of these models are demonstrated in the pre-treatment parts on the diagrams showing model output.

The aluminium chemistry in the SLiM model is based on the concept of gibbsite equilibrium. This allows the aluminium concentration to be calculated from one equilibrium constant and from the pH in solution. Although highly disputed, the model give a good correlation between pH and ANC. With available stream chemistry data, the gibbsite equilibrium constant approximated 7·7, a relatively low value.

The reliability of the output from a model calculation reflects how well input data are selected. One may choose some parameters relatively carelessly, while the simulation results may be very sensitive to the magnitude of others. One important input is clearly the fraction of baseflow.

The two input parameters that introduce most uncertainty to the SLiM soil chemistry calculations are soil carbon dioxide pressure and soil water content. Both these conditions influence the rate of limestone dissolution. Soil water was derived from density measurements, but values for carbon dioxide pressure presented larger problems because the pressure varies both temporally and spatially (Castelle and Galloway, 1990). Values were therefore chosen on an empirical basis. The maximum ANC obtained after treatment was used to select the value, since increased carbon dioxide pressure increases the potential solution of limestone.

Among other soil chemistry factors, the most important are the limestone dose in relation to the neutralization requirement of the uppermost soil layer. Good data were available for all parameters related to this ratio (Wilson, 1989).

Input data supplied to the model appear in Table 16.2. In addition, rate constants for the dissolution and exchange reactions and equilibrium constants for the aqueous phase are needed. Available relationships describing equilibrium constants as a function of temperature were used. The modified Oliver equation presented above was used to provide pH–alkalinity relationships. Rate constants and the temperature dependence of dissolution were taken from available literature and laboratory data. The soil temperature was approximated by stream water temperatures.

An important condition for validity of the model simulations was that the limestone powder be disaggregated according to the particle size distribution. This can be achieved by application of the material as a slurry, while dry application of ground material results in a less favourable disaggregation and, therefore, dissolution rate. Figure 16.7 shows the particle size distributions of the material considered. In the calculations, each powder could be distinguished as five fractions, each being treated individually in the submodel for limestone dissolution.

16.4.3 SLiM Results

Figure 16.8 shows output from calculations using SLiM simulations for sector VII. This shows the modelled stream chemistry of the Altiwhat (sector VII) at the point where it reaches the loch, together with weekly average data. Based only on measured data such as catchment characteristics, liming material properties, and soil chemistry, the model reproduces the general response to limestone application. SLiM describes the immediate increase in stream water calcium concentration, alkalinity, and pH following liming, as well as short-term variations. The reacidification of the soil system apparently begins as early as the first year following treatment. There is a tendency for the model to over-predict the calcium concentration and alkalinity leached during the first winter following treatment. This also occurs in calculations for the other sectors modelled.

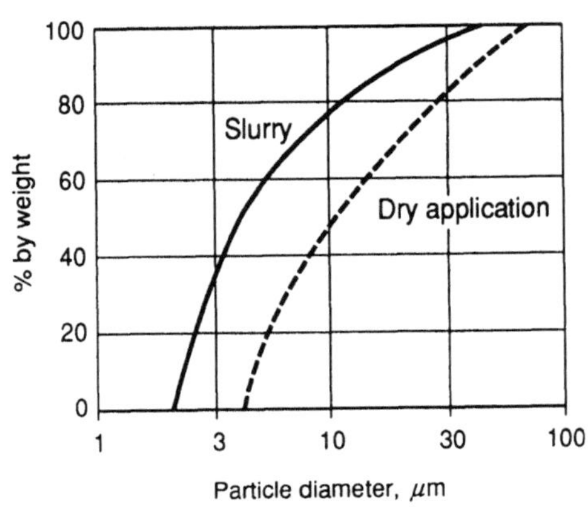

FIG. 16.7 Particle size distributions used in the SLiM model calculations; —— sector IV, - - - sectors VI and VII.

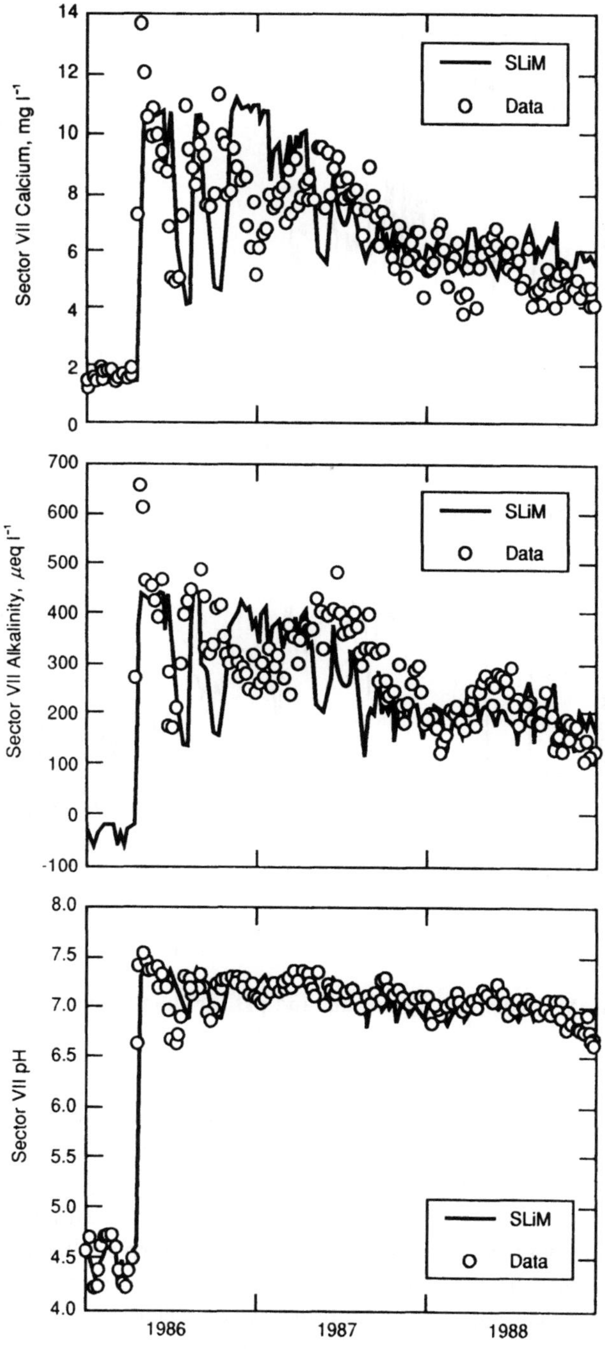

FIG. 16.8 SLiM calculations and measured values for Altiwhat stream, sector VII, 1986–88.

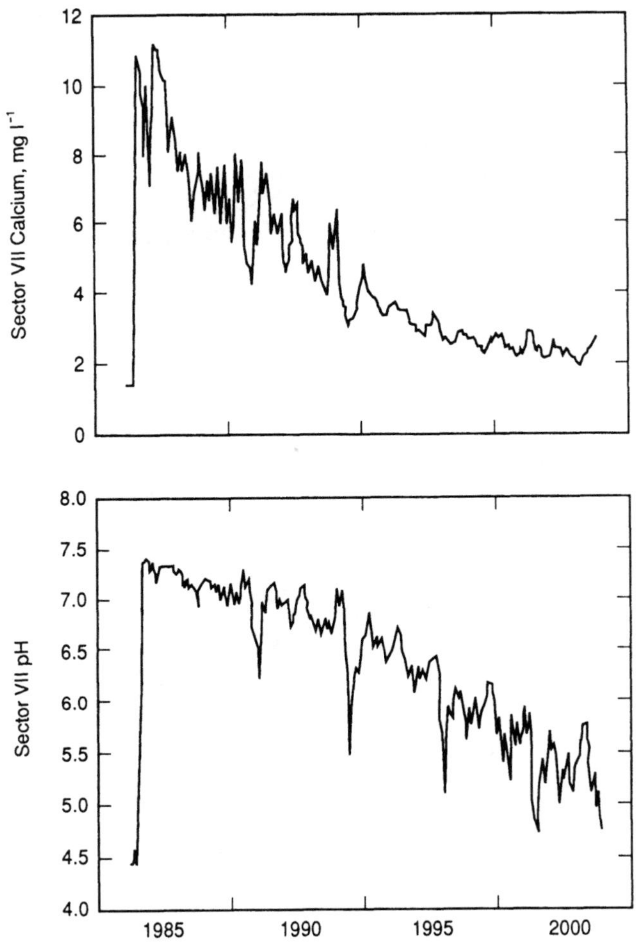

FIG. 16.9 SLiM predictions of calcium and pH for runoff from sector VII to the year 2000.

Long-term simulations for sector VII indicate that the stream pH may drop below pH 6 during episodic events in 1992, but for most conditions they stay above pH 6 until 1995 (Fig. 16.9). At this stage the buffering of the stream water will be very weak, and one could anticipate a rapid decline to pH values in the range 5·1–5·3. This sharp decline is commonly observed in lake liming, but in the case of catchment liming it should not be, since these systems will be buffered by soil reaction, not by the stream equilibria. By the end of the century, values will have reached levels critical for fish reproduction.

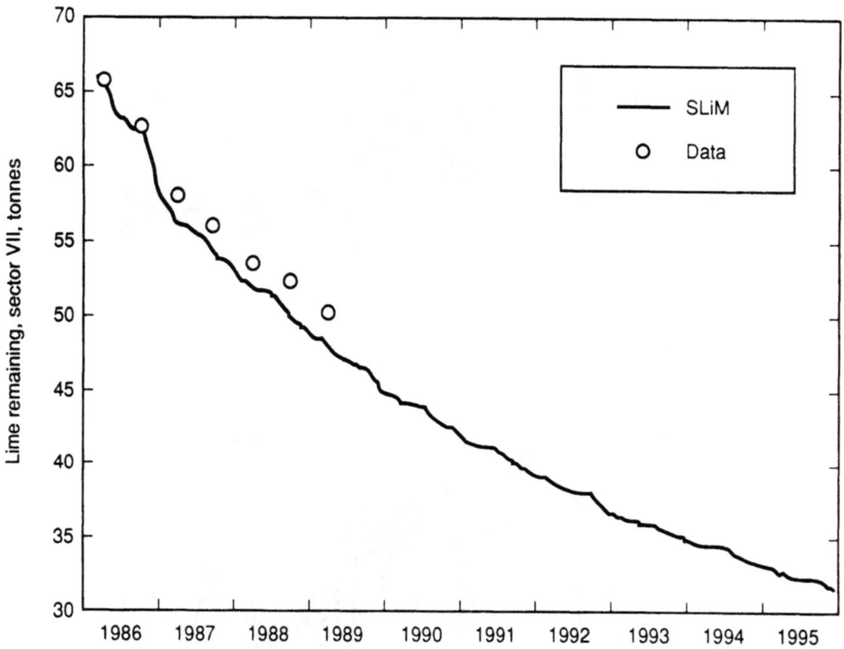

FIG. 16.10 SLiM predictions for amount of calcium retained, and measured values, on sector VII to 1995.

One may also compare simulations with data by studying the amount of limestone leached from a catchment. This method tends to put emphasis on periods of high flow rate where discrepancies will have a larger influence. Figure 16.10 shows the rate of leaching expressed in terms of the amount remaining on the catchment. SLiM reproduces the leaching during the first year, but during the second 6-month period after liming, calculations suggest a higher leaching rate. After this period, the calculated leaching rate again follows the measured rate quite closely.

Compared with the alternative method of predicting the water quality by extrapolation of stream water chemistry and calcium flux, SLiM predicts a higher leaching rate, but a slower reduction in stream water pH for sector VII. These differences are due to the properties of the soil cation exchange buffering system. In SLiM this controls the runoff composition during reacidification.

Sectors IV and VI were also modelled with equal agreement, using the data as shown in Fig. 16.11 and Fig. 16.12. It should be noted that

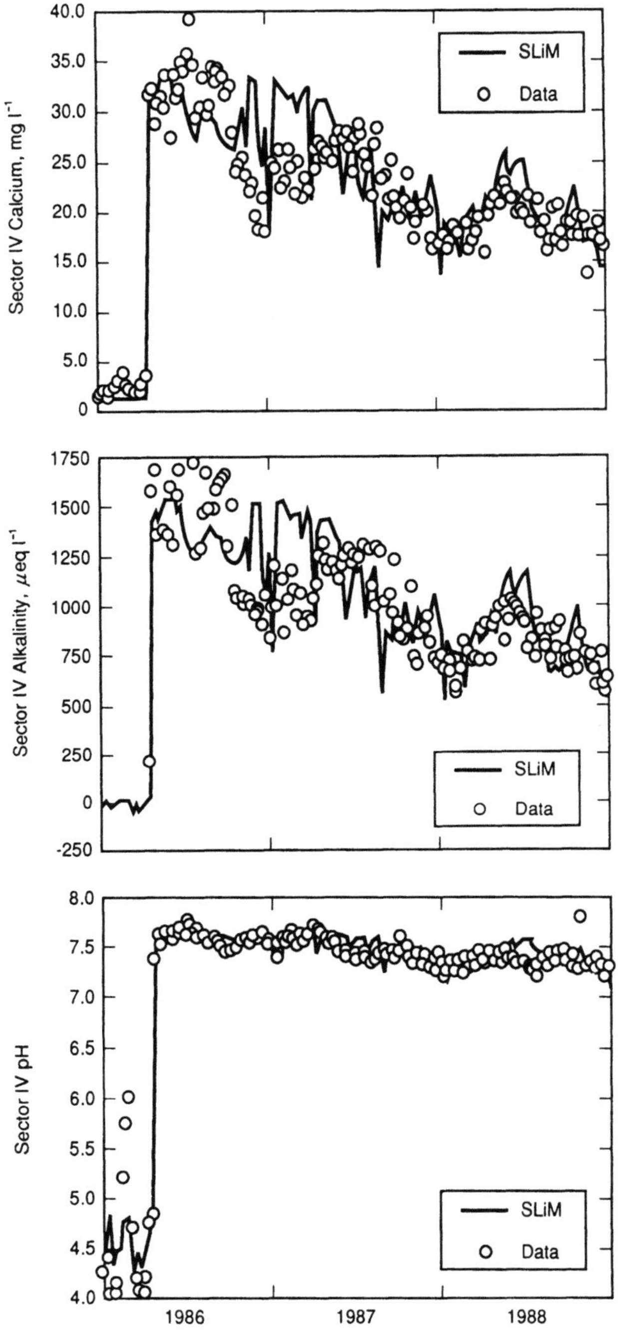

FIG. 16.11 SLiM predictions and measured values for runoff from sector IV, 1986–88.

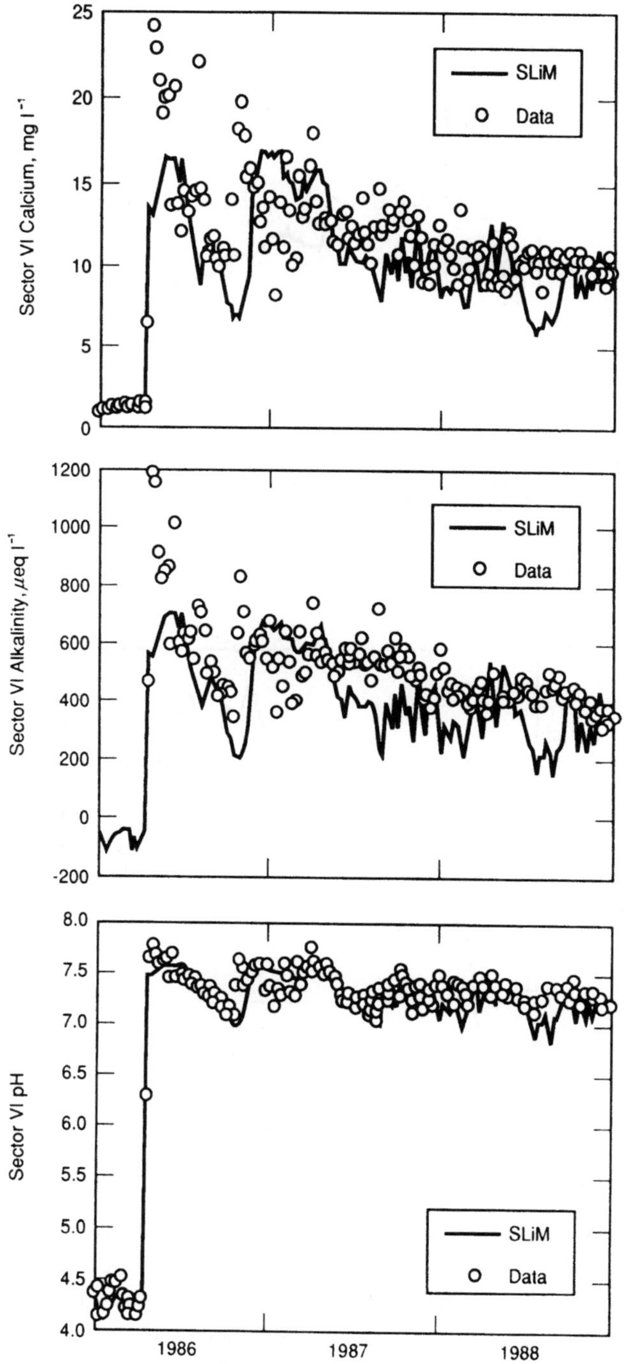

FIG. 16.12 SLiM predictions and measured values for runoff from sector VI, 1986–88.

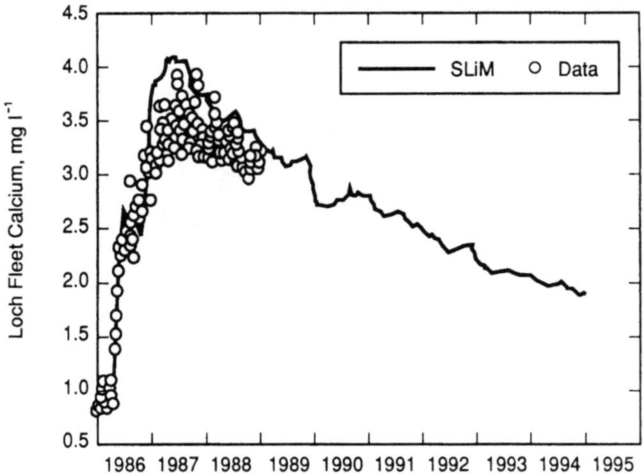

(a)

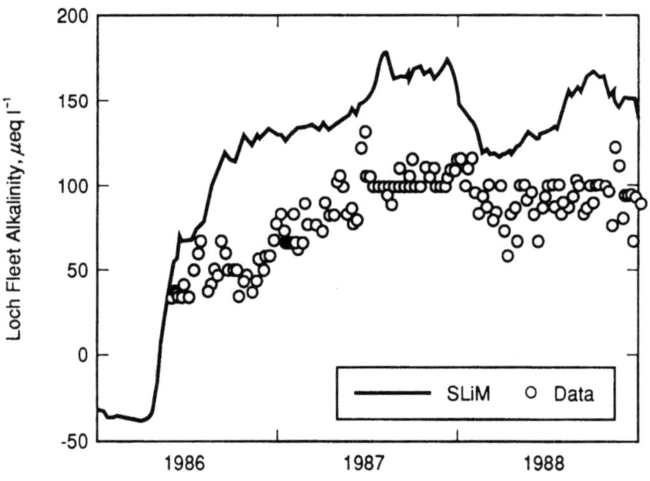

(b)

FIG. 16.13 SLiM predictions and measured values for Loch outlet 1986–88.

although the modelled short-term increase in alkalinity is large, all data sets show additional peaks in alkalinity and, with the exception of sector VI, also in calcium concentration. These peaks may be explained by limestone particles being washed into drainage streams.

SLiM predicts the calcium concentration from sector IV to decrease to about 7 mgl^{-1} (350 μeq $litre^{-1}$) and pH 6·8 by 1996. Sector VI will reacidify faster because of a somewhat lower application rate and greater runoff, with a predicted calcium concentration in 1996 typically $4\,mg\,l^{-1}$ ($200\,\mu eq\,litre^{-1}$), in parallel with a pH value of 6·5.

On the basis of these results, the chemistry of Loch Fleet itself was modelled. The loch receives input from three sources: tributaries from limed areas; discharge from untreated zones; and as direct rainfall on the lake surface. The contribution from groundwater is assumed to be constant. The loch responds fairly rapidly to the treatment. As shown in Fig. 16.13(a), SLiM gives a good reproduction of the increase in calcium concentration. For the end of 1986, however, the input of calcium is over-predicted. While this shows in stream data as an over-prediction in calcium concentration, this error becomes more apparent when simulating the loch response since it coincides with high flow. This offset remains in the calculations, showing as an over-prediction of the loch water calcium concentration of about $10\,\mu eq\,litre^{-1}$. The calculated loch alkalinity does not agree with data; consistently, SLiM over-predicts the alkalinity (Fig. 16.13(b)), although the patterns are identical. This may be caused by an internal sink of alkalinity, such as nitrification. Sediment exchange reactions are not likely to be the source of discrepancy since this would cause corresponding errors in calcium simulations. In terms of treatment duration, SLiM calculations covering a longer period of time show that Loch Fleet as a whole may stay above pH 6 towards the end of this century. This is mainly because of the acid neutralising capacity supplied from sectors IV and VI.

16.5 DISCUSSION AND CONCLUSIONS

Both the extrapolation and modelling approaches to estimate the times over which the catchment liming treatments at Loch Fleet are likely to remain effective give similar predictions.

From an economic standpoint, the length of time that a liming application continues to ensure acceptable water quality for fish survival is paramount. The predictions of water quality, using data from the 3·5 years since liming at Loch Fleet, suggest that overall coverage of limestone in this catchment, either as a slurry or as a dry powder, at application rates greater than $20\,t\,ha^{-1}$, will maintain acceptable drainage water quality conditions for more than 15 years. The application rate on sector VII was approximately half that applied to sectors IV and VI and is estimated to remain effective for only about 7 years. It should be noted, however, that careful direction of liming treatments to parts of a catchment, for example to wetland or "water source" areas, may be more economic than complete catchment coverage. This approach may also minimise the ecological impacts of liming areas that support calcifuge vegetation types, especially if only a proportion of the wetter areas is sufficient for treatment.

In contrast to these predicted treatment effectiveness times, direct application of limestone to the loch would need to be repeated at regular intervals to maintain satisfactory loch water quality since its retention time is only 173 days. This method would also provide no water quality improvement in the main spawning stream.

The calcium release calculations and the input data for the soil liming model and the predictions based on the two methods relate to the Loch Fleet catchment with its specific soil and vegetation types, rainfall conditions, and hydrology. Extrapolation to other sites will need to be made with care. In addition, the importance of rainfall quantity and pattern of incidence and hence stream water volumes is crucial in determining the effectiveness times of catchment liming treatments. The predictions made for Loch Fleet rely on recent rainfall and runoff conditions, and if these were to change in the future, then the treatment effectiveness times will be affected. Continued monitoring of stream water quality and quantity and a continued programme of hydrochemical flux calculations and modelling for the Loch Fleet catchment will serve to verify predictions made in this chapter.

At this stage the few catchment liming projects reported supply too little information from which similar predictions could be made. In addition, the variety of catchment conditions and of limestone applications make such comparisons uncertain.

REFERENCES

Allison, A.J. (1989). Report on the Examination of Peat Sample ex Loch Fleet. British Coal Report.

Brown, D.J.A. (1983). Effect of calcium and aluminium concentrations on the survival of brown trout (*Salmo trutta*) at low pH. *Bull. Env. Contam. Toxicol.* **30**: 582–87.

Castelle, A.J. and Galloway, J.N. (1990). Carbon dioxide dynamics in acid forest soils in Shenandoah National Park, Virginia. *J. Soil Sci. Amer.* **54**: 252–57.

Chester, P.F. (1984). General Discussion, pp. 564–65 in *Acid Deposition: Effects of Sulphur and Nitrogen Compounds. Phil. Trans. Roy. Soc. B* **305**, 319 pp.

Howells, G. (1990). *Acid Rain and Acid Waters.* Ellis Horwood, Chichester, 215 pp.

Howells, G., Brown, D.J.A. and Sadler, K. (1983). Effects of acidity, calcium and aluminium on fish survival and productivity – a review. *J. Sci. Food Agric.* **34**: 559–70.

Howells, G., Dalziel, T.R.K., Reader, J.P. and Solbe, J.F. (1990). EIFAC water quality criteria for European freshwater fish: report on aluminium. *Chem. Ecol.* **4**: 117–73.

Magnuson, J.J., Baker, J.P. and Rahel, E.J. (1984). A critical assessment of effects of acidification on fisheries in North America. *Phil. Trans. Roy. Soc. Lond. B* **305**: 501–16.

Oliver, B.G., Thurman, E.M. and Malcolm, R.L. (1983). The contribution of humic substances to the acidity of colored natural waters. *Geochem. Cosmochim. Acta* **47**: 2031–35.

Stumm, W. and Morgan, J.J. (1983). *Aquatic Chemistry*. John Wiley and Sons, New York, 780 pp.

Warfvinge, P. (1988). Modeling acidification mitigation in watersheds. Doctoral Thesis, Department of Chemical Engineering, Lund Institute of Technology.

Warfvinge, P. and Sverdrup, H. (1988a). Soil liming as a measure to mitigate acid runoff. *Water Res. Res.* **24**: 701–12.

Warfvinge, P. and Sverdrup, H. (1988b). Watershed liming. *Lake and Reservoir Manag.* **4**: 99–106.

Warfvinge, P. and Sverdrup, H. (1989). Modeling limestone dissolution in soils. *J. Soil Sci. Amer.* **53**: 44–51.

Warfvinge, P., Dickson, A. and Sverdrup, H. (1990). Modelling acidification reversion at Loch Fleet (UK). Proc. Internat. Conference Acidic Deposition: Its Nature and Impacts. Roy. Soc. Edin., Conference Abstracts, p. 589.

Wilson, E.J. (1989). The Effect of Lime Application on Soils at Loch Fleet – 1987 Sampling. CERL Research Report ESTD/0010/R89, 18 pp.

Wilson, E.J., Roberts, D.J. and Skeffington, R.A. (1988). The Effect of Lime Application to the Loch Fleet Catchment on Tree and Soil Chemical Properties – An Initial Report. CERL Research Report TPRD/L/3283/R88, 25 pp.

CHAPTER 17

The Loch Fleet Project and Catchment Liming in Perspective

G. Howells
University of Cambridge, Cambridge, UK

T.R.K. Dalziel
PowerGen plc, Nottingham, UK

17.1 INTRODUCTION

It has been the aim, in this book, to deal with the principles and practice of catchment liming based on what has been learned from the Loch Fleet project and by reference to experience reported for other liming studies. Source material for the preceding chapters is derived from diverse disciplinary fields. This chapter aims to integrate and assess the results of all these areas of investigation and to provide an appraisal of the Loch Fleet project as a catchment liming experiment, placing what has been learnt at Loch Fleet in the context of other projects aimed at mitigating acidification.

17.2 THE LOCH FLEET CATCHMENT: A REPRESENTATIVE SITE?

The general principles and practical considerations involved in identifying a suitable site for catchment liming have been mentioned earlier and are set out in Chapter 4. It should be remembered that Loch Fleet was identified not only as a suitable site where catchment liming could be applied, but also as a suitable experimental site. From a practical point of view, features like the size of a lake and its catchment, and a single outlet stream, are obviously important. The fishery history of the loch and the land-use history of the catchment were both known. Since from the outset it was the intention of the project to test different land-based methods of surface water acidity neutralisation, and the expectation that the results obtained might be applicable elsewhere, it was important, too, that the site was representative of the types of area, certainly within the UK, that are perceived to have acidification problems.

Early in the planning stages of the project consideration was given to the possibility of identifying a number of separate lakes and associated catchments each as similar to one another as possible and to test a different catchment liming strategy in each one. The difficulties of finding a range of such lakes and catchments in the same area that were sufficiently similar so as to allow comparisons of different liming strategies to be made led to the decision to focus on a single catchment in which separation into discrete sub-catchments would allow the testing of a number of treatments within one catchment. This conclusion alone emphasises that the heterogeneity and "uniqueness" of individual sites is often their most common feature. The question, "How applicable are the findings from Loch Fleet to other potential sites for restoration/protection by catchment liming?", and arising from this, "How justified are we in comparing data from different sites given the heterogeneity between sites?", are challenging and not completely resolved at this stage, but the intensive study at Loch Fleet will help to make an informed guess, and will identify those components that are paramount.

Figure 17.1 shows the areas within the UK in which surface water acidification is considered a problem. In general, these lie to the west of Britain and are subject to similar rainfall quantity and quality. Their susceptibility to acidification relates both to the rainfall quantity and hence acid loading experienced (rather than the acid concentration in rain), and to the nature of their underlying bedrock geology, which in general is granitic, gneissic, or shaly, providing little inorganic weathered material, with overlying soils of low mineral content and consequently little ability to neutralise acid inputs. These predominantly organic peaty soils are also, by their very nature, acid because of proton-generating processes within the peat (see Chapter 2).

Sites in the UK, in addition to Loch Fleet, where catchment liming has been tried, sometimes in association with other management practices, are

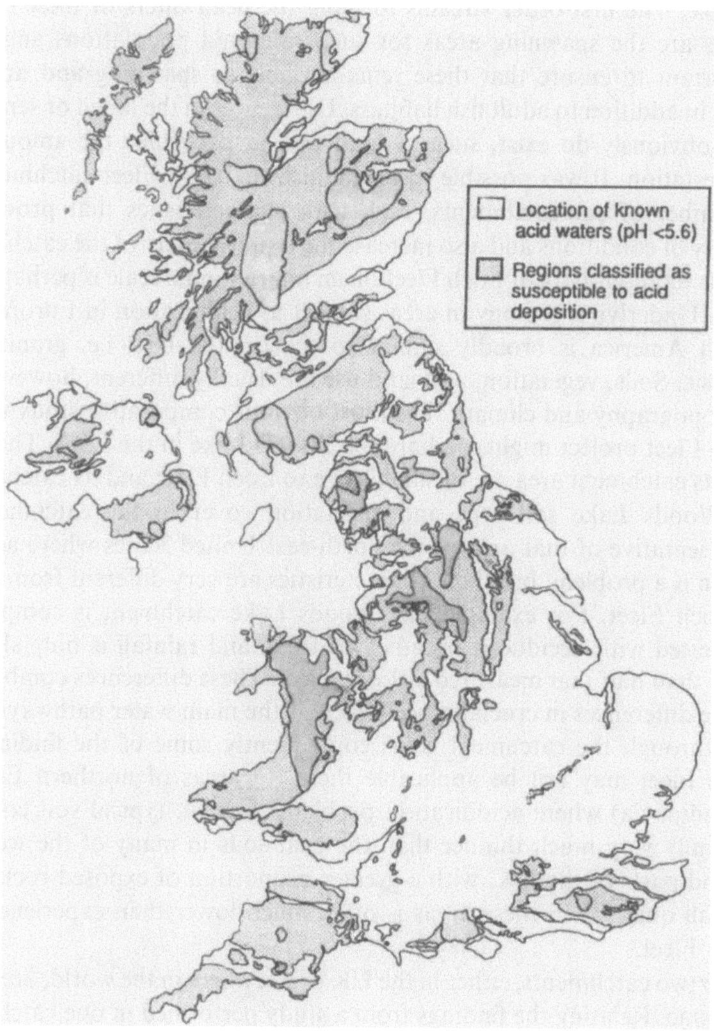

FIG. 17.1 Areas in the UK where acid waters are known, and those considered to be sensitive to acidification.

given in Chapter 1, Table 1.1. They are all westerly sites with broadly similar orographic conditions and mineral-depleted soil characteristics. In these areas it is usually salmonid fish that are the main concern and the subject of protection or restoration. Historic records of trout catch in Loch Fleet prior to acidification suggested that restoration could be feasible. The first-order streams forming the headwaters of these catchments are the spawning areas for such salmonid populations and it is important to ensure that these remain viable as spawning and nursery areas in addition to adult fish habitats. Differences in the limed or sensitive sites obviously do exist, such as land use, in particular the amount of afforestation. It was possible to distinguish the Loch Fleet catchment as a number of sub-catchments, each with characteristics that provide a variety of conditions and also increase the representivity of the catchment.

The representivity of Loch Fleet on an international scale is perhaps less clear. Underlying geology in areas subject to acidification in Europe and North America is broadly similar to that in the UK, i.e. granitic or gneissic. Soils, vegetation, and land use are usually different, however, as are topography and climate. The most obvious comparative study to the Loch Fleet project might, perhaps, be Woods Lake in the USA. The lake and its catchment area are of similar size to Loch Fleet and its catchment. At Woods Lake soil type and vegetation cover in the catchment is representative of that area of the north-east United States where acidification is a problem, but these characteristics are very different from those at Loch Fleet. For example, the Woods Lake catchment is completely afforested with deciduous broadleafed trees, and rainfall is only slightly more than half that measured at Loch Fleet. These differences combine to create differences in crucial aspects such as the main water pathways over and through the catchment soils; consequently some of the findings at Loch Fleet may not be applicable there. In areas of northern Europe (Scandinavia) where acidification problems prevail, typical soil cover is generally very much thinner than the peat soils in many of the western upland parts of the UK, with a greater proportion of exposed rock, and rainfall quantity in these areas is often much lower than experienced at Loch Fleet.

No two catchments, either in the UK or elsewhere in the world, are quite the same. Relating the findings from a study performed in one catchment to other sites must be done with care, with a careful consideration of the differences between sites. Generally, the representivity of Loch Fleet, at least in a UK context, must be considered good, and even in comparison with areas overseas the main findings should prove valid and useful.

17.3 THE LOCH FLEET PROJECT: THE IMPORTANT MESSAGES

17.3.1 Hydrology: A Crucial Feature

The Loch Fleet project was not the first attempt to rectify the acidity of surface waters by applying limestone to parts of a catchment area. Such has been the extent and range of studies within the project, however, that it must be regarded as important from the point of view of refining the techniques of catchment liming. The results gained from these various studies have yielded a number of important messages.

Perhaps the investigation that has proved to be most important in the consideration of a programme of catchment liming is that of hydrology. A detailed hydrological analysis of any site is critical since the maximum efficacy of any applied neutralising material will only be achieved if the application takes into account the major hydrological pathways. Reliable hydrological measurements, coupled to a programme of chemical analysis, are thus essential for calculating the amount of applied material that has become mobilised and that remaining on the catchment and within the soil. Different techniques exist with which to assess the detailed hydrology of a catchment. At Loch Fleet, several approaches were employed and some findings unresolved and not entirely consistent. It is worthwhile to consider reasons for this and to invoke data from other related studies, in particular the soil sampling and analysis (see Chapter 11) and the hydrochemical budget calculations (see Chapter 12).

Chapters 6–8 provide alternative models of hydrological pathways and time-scales. Conventional hydrological records of rainfall, aerial measurements, and streamflow, as well as estimates of evapotranspiration, are used to interpret the transfer of deposition input to stream and lake outflows. The stream hydrographs recorded through variable conditions from drought to spate are used to interpret the flows in terms of baseflow and more superficial flow shed from the catchment during heavy rain. In the case of Loch Fleet, monthly mean data were employed in an accepted "catchment model" to calibrate the flow records with a view to infilling missing data records when required. This approach estimated that 70%–80% of drainage could be considered as "fast runoff" through and over a surface soil horizon of about 20 cm. This model was reasonably consistent with the measured ratio of stream flow to rainfall, with 20%–30% due to evapotranspiration (in line with that estimated for similar moorland catchments in Scotland (Hall, 1987)). Criticisms of this conventional approach are that interpretation of the usual hydrograph is largely intuitive, nor does it represent conditions in low order streams, and that it depends on detailed and accurate measurement of stream flows.

The two main methods employed at Loch Fleet to gain an understanding of the catchment hydrology were the oxygen isotope approach (see Chapter 7) and the soil pit approach (see Chapter 8). The oxygen isotope approach, involving ^{18}O isotope tracer separation, provided estimates of the rates and routes of runoff from moorland through individual storm events. This indicated that the greater part (66%–75%) of the storm hydrograph represents pre-event water, in other words that displaced from a reserve within catchment soils (down to 20 cm depth) by the rainfall during the monitored event. The soil pit approach, however, yielded a different picture. This involved the digging of plots in moorland and forest terrain draining at different soil depths, thus providing measurements through the year of flow within the soil horizons. In the moorland pits, runoff responded rapidly to rain inputs, with throughflow within the upper 5 cm accounting for 87%–99% of the total. Measurements made in the forest suggested that even there the throughflow in the litter layer and upper few centimetres was greater than 95%. These data were consistent with chemical analyses, which indicated relatively little pick-up of soil-derived elements. They were also seen to reflect the non-porous nature of the saturated peaty soils characteristic at Loch Fleet.

The oxygen isotope approach is undoubtedly a precise technique (certainly in terms of the oxygen isotope ratios) but it may be that uncertainties in interpretation remain. The approach was used to provide estimates of response to individual storm events, in contrast to the pit studies which provided regular data through a 2-year period. Few studies have been made on sites elsewhere at which conventional hydrograph interpretations have been matched by either plot studies or isotope separations, or both.

The criticism levelled against the soil pit approach is that the construction of the pits inevitably changes the "natural" drainage pathways and that in fact aluminium, almost absent in rainfall, was present in runoff waters, indicating soil contact. The discrepancies of flow paths and timing might be resolved by considering the real-time response to events – if the soil reserve is mobilised quickly (and replaced equally quickly) the recharge water flow could simulate surface throughflow, although the apparent non-permeable nature of the peaty soils militates against this.

The results of the soil sampling programme and of hydrochemical budget calculations provide an alternative, although direct and pragmatic, approach to understanding the major flow pathways. The flux of applied calcium from the treated sectors has declined since the time of application; the same volume of runoff now will not yield the same export of applied calcium as the equivalent volume of runoff in the first months following liming. The soil studies show that since application there has been a downward movement of free lime in the soil. In the absence of any decrease in particle dissolution through coating, the calcium flux and the soil data argue in favour of surficial runoff predominating, broadly in agreement with the results of the soil pit approach. While not a rigorous test of the assumptions, this does lead to a pragmatic adoption of the concept of rapid and superficial runoff. It also implies that applied calcium reaching the lower soil horizons is not freely accessible to runoff, until perhaps dry conditions allow a true baseflow to mobilise the reserves.

A sound knowledge of hydrology would allow liming treatments to be targeted at the main flow pathways, thus ensuring effective export or flux of the applied material to the target lake or stream.

At some sites, areas recognised as "discharge zones" can be identified where water, having entered the soil higher up-slope, re-emerges. Such areas are often found at the base of concave slopes adjacent to streams and lakes. Where such areas can be identified they should be considered as appropriate for liming. The boggy water-source area at the top of sector VII at Loch Fleet was identified as a suitable target area for liming treatment since it drains directly to the main inlet stream, the Altiwhat, and its delayed release of limed water would damp out high flow acid episodes. This makes a proportionate contribution to the loch water chemistry, but is also the main spawning area for fish and thus ensures that water quality is maintained at a level where population recruitment can be successful.

17.3.2 Role of Terrestrial Vegetation, Land Use, and Management

It is clear that collected rain does not represent the total input of material from the atmosphere; dry deposition of particulate and aerosol materials are thought to be an important component, and wet deposition of fog and mist is also not sampled efficiently by standard rainfall collectors. At other sites in the district, rainfall is greater with higher altitude. Because of the practical difficulty of collecting a true total deposition, it has been claimed that the composition and flow of first-order streams are better estimates than bulk rain samples.

At Loch Fleet, where measured rainfall totals nearly 2300 mm yr^{-1}, and the frequency of wet conditions is high, wet deposition is estimated at about 70% of the total deposition from atmosphere. The ratio of influx of measured rainwater to measured water flow from the lake is 1·25 (range 1·17–1·60 over 5 years) with evapotranspiration estimated to be 25%, assuming no net retention in catchment soils, consistent with estimates of 20%–30% for similar upland terrain in Scotland (Hall, 1987). Similar ratios of influx to outflow calculated for the different sectors indicate that evapotranspiration is as much as 45% for sector IV (forest), 12% for sector VI (moorland), and 21% for sector VII, where standing bog pools may increase the loss by evaporation. However, the interception gauge at 6 m height between sectors II and III, operating for the two years 1985 and 1986, collected about twice the volume of bulk rainfall, whereas throughfall volumes were 66%–85% of the bulk volumes. In the absence of validated data for mist and fog interception, or for throughfall on more than a plot basis, bulk rainfall is used for water and ion budgets, with allowance made for estimated evapotranspiration.

Sulphate (total of sea salt and anthropogenic sources) provides an influx-to-efflux ratio of 0·64 for the whole catchment. Similar sectoral ratios for sector IV and sector VI do not differ significantly (0·56 and 0·55) but are less than for sector VII and the loch outflow. This is inconsistent with the widely held view (based on single tree or plot studies) that tree canopies effectively scavenge the atmosphere for aerosol or mist sulphate. The interception sampler appeared to increase the levels of ions of marine origin as much as fourfold, but throughfall below Sitka spruce was only about twice that of rain. It is difficult to reconcile these data with the scavenging hypothesis unless the sulphate influx in rain is underestimated or that sulphate efflux is enhanced by mobilisation of sulphur from canopy soils or lake sediments; none of these measurements is available. The behaviour of sectors IV and VI also suggests that both forest and rough moorland behave in a similar fashion. This conclusion, and the fact that the forested area of the catchment is only 10% of the total, suggests that afforestation has little influence on acidification at this site. However, the hydrological consequences of ploughing prior to planting the forested area (Section 17.2) should be kept in mind.

Nitrogen budgets and ratios are more complex, with influx-to-efflux ratios for ammonia varying from 6 (sector IV) to 43 (loch outlet), whereas nitrate in bulk rain, throughfall, and in efflux are little different. Ammonia in rain appears to be taken up by the nitrogen-deficient forest, but is also subject to oxidation to nitrate as it is transferred through the catchment. The nitrate influx-to-efflux ratio for sector IV, 1·26, suggests that it may be taken up by the forest, and to a lesser extent in the wetland mires of sector VII. In contrast, in the moorland sector VI it has a ratio of only 0·79.

Chloride is generally regarded as a mobile anion, is seldom generated within a catchment, and in general not retained. The throughfall studies indicate that the forest canopy scavenges chloride with twice the bulk rain flux in throughfall below Sitka spruce. The greater flux (nearly six times greater) in interception samples is difficult to explain. The flux ratio for the whole catchment is 0·84 (and smaller values for sectoral ratios) and suggests that either input has been grossly underestimated, or that chloride is retained. A case for chloride retention in the adjacent Loch Dee has been argued (Farley and Werritty, 1990). Since no mechanism for chloride retention has been proposed, however, the former explanation appears more realistic.

17.4 TARGET RESPONSES

It is clear that the most usual target of liming projects (either direct or to the catchment) has been to improve or recover fisheries in acidified waters. Overall, liming has been followed by expected improvements in water quality, specifically in raised pH and alkalinity, calcium concentration, and a reduction in toxic forms of aluminium and the toxicity of other trace metals. This is associated with rapid recovery of the fishery in most cases, although at some sites natural recolonisation has had to be encouraged by restocking programmes.

On the other hand, there are records of fishery recovery in acidified waters where liming has not improved water quality. Two examples are the Loch Dee project in south-west Scotland and the River Esk in Cumbria, UK. In the former, a limited (mostly stream-side) liming programme has provided limited water quality improvement; indeed, the pH declined from 5·58 in 1985 to 5·23 in 1989, calcium about 60 μeq litre^{-1} in 1985 declining to less than 50 μeq litre^{-1} in 1987–88, but returning to 60 μeq litre^{-1} in 1989. Notwithstanding this lack of improved water quality, the healthy trout population is now estimated at 9000–14 000, even though fishing effort there has probably increased. It has been suggested that changes in fishery management have led to the improved status of the trout (Loch Dee Project, pers. comm.). In the River Esk, liming of riparian areas also led to some increase in pH and reduction in aluminium, but no increase in calcium concentrations, but with a concurrent increase in the distribution and abundance of salmonids in the river system, and no evidence of the spawning mortalities seen prior to liming in the lower reaches (Diamond *et al.*, 1992).

There are also reports (e.g. Renberg and Grahn, 1990) where natural de-acidification has occurred with the decline of *Sphagnum* and recovery of the perch population, although water quality changes are not reported. Reduced acid deposition is thought unlikely, but forestry and weather changes are suggested as possible factors. It has to be acknowledged that uncontrolled changes in both biotic and abiotic conditions make interpretations of change difficult in some cases. Nonetheless, there is now abundant evidence that liming procedures at many sites have improved fisheries in a variety of water conditions.

17.5 COSTS AND CHOICES

17.5.1 Liming and Other Options

A range of treatments were applied at Loch Fleet, selected on the basis of experience elsewhere and from laboratory and field trials at Fleet. Overall, these treatments involved limestone applications at various rates and by various methods. Limestone application ranged from 5 t ha^{-1} (prior to 1976 a familiar application rate for upland grazing), to 25 t ha^{-1} to moorland and afforested areas. Evaluation of other catchment liming projects had suggested that rates of less than 5 t ha^{-1} are ineffective, possibly because acid soils must overcome their inherent soil acidity as well as that deposited in rain.

In the flow-gauged sectors and the loch outflow, calculations based on calcium and water budgets estimate a time-scale of effectiveness from about 5 years to greater than 15 years, varying with the dose applied and the fraction of area of catchment treated. It is evident that the yield of calcium from the treated sectors is not linear, but exponential in character; when the response curve flattens, no further export of calcium or alkalinity is likely. This suggests that a substantial fraction of the applied lime is retained in the soil and is ineffective, possibly because soil drainage is predominantly through the upper soils at Loch Fleet. However, the soil base reserve, which has been restored by treatment, may continue to be an effective source of calcium if soil pH falls back to more acidic pre-liming values.

These estimates of effectiveness are based on mean values over 6-month periods, thus ignoring short-term stream events that might be critical to fry and juvenile fish. A possible increased frequency of such events as the catchment base reserve is reduced is a matter of concern if the critical question is to maintain spawning and nursery conditions suitable for salmonids even at high flows.

The advantages of catchment liming over direct lake liming or stream dosing systems lies for the most part in its longer time-scale of effectiveness and sustained response to rainfall events. It does involve greater initial costs and the potential of some unwelcome non-target effects (see below). The alternative of reducing acid deposition by emission reduction is one which may well bring biological improvements to sensitive or acidified waters, but questions remain as to the critical level of emission reduction, and how quickly this improvement will come. Field experiments (e.g. RAIN, Wright *et al.*, 1988) and modelling approaches (e.g. MAGIC, Cosby *et al.*, 1985; Whitehead, 1989; Skeffington and Brown, 1992) suggest that recovery would take decades, possibly a century (see Chapter 2).

The development of equilibrium invertebrate communities may take several years and the attainment of a stable age structure of fish may take even longer. Both depend on a stable abiotic regime. The longer time-scale of catchment vs direct water liming promotes this abiotic stability and should favour the establishment of a more stable ecosystem. A shallow, prolonged, lime decay curve offers a better prospect of controlling water chemistry within target limits.

However, the choice of emission control, direct lake or stream liming, catchment liming, or land-use changes to mitigate effects of acidifying coniferous forests will be determined by the conditions at the site to be restored.

17.5.2 Practical Aspects

The decision to mitigate acidic conditions at any site will depend on a variety of factors, amongst which are that this objective is paramount, that physical conditions are favourable, and that the task is within acceptable economic limits. Experience at Loch Fleet provides some basis for judgement in making such decisions.

The physical conditions favouring lime applications, and the choice of application and distribution methods, will inevitably vary according to each site of interest. The findings at Loch Fleet, while governed by the characteristics there – heavy rainfall, predominantly peaty soils, high altitude, rough terrain – are characteristic of western Britain where most identified acid waters are found. Hydrological and drainage pathways in these conditions are important – direct lake liming is effective only if lake

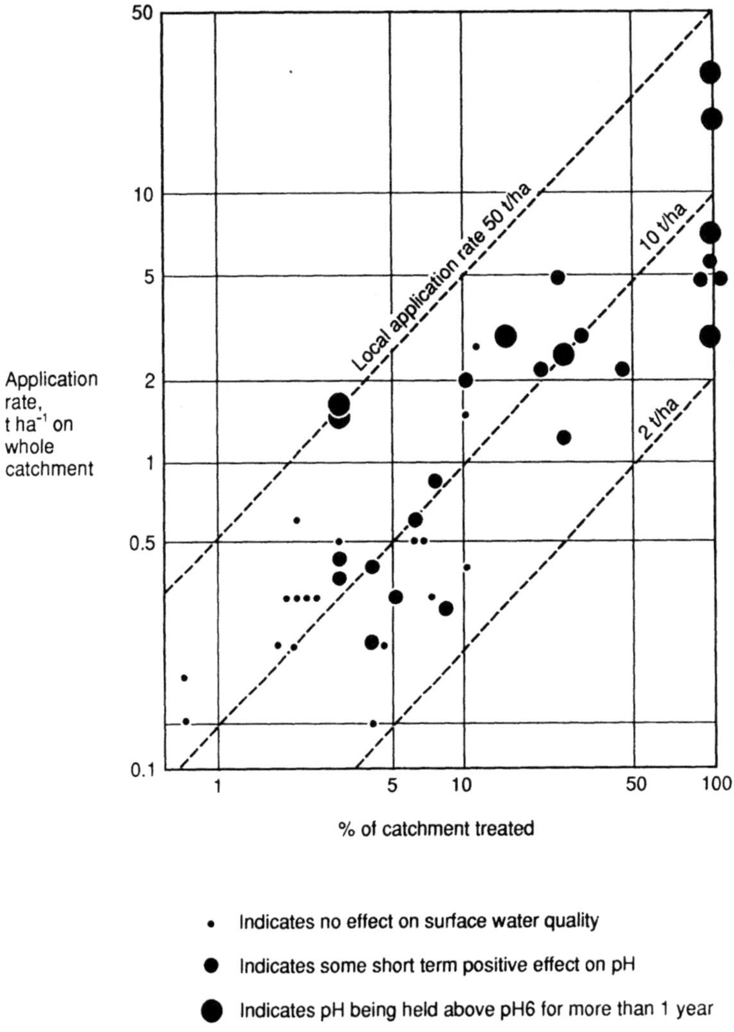

FIG. 17.2 Effectiveness of catchment liming experiments, world-wide.

turnover is slow and in areas such as Galloway with rapid lake water replacement, catchment liming is favoured.

Experience has shown that the fraction of the catchment treated also influences the effectiveness of the lime application (Fig. 17.2), with 50% or more useful as a general guide, although a smaller proportion will suffice if this generates a larger fraction of the runoff, as via an upland wetland.

Since acid upland catchments provide sites for wildlife, partly due to their acidic nature and to their typical lack of farming activity because of low productivity, lime application may be seen as an adverse action by some interests. In this case there is a need to choose what is desired by the landowners or the local community, or a wider group, bearing in mind any unique features of a site.

17.5.3 Economic Aspects

In the context of the experimental constraints of the Loch Fleet project it is difficult to specify the costs of a practical exercise; it should be noted that many projects reported include the costs of research investigations. The distance of a candidate site from a suitable limestone source, cost of formulation of the desired product, alternative means of transport to the site, and methods of distribution on site are all pertinent. In addition, the application dose will depend on the target water quality for a fishery, i.e. self-sustaining or annually restocked, and the expected or acceptable frequency of reapplication. For these reasons, costs are inevitably specific to each site and to the specified objectives.

The "lime requirement" for a catchment, analogous to that of upland soils, can be calculated from the product of the deposited acid "load", the soil characteristics (percentage of base reserve), rate of lime dissolution, and the rate of catchment runoff. This calculation will estimate a minimal dose since it ignores any fraction that is unavailable because it is irretrievably bound in the soil, or permeates to depths not usually contributing to flow, or becomes chemically inactivated. It also ignores that part of the calcium contributed from deposition or soil–rock weathering.

Provisional estimates of costs of liming to meet this "requirement" can be made, with reservations, on the basis of studies at Loch Fleet and elsewhere. Taking, as an example, the conditions in the area of Galloway in which Loch Fleet lies, six fishless lochs of pH less than 4·6 have been identified, while a further six lochs have a pH of 4·6–4·8, but still contain fish. This region contains the larger part of acidified or fishless lakes in Scotland (Fig. 17.3).

For these 12 Galloway lochs, with a total catchment area of about 4000 ha, a lime application based on that applied overall at Loch Fleet, i.e. about 3·5 t ha^{-1}, would amount to 14 000 t, possibly sufficient for an estimated 10 years. The costs of applied lime are estimated (1989 values) at about £15 per tonne (agricultural rates) to about £60 per tonne in the

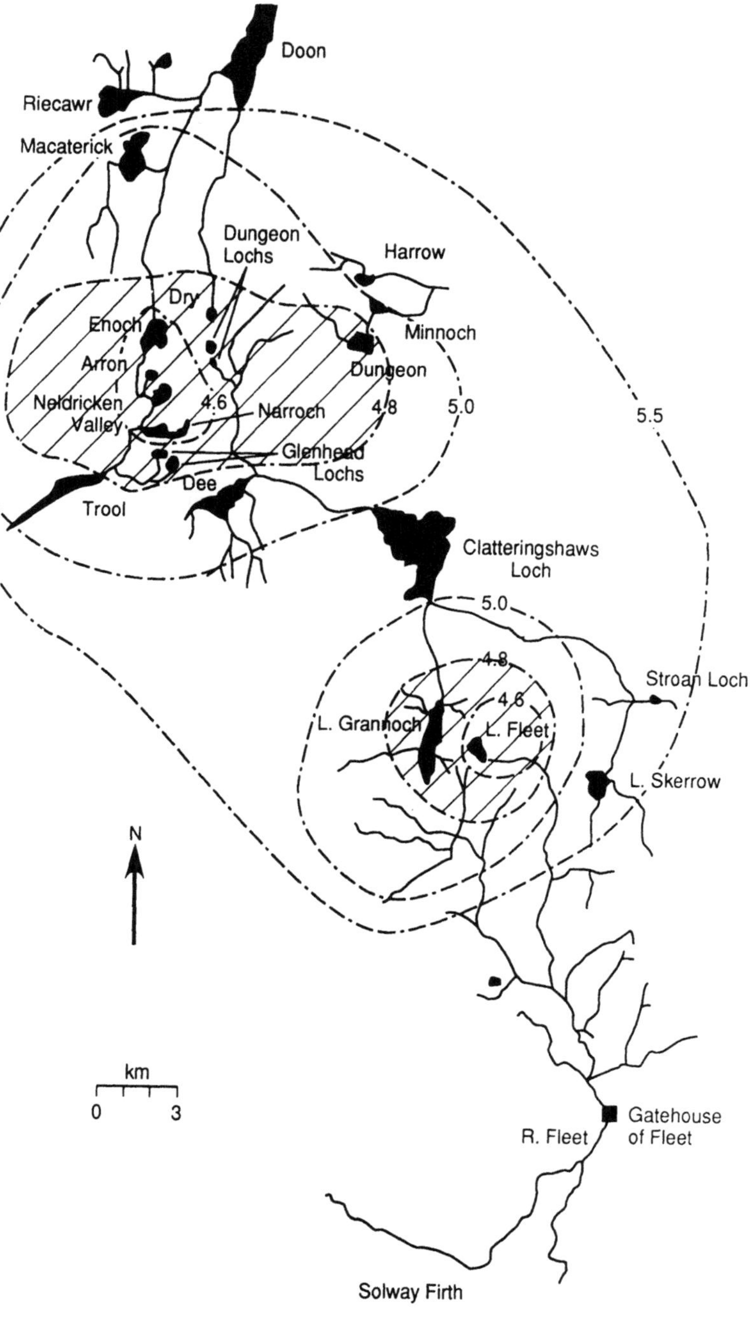

FIG. 17.3 Distribution of 12 acid lakes (pH < 4·8) in Galloway, with potential for mitigation by liming.

PLATE V Limed wetland source area of the Altiwhat, sector VII.

rough terrain characteristic of Loch Fleet. The total cost at 1989 prices would be £240 000 to £960 000.

The Loch Fleet applications are certainly effective to restore the former catch in this 17 ha lake to 100 fish per year, but potentially even greater. For the 12 Galloway lochs, with a total lake area of 350 ha, catch might be reasonably estimated at more than 2000 fish per year, i.e. 20 000 fish per 10 years after treatment, amounting to £10–£50 for each fish restored.

However, this is pessimistic since production will be improved, not only in the limed headwaters but also in waters downstream. Based on comparable waters in the UK, the carrying capacity of Loch Fleet is estimated to be as much as 800 kg ha^{-1}, with a maximum sustainable yield, if present water conditions are maintained, of almost 1000 fish (or more than 500 kg total), assuming that spawning areas were sufficient. The expectation of improved fisheries downstream suggests an estimated "standing crop" of nearly 1000 fish (or 200 kg) of angling potential. Thus the cost per fish would reduce to £1–£5.

An alternative hydrological management scheme for improving loch and stream water quality at Loch Fleet by increasing the contribution of alkaline groundwater has been proposed (Cook *et al.*, 1991). To meet the target water quality specified for maintenance of trout populations would require only a 1·5% increase of current flows, or a 3·5% increase to return to conditions believed to exist in about 1800. This is estimated to cost about £6000 capital plus maintenance, or £6+ per fish. However, it is difficult to judge whether this approach could be applied to lakes elsewhere without first ascertaining groundwater supply and quality. Some desk studies have been reported in the USA (Olem, 1990); it is not known whether the opportunities for groundwater use are widely available, nor how the costs of treatment compare there with those of liming or other options.

Whether liming costs are considered acceptable depends on what other options are available and the perceived value of the fisheries. For the fisherman, if the affected lochs have only limited potential (e.g. 100 trout per year as in pre-acidified Loch Fleet) owing to their innate characteristics, the cost of regaining only the past fisheries by liming may appear exorbitant (about £100 per fish) and it would be more economic to promote productive fisheries for angling elsewhere. If the predicted improvements overall result in restoration and/or greater production of a fishery in the limed waters and similar benefits downstream, the liming option is clearly economic.

17.6 NON-TARGET EFFECTS

Changes in Other Flora and Fauna

It is sometimes claimed that liming adversely changes the character of uplands and oligotrophic lakes which are sensitive to acidification. There are few long-term investigations reported which justify this concern, but an opportunity was provided at Loch Fleet to study the effects of liming treatments on non-target fauna and flora of both the lake and its catchment. Data from the extensive programme of biological monitoring at Loch Fleet show that any non-target effects of catchment liming are minimal. The changes recorded following liming of open moorland were minor, with the observation that flying bent grass (*Molinia caerulea*) at one site was being favoured at the expense of heather (*Calluna vulgaris*). Studies of the 25-year-old stands of lodgepole pine and Sitka spruce in limed and unlimed areas showed no measurable significant or predictable change. The only terrestrial vegetation that was seen to suffer as a result of limestone application were two of several *Sphagnum* species in the wetland area at the head of the Altiwhat, where direct contact with limestone powder or runoff from limed parts nearby caused the death of *Sphagnum*. It is worth emphasising that no effects of liming on other species typical of acidophilic mires in the area were recorded (Plate V).

Independent studies were also carried out to establish any effects of liming on terrestrial invertebrate and small mammal populations. These showed no differences in the numbers of animals recovered from limed compared with unlimed parts of the catchment. Some short-term (less than 6 months) changes in wetland invertebrates were reported in an experimental plot study elsewhere (see Chapter 14).

Following the water quality changes caused by catchment liming, the diatom flora of the loch showed some expected changes, such as the return of less acidophilic species. Biochemical studies of the microbial flora, however, showed little change other than seasonal and year-by-year differences, which were paralleled in adjacent lochs with a wide range of pH. Primary production showed a progressive increase, sustained for at least 3 years following liming, even though additional nutrients were not detected. The most marked changes in the aquatic environment following liming were in the macrophyte flora. Submerged *Sphagnum auriculatum* decreased markedly, while less acidophilic species such as *Isoetes echinospora*, *Juncus fluitans*, and *Ultricularia intermedia* increased in abundance. Similar studies are reported elsewhere (e.g. Renberg and Grahn, 1990).

The programme of invertebrate sampling of the loch benthos, but not in shore samples and tributary streams, has shown that in the year following liming there were significant reductions in the numbers of individuals in most groups. Numbers of some groups recovered, thereafter, to levels similar to those before liming; numbers of caddis fly larvae and mayfly nymphs actually increased. It is difficult to explain these findings, even with the introduction of trout into the loch approximately 1 year and 2 years after liming. Mayfly (*Baetis rhodani*) nymphs colonised the Altiwhat and the outlet stream after liming. This species had not previously been recorded there.

In summary, the extensive programme of biological monitoring at Loch Fleet has provided an assessment of the overall ecological consequences of adding limestone to naturally acidic and calcium-deficient ecosystems. This monitoring programme is set to continue at Loch Fleet even though, so far, the results indicate only minimal non-target effects of catchment liming.

17.7 RESTORING ACID WATERS: CATCHMENT LIMING IN PERSPECTIVE

The various possible methods of restoring acid waters are discussed in Chapter 1. Recognising that sulphur and nitrogen emission reductions will not improve surface water quality in the short, or, in some cases, even in the long term, in that such reductions do nothing to re-establish lost base materials from soils and surface waters (Skeffington and Brown, 1992), one is led to conclude that liming is a practical method by which to restore acid waters.

Liming acid streams and lakes directly brings about a rapid change in water quality with decreased acidity and increased alkalinity and calcium concentration. There are difficulties, however, associated with this approach. Liming the water directly does not neutralise acid soils in the catchment from where the acidity and elevated aluminium concentrations of the surface waters are derived. The length of time over which direct lake liming will remain effective is governed by the hydrological turnover time. In many cases (certainly in the UK) lakes subject to acidification are often those with rapid hydrological turnover times. Where improvement or protection of water quality in tributary streams is required, liming only the lake water will not be effective in improving or protecting the water quality of tributary streams that provide the nursery areas for spawning and for juvenile growth. Catchment liming, properly targeted, can provide a rapid water quality improvement in these areas and can overcome the difficulties associated with direct water liming.

Decisions regarding whether or not catchment liming would be an appropriate measure need to take into account a range of questions already described in Chapter 4 (Fig. 4.1.). At the outset it is assumed that a rapid response to treatment is required and that the cost of the treatment can be recovered in time by the benefits of an improved water quality (i.e. an improved fishery). If catchment liming is perceived as being an appropriate course of action, consideration must be given to any adverse ecological side-effects. As explained earlier, in the case of Loch Fleet such side-effects are limited, but it should be remembered that there are some species of acidophilic *Sphagnum* killed by contact with limestone. This may be regarded as important, especially if it could lead to undesirable effects, such as erosion.

In summary, then, a balance needs to be established in terms of whether the higher cost of catchment liming can be offset by the longer effectiveness time of treatment when compared with direct water application. A balance should be sought, too, between the perceived benefits and any possible ecological perturbation.

The Loch Fleet project has been important in furthering an understanding of the ways in which catchment liming can be carried out to achieve specific objectives. It has set lower ($5\ t\ ha^{-1}$) and higher ($> 25\ t\ ha^{-1}$) limits of effective "dose", has identified preferred areas for lime application, and has explored alternative modes of dry, wet, and pellet application. Whether the techniques successful at Loch Fleet can be applied with similar effectiveness elsewhere, remains to be tested. The findings at Loch Fleet relating to the type and particle size of the material used, targeting particular areas of a catchment for application, and involving a sound knowledge of the local hydrology, as well as possible adverse effects (or lack of them), are likely to apply to most situations, however.

The collaborative project undertaken at Loch Fleet since 1984 provides a practical demonstration, underpinned by extensive laboratory and intensive field studies, of the improvements that can be achieved in the short term by various catchment liming techniques. The project has also shown how organisational and disciplinary coordination can enhance the interpretation and assessment of such a programme, with continuity and consistency of studies extending from deposition measurements through to benefits for targeted fish populations and possible effects on other biological communities. The scientific outcome of the project will, we believe, be exemplary; it will also help to guide practical measures to mitigate surface water acidification elsewhere.

REFERENCES

Cook, J.M., Edmunds, W.M. and Robins, N.S. (1991). Groundwater contributions to an acid upland lake (Loch Fleet, Scotland) and the possibilities for amelioration. *J. Hydrol.*, **125**: 111–28.

Cosby, B.J., Hornberger, G.M., Galloway, J.N. and Wright, R.F. (1985). Timescales of catchment acidification. *Env. Sci. Technol.* **19**: 1144–49.

Diamond, M., Hirst, D., Winder, L., Craig-Hine, D. and Crawshaw, D.L. (1992). The effect of liming agricultural land on the chemistry and biology of the River Esk, N.W. England. Internat. Conference Acidic Deposition: Its Nature and Impacts, Glasgow, UK. *Env. Poll.* (in press).

Farley, W.A. and Werritty, A. (1990). Is chloride always chemically conservative? Results from monthly budgets derived for acidifying catchments in S.W. Scotland. Internat. Conference Acidic Deposition: Its Nature and Impacts, Glasgow, UK.

Hall, R.T. (1987). Processes of evaporation from vegetation of the uplands of Scotland. *Trans. Roy. Soc. Edin. Earth Sci.*, **78**, 327–34.

Olem, H. (1990). *Liming Acidic Surface Waters.* Lewis Publishers. Michigan, USA, 331 pp.

Renberg, I. and Grahn, O. (1990). Recovery of the acidified lake Orvattnet, Sweden; but why? (poster). Internat. Conference Acidic Deposition: Its Nature and Impacts, Glasgow, UK, Roy. Soc. Edinburgh, Abstract p. 612.

Skeffington, R.A. and Brown, D.J.A. (1992). Timescales for recovery from acidification: implications of current knowledge for aquatic organisms. Internat. Conference Acidic Deposition: Its Nature and Impacts, Glasgow, UK. *Env. Poll.* in press.

Whitehead, P. (1989). Future trends in acidification, pp. 114–21 in *Acidification in Scotland.* Scottish Development Dept, Edinburgh, 128 pp.

Wright, R.F., Lotse, E. and Semb, A. (1988). Reversibility of acidification shown by whole-catchment experiments. *Nature* **334**, 670–75.

Index

MIX
Papier aus verantwortungsvollen Quellen
Paper from responsible sources
FSC® C105338
FSC
www.fsc.org

If you have any concerns about our products,
you can contact us on
ProductSafety@springernature.com

In case Publisher is established outside the EU,
the EU authorized representative is:
Springer Nature Customer Service Center GmbH
Europaplatz 3, 69115 Heidelberg, Germany

Printed by Libri Plureos GmbH
in Hamburg, Germany